CHAPMAN & HALL/CRC
Monographs and Surveys in
Pure and Applied Mathematics   **120**

ABSTRACT

CAUCHY

PROBLEMS:

Three Approaches

CHAPMAN & HALL/CRC
Monographs and Surveys in
Pure and Applied Mathematics   **120**

ABSTRACT

CAUCHY

PROBLEMS:

Three Approaches

IRINA V. MELNIKOVA
ALEXEI FILINKOV

CRC Press
Taylor & Francis Group
Boca Raton London New York

CRC Press is an imprint of the
Taylor & Francis Group, an **informa** business
A CHAPMAN & HALL BOOK

CRC Press
Taylor & Francis Group
6000 Broken Sound Parkway NW, Suite 300
Boca Raton, FL 33487-2742

First issued in paperback 2019

© 2001 by Taylor & Francis Group, LLC
CRC Press is an imprint of Taylor & Francis Group, an Informa business

No claim to original U.S. Government works

ISBN-13: 978-1-58488-203-9 (hbk)
ISBN-13: 978-0-367-39747-0 (pbk)

Library of Congress Cataloging-in-Publication Data

Filinkov, A.I. (Aleksei Ivanovich)
    Abstract Cauchy problems : three approaches / Alexei Filinkov, Irina
V. Melnikova
        p.   cm. (Chapman & Hall/CRC monographs and surveys in pure and
applied mathematics ; 120)
    Includes bibliographical references and index.
    ISBN 1-58488-250-6 (alk. paper)
    1. Cauchy problem. I. Melnikova, I.V. (Irina Valerianovna) II.
Title. III. Series.
[DNLM: 1. Hepatitis B virus. QW 710 G289h]
QA377 .F48 2001
515′.35—dc21   2001017069
                                                                CIP

Library of Congress Card Number 2001017069

Visit the Taylor & Francis Web site at
http://www.taylorandfrancis.com

and the CRC Press Web site at
http://www.crcpress.com

To our teacher

V.K. Ivanov

Contents

Preface

Many mathematical models in physics, engineering, finance, biology, etc., involve studying the Cauchy problem

$$u'(t) = Au(t) + F(t), \quad t \in [0, T), \ T \leq \infty, \ u(0) = x,$$

where A is a linear operator on a Banach space X and F is an X-valued function that represents the deterministic or stochastic influence of a medium. The first and main step in solving such problems consists of studying the homogeneous Cauchy problem

$$u'(t) = Au(t), \quad t \in [0, T), \ T \leq \infty, \ u(0) = x. \tag{CP}$$

The problem (CP) has been comprehensively studied in the case when the operator A generates a C_0-semigroup. A C_0-semigroup generated by a bounded operator A is nothing but an exponential operator-function

$$e^{At} = \sum_{k=1}^{\infty} \frac{A^k t^k}{k!}.$$

In general, C_0-semigroups inherit some main properties of exponential functions. As it turns out, the generation of a C_0-semigroup by the operator A is closely related to the uniform well-posedness of the Cauchy problem (CP). However, many operators that are important in applications do not generate C_0-semigroups.

The focus of this book is the Cauchy problem (CP) with operators A, which do not generate C_0-semigroups. We focus on three approaches to treating such problems:

- semigroup methods

- abstract distribution methods

- regularization methods

xi

The strategic concept of the first two approaches is the relaxation of the notion of 'well-posedness' so that a Cauchy problem that is not well-posed in the classical sense becomes well-posed in some other sense. In the first approach, using semigroups more general than C_0-semigroups – integrated, C-regularized, and κ-convoluted semigroups – one can construct solution operators on some subsets of $D(A)$. The corresponding solutions are stable with respect to norms that are stronger than the norm of X. Using distribution semigroups in the second approach, one can construct a family of generalized solution operators on X. The essence of the third approach is the following: assuming that for some x there exists a solution $u(\cdot)$ of (CP), for given x_δ ($\|x - x_\delta\| \leq \delta$) we construct a solution $u_\varepsilon(\cdot)$ of a well-posed (regularized) problem depending on a regularization parameter ε and choose $\varepsilon = \varepsilon(\delta)$ in a way that $u_\varepsilon \to u$ as $\delta \to 0$. So $u_\varepsilon(\cdot)$ can be taken as an approximate solution and the operator $\mathbf{R}_{\varepsilon,t}$ defined by $\mathbf{R}_{\varepsilon,t} x_\delta := u_\varepsilon(t)$ is a regularizing operator.

The motivation for writing this book was twofold: first, to give a self-contained account of the above-mentioned three approaches, which is accessible to nonspecialists. Second, to demonstrate the profound connections between these seemingly quite different methods. In particular, we demonstrate that integrated semigroups are primitives of generalized solution operators, and many regularizing operators coincide with C_ε-regularized semigroups.

The book's three chapters are devoted respectively to semigroup methods, abstract distribution methods, and regularization methods. We discuss not only the Cauchy problem (CP), but also the following important generalizations: the degenerate Cauchy problem, the Cauchy problem for inclusion and the Cauchy problem for the second order equation.

Acknowledgments

This project was supported by the Australian Research Council. The first author is grateful to the Department of Pure Mathematics of Adelaide University for their hospitality during her visits to Adelaide. The first author was also partially supported by grant RFBR N 99-01-00142 (Russian Federation).

The authors thank Professor Alan Carey, Professor Mike Eastwood, Dr. John van der Hoek and Dr. Nick Buchdahl for their support and encouragement.

The authors thank Dr. Maxim Alshansky, Dr. Uljana Anufrieva, Dr. Alexander Freyberg and Dr. Isna Maizurna for participation in some parts of this project

The authors also thank Cris Carey for editing the manuscript and Ann Ross for technical support.

Irina Melnikova
Alexei Filinkov

Introduction

We aim to present semigroup methods, abstract distribution methods, and regularization methods for the abstract Cauchy problem

$$u'(t) = Au(t), \quad t \in [0, T), \ T \leq \infty, \ u(0) = x, \qquad \text{(CP)}$$

where A is a linear operator on a Banach space X, $u(\cdot)$ is an X-valued function and $x \in X$. We also demonstrate the connections among these methods.

In Chapter 0 we use Heat and Wave equations to illustrate some of the main ideas, notions, and connections from the discussion below.

In Section 1.1 we give a brief account of the theory of C_0-semigroups. Here the reader will find a proof of the following fundamental result of this theory.

Theorem 1.1.1 *Suppose that A is a closed densely defined linear operator on X. Then the following statements are equivalent:*

(I) *the Cauchy problem (CP) is uniformly well-posed on $D(A)$: for any $x \in D(A)$ there exists a unique solution, which is uniformly stable with respect to the initial data;*

(II) *the operator A is the generator of a C_0-semigroup $\{U(t), \ t \geq 0\}$;*

(III) *Miyadera-Feller-Phillips-Hille-Yosida (MFPHY) condition holds for the resolvent of operator A: there exist $K > 0$, $\omega \in \mathbb{R}$ so that*

$$\left\| R_A^{(k)}(\lambda) \right\| \leq \frac{K \, k!}{(\operatorname{Re} \lambda - \omega)^{k+1}},$$

for all $\lambda \in \mathbb{C}$ with $\operatorname{Re} \lambda > \omega$, and all $k = 0, 1, \dots$.

In this case the solution of (CP) has the form

$$u(\cdot) = U(\cdot)x, \qquad x \in D(A),$$

and $\sup_{t \in [0,T]} \|u(t)\| \leq K \|x\|$ for some constant $K = K(T)$.

Condition (III) is usually used as the criterion for the uniform well-posedness of (CP).

Section 1.2 is devoted to n-times integrated semigroups connected with the well-posedness of (CP) on subsets from $D(A^{n+1})$. Here is the main result for exponentially bounded semigroups.

Theorem 1.2.4 *Let A be a densely defined linear operator on X with nonempty resolvent set. Then the following statements are equivalent:*

(I) *A is the generator of an exponentially bounded n-times integrated semigroup $\{V(t),\ t \geq 0\}$;*

(II) *the Cauchy problem (CP) is (n, ω)-well-posed: for any $x \in D(A^{n+1})$ there exists a unique solution such that*

$$\exists K > 0,\ \omega \in \mathbb{R}: \qquad \|u(t)\| \leq K e^{\omega t} \|x\|_{A^n},$$

where $\|x\|_{A^n} = \|x\| + \|Ax\| + \ldots + \|A^n x\|$.

Statements (I), (II) of this theorem are equivalent to the MFPHY-type condition

$$\left\| \left[\frac{R_A(\lambda)}{\lambda^n} \right]^{(k)} \right\| \leq \frac{K\,k!}{(\lambda - \omega)^{k+1}}$$

for all $\lambda > \omega$, and $k = 0, 1, \ldots$.

In the particular case when A generates a C_0-semigroup U, the n-times integrated semigroup generated by A is the n-th order primitive of U. In general (and in contrast to C_0-semigroups), integrated semigroups may be not exponentially bounded, may be locally defined, degenerate, and their generators may be not densely defined. In Section 1.2 we consider non-degenerate exponentially bounded and local integrated semigroups, their connections with the Cauchy problem, and discuss several examples. We also show that an n-times integrated semigroup with a densely defined generator A can be defined as a family of bounded linear operators $V(t)$ satisfying

$$
\begin{aligned}
V(t)x - \frac{t^n}{n!}x &= \int_0^t V(s)Ax\,ds \\
&= \int_0^t AV(s)x\,ds,\quad 0 \leq t < T,\quad x \in D(A),
\end{aligned}
$$

and

$$V(t)x - \frac{t^n}{n!}x = A\int_0^t V(s)x\,ds,\quad x \in X.$$

Replacing function $t^n/n!$ by a continuously differentiable function $\Theta(t)$, one arrives at the notion of κ-convoluted semigroup, where $\kappa = \Theta'$.

Definition 1.3.1 *Let A be a closed operator and $\kappa(t)$ be a continuous function on $[0, T)$, $T \leq \infty$. If there exists a strongly continuous operator-family $\{S_\kappa(t), \ 0 \leq t < T\}$ such that $S_\kappa(t)Ax = AS_\kappa(t)x$ for $x \in D(A)$, $t \in [0, T)$, and for all $x \in X$*

$$S_\kappa(t)x = A \int_0^t S_\kappa(s)x \, ds + \Theta(t)x, \ 0 \leq t < T,$$

where $\Theta(t) = \int_0^t \kappa(s)ds$, then S_κ is called a κ-convoluted semigroup generated by A, and A is called the generator of S_κ.

In the particular case $\Theta(t) = t^n/n!$, the κ-convoluted semigroup S_κ is an n-times integrated semigroup. If A is the generator of a C_0-semigroup U, then

$$S_\kappa(t)x = \int_0^t \frac{(t-s)^{n-1}}{(n-1)!} U(s)x \, ds$$

is an n-times integrated semigroup and a κ-convoluted semigroup. This is the reason that S_κ is called a κ-convoluted semigroup, but not a Θ-convoluted semigroup.

The corresponding Cauchy problem

$$v'(t) = Av(t) + \Theta(t)x, \quad 0 \leq t < T, \quad v(0) = 0,$$

is called the Θ-convoluted Cauchy problem. If there exists a solution of the Cauchy problem

$$u'(t) = Au(t), \quad 0 \leq t < T, \quad u(0) = x,$$

then, as usual for a nonhomogeneous equation, we have $v = \Theta * u$.

The results of Section 1.3 connect the existence of a κ-convoluted semigroup and well-posedness of a Θ-convoluted Cauchy problem with the behaviour of the resolvent $R_A(\lambda)$. In contrast to the case of an exponentially bounded and local n-times integrated semigroup, where the resolvent has polynomial estimates in a half-plane or in a logarithmic region respectively, in the 'convoluted' case the resolvent exists in some smaller region and is allowed to increase exponentially:

$$\|R_A(\lambda)\| \leq K e^{M(\lambda)}.$$

Here a real-valued function $M(\lambda)$, $\lambda \in \mathbb{C}$, grows not faster than λ and is defined by the order of decreasing of $\tilde{\kappa}(\lambda)$, the Laplace transform of $\kappa(t)$. In Chapter 3 we demonstrate that a (CP) with a generator of a κ-convoluted semigroup takes an intermediate place between 'slightly' ill-posed problems (with a generator of an integrated semigroup) and 'essentially' ill-posed problems (with $-A$ generating a C_0-semigroup).

In Section 1.4 we consider the Cauchy problem (CP) with A being the generator of an exponentially bounded C-regularized semigroup $\{S(t),\ t \geq 0\}$, where C is some bounded invertible operator on X. It is shown that for $x \in CD(A)$ there exists a solution of this problem:

$$u(\cdot) = U(\cdot)x = C^{-1}S(\cdot)x.$$

Solution operators $U(t)$ are not necessarily bounded on X, therefore in general the solution $u(\cdot)$ is not stable in X with respect to variation of x. The well-posedness of the Cauchy problem can be restored in a certain subspace of X with a norm stronger than the norm of X. Namely, if A is the generator of a C-regularized semigroup, then for any $x \in CD(A),\ u(\cdot) = C^{-1}S(\cdot)x$ is a unique solution of (CP), and

$$\sup_{0 \leq t \leq T} \|u(t)\| \leq K\|C^{-1}x\|$$

for some constant K. There are operators A (for example, those having arbitrary large positive eigenvalues) that generate C-regularized semigroups, defined only for t from some bounded subset of $\mathbb{R}$. Such semigroups correspond to the local C-well-posedness of the Cauchy problem, and are also considered in Section 1.4.

In Section 1.5 we study degenerate C_0-semigroups and integrated semigroups connected with the degenerate Cauchy problem

$$Bu'(t) = Au(t), \quad t \geq 0,\ u(0) = x, \tag{DP}$$

where $B, A : X \to Y$ are linear operators in Banach spaces X, Y, and $\ker B \neq \{0\}$. We assume that the set

$$\rho_B(A) :=$$
$$\left\{\lambda \in \mathbb{C} \mid R(\lambda) := (\lambda B - A)^{-1}B \text{ is a bounded operator on } X\right\}$$

is not empty. This set is called the B-resolvent set of operator A and the operator $R(\lambda)$ is called the B-resolvent of operator A.

We prove the equivalence of the uniform well-posedness of (DP) on the maximal correctness class, the existence of a degenerate C_0-semigroup with A and B being the pair of generators, and MFPHY-type estimates together with the decomposition of X:

$$X = X_1 \oplus \ker B, \quad X_1 := \overline{D_1}, \ D_1 := R(\lambda)X, \ \lambda \in \rho_B(A).$$

This decomposition plays a role similar to the density condition for the generator of a C_0-semigroup. We also prove results about (n, ω)- and n-well-posedness of the degenerate Cauchy problem.

Both degenerate Cauchy problems

$$Bu'(t) = Au(t), \quad t \in [0, T), \ T \leq \infty, \tag{DP}$$
$$u(0) = x, \quad \ker B \neq \{0\}$$

and

$$\frac{d}{dt} Bv(t) = Av(t), \quad t \in [0, T), \ T \leq \infty, \tag{DP1}$$
$$Bv(0) = x,$$

can be considered as particular cases of the Cauchy problem for an inclusion with a multivalued linear operator $\mathcal{A}$

$$u'(t) \in \mathcal{A}u(t), \quad t \in [0, T), \ T \leq \infty, \ u(0) = x. \tag{IP}$$

If we set $\mathcal{A} = B^{-1}A$ for problem (DP), or $\mathcal{A} = AB^{-1}$ and $u(\cdot) := Bv(\cdot)$ for (DP1), then $u(\cdot)$ is a solution of the Cauchy problem (IP). Conversely, if $u(\cdot)$ is a solution of (IP) with $\mathcal{A} = B^{-1}A$ or $\mathcal{A} = AB^{-1}$, then $u(\cdot)$ is the solution of (DP) or any $v(\cdot)$ from the set $Bv(\cdot) = u(\cdot)$ is the solution of (DP1), respectively.

In Section 1.6 we use the technique of degenerate semigroups with multivalued generators to study the uniform well-posedness and the (n, ω)-well-posedness of the Cauchy problem (IP). As a consequence, we obtain MFPHY-type necessary and sufficient conditions for the well-posedness of (DP) and for the existence of a solution of (DP1).

Section 1.7 is devoted to semigroup methods for abstract Cauchy problems

$$u''(t) = Au'(t) + Bu(t), \quad t \geq 0, \ u(0) = x, \ u'(0) = y, \tag{CP2}$$

where $A, B : X \to X$, and

$$Qu''(t) = Au'(t) + Bu(t), \quad t \geq 0, \ u(0) = x, \ u'(0) = y, \tag{DP2}$$

where $Q, A, B : X \to Y$ are linear operators, $\ker Q \neq \{0\}$, and X, Y are Banach spaces. In this section we use semigroup methods for studying the well-posedness of these problems. First, we construct $\mathcal{M}, \mathcal{N}$-functions, the solution operators for problem (CP2), which generalize semigroups and cosine/sine operator-functions. Second, we study (CP2), (DP2) by reducing them to a first order Cauchy problem in a product space. Using the results of Sections 1.2 and 1.6 on integrated semigroups we obtain necessary and sufficient conditions for the well-posedness of (CP2), (DP2) in terms of resolvent operators

$$R_r(\lambda^2) := (\lambda^2 I - \lambda A - B)^{-1} \quad \text{and} \quad R_d(\lambda^2) := (\lambda Q^2 - \lambda A - B)^{-1},$$

respectively.

The relation between the operators A, B plays a very important role in studying well-posedness of second order problems. We show that for the equivalence of the well-posedness of (CP2) and the existence of an integrated semigroup for the Cauchy problem

$$v'(t) = \Phi v(t), \quad t \geq 0, \quad v(0) = v_0,$$

$$\Phi = \begin{pmatrix} 0 & I \\ B & A \end{pmatrix}, \quad v(t) = \begin{pmatrix} u(t) \\ u'(t) \end{pmatrix}, \quad v_0 = \begin{pmatrix} x \\ y \end{pmatrix},$$

operators A, B have to be biclosed. With the help of the theory of integrated semigroups we prove that the MFPHY-type condition:

$\exists K > 0, \ \omega \geq 0:$

$$\left\| \frac{d^k}{d\lambda^k} R_r(\lambda^2) \right\|, \ \left\| \frac{d^k}{d\lambda^k} \overline{R_r(\lambda^2) \left(I - \frac{A}{\lambda} \right)} \right\| \leq \frac{K \, k!}{(\operatorname{Re} \lambda - \omega)^{k+1}},$$

for all $\lambda \in \mathbb{C}$ with $\operatorname{Re} \lambda > \omega$ and all $k = 0, 1, \ldots,$

is necessary and sufficient for well-posedness of (CP2) with biclosed operators A, B. Using the $\mathcal{M}, \mathcal{N}$-functions theory, we obtain the following MFPHY-type well-posedness condition

$\exists K > 0, \ \omega \geq 0:$

$$\left\| \frac{d^k}{d\lambda^k} R_r(\lambda^2) \right\|, \ \left\| \frac{d^k}{d\lambda^k} \left(I - \frac{A}{\lambda} \right) R_r(\lambda^2) \right\| \leq \frac{K \, k!}{(\operatorname{Re} \lambda - \omega)^{k+1}},$$

for all $\lambda \in \mathbb{C}$ with $\operatorname{Re} \lambda > \omega$ and all $k = 0, 1, \ldots,$

for problem (CP2) with commuting operators A, B. Using the theory of degenerate integrated semigroups we prove MFPHY-type necessary and sufficient conditions for well-posedness of the degenerate Cauchy problem (DP2).

Chapter 2 is devoted to the abstract distributions methods. In Section 2.1 for any $x \in X$, we consider the Cauchy problem (CP) with $A : D(A) \subset X \to X$, in the space of (Schwartz) distributions. Our main aim here is to obtain necessary and sufficient conditions for well-posedness in the space of distributions in terms of the resolvent of A. We show that they are the same as in the case when A is the generator of a local integrated semigroup. Thus, only the Cauchy problem with a generator of such a semigroup can be well-posed in the space of distributions. In this case, a distribution semigroup is the solution operator distribution, and an integrated semigroup is the continuous function whose existence is guaranteed by the abstract structure theorem for this distribution.

In Section 2.2, we prove the necessary and sufficient conditions for well-posedness of (DP) and (IP) in spaces of distributions, generalizing the conditions obtained in Section 2.1. Naturally, the semigroups connected with these problems are degenerate. The important fact is that an n-times integrated semigroup with a pair of generators A, B or with the multivalued generator $\mathcal{A}$ is degenerate on $\ker B = \ker R(\lambda)$, but the corresponding solution operator distribution and the distribution semigroup are degenerate on $\ker R^{n+1} := \ker R^{n+1}(\lambda)$. Here $R(\lambda)$ is the pseudoresolvent that is equal to $(\lambda B - A)^{-1} B$ or $(\lambda I - \mathcal{A})^{-1}$, respectively. In this section we assume that X admits the decomposition

$$ X = X_{n+1} \oplus \ker R^{n+1}, \quad X_{n+1} := \overline{D}_{n+1}, \quad D_{n+1} := R^{n+1}(\lambda)X. $$

First, we describe the structure of $\ker R^{n+1}$, which is used in the main theorem, where we discuss the structure of solution operator distributions and distribution semigroups on $\ker R^{n+1}$. We prove that (DP) is well-posed in the space of distributions if and only if $\hat{A}$ (the part of operator $\mathcal{A}$ in X_{n+1}) is the generator of a distribution semigroup, or equivalently, $\hat{A}$ is the generator of a local k-times integrated semigroup for some k. Finally, we consider (DP) in the space of exponential distributions. The distinctive feature of the exponential case is that the parameter k of the integrated semigroup coincides with the parameter in the estimate for pseudoresolvent $R(\lambda)$.

Studying the Cauchy problem in the spaces of distributions we prove that polynomial estimates for the resolvent of A (or for the pseudoresolvent of a multivalued operator $\mathcal{A}$) in a certain logarithmic region of a complex plane is the criterion for well-posedness of the Cauchy problem with such an operator A (or $\mathcal{A}$, respectively) in the sense of distributions. This domain and resolvent estimates are connected with the order of a corresponding distribution semigroup and solution operator distribution.

In Section 2.3 we consider the Cauchy problem with operator A having the resolvent in a certain region Λ smaller than logarithmic. We investigate well-posedness of such (CP) in the space of (Beurling) ultradistributions. Utradistributions are defined as elements of dual spaces for spaces of infinitely differentiable functions with a locally convex topology. In these spaces well-posedness of the Cauchy problem is studied in a more general sense than the spaces of Schwartz distributions. We show that the existence of a solution operator ultradistribution is equivalent to exponential estimates for the resolvent in some domain Λ. As is proved in Section 1.3, the latter fact is equivalent to the existence of a unique solution of the convoluted equation

$$ v(t)' = Av(t) + \Theta(t)x, \quad t \geq 0, \ v(0) = 0, \quad \cdot $$

where function $\Theta(t)$ is defined by the resolvent of A.

Spaces of test functions for ultradistributions are spaces of infinitely differentiable functions defined in terms of estimates on their derivatives, depending on some sequence M_n. The same sequence allows one to estimate coefficients of differential operators of infinite order, which are called ultradifferential operators. The spaces of ultradistributions are invariant under ultradifferential operators just as spaces of distributions are invariant under differential operators. In this section, firstly we discuss some properties of ultradistributions. Then we investigate the well-posedness of (CP) in the spaces of ultradistributions. The limit case for (CP) well-posed in the spaces of ultradistributions is the case of the Cauchy problem with operator A having no regular points in a right semiplane. For example, the reversed Cauchy problem for the heat equation is equivalent to the local Cauchy problem (CP) with $A = -d^2/dx^2$ in the corresponding space X. It is proved that such a (CP) is well-posed in spaces of new distributions. Spaces of test elements $x \in X$ for construction of spaces of new distributions are defined in terms of the behaviour of $A^n x$ for any n, and their dual spaces of new distributions are subspaces of X^*. Section 2.3, together with Section 1.6 on κ-convoluted semigroups, demonstrates the transition from uniformly well-posed problems to 'essentially' ill-posed problems via 'slightly' ill-posed problems.

Thus, in Chapters 1 and 2 we investigate Cauchy problems that are not uniformly well-posed. The technique of integrated, κ-convoluted, and C-regularized semigroups presented in Chapter 1 allows one to construct a solution of (CP) and local (CP) for initial values x from various subsets of $D(A)$, stable in X with respect to x in corresponding graph-norms. The technique of distributions, ultradistributions, and new distributions presented in Chapter 2 allows the construction of a generalized solution for any $x \in X$, stable in a space of distributions.

In Chapter 3 we consider regularizing operators which allow one to construct an approximate solution of (CP), which is stable in X for concordant parameters of regularization and error. The general theory of ill-posed problems usually treats ill-posed problems in the form of the equation of the first kind

$$\Psi u = f, \quad \Psi : U \to F, \tag{OP}$$

where, in general, operator Ψ^{-1} either does not exist or is not bounded. The main regularization methods for (OP) are variational:

    (a) Ivanov's quasivalues method,
    (b) the residual method,
    (c) Tikhonov's method.

In Section 3.1 we study regularization methods for the ill-posed local Cauchy problem

$$u'(t) = Au(t), \quad 0 \le t < T, \ T < \infty, \ u(0) = x,$$

with operator $-A : D(A) \subset X \to X$ generating a C_0-semigroup or satisfying some similar conditions.

These methods employ the differential nature of this problem and allow us to construct a regularized solution of local (CP) as a solution of a well-posed regularized problem, without reducing it to the form (OP). They are the following:

1) the **quasi-reversibility method**, where the regularized solution is a solution of the Cauchy problem

$$u'_\varepsilon(t) = (A - \varepsilon A^2)u_\varepsilon(t), \quad 0 < t \leq T,$$
$$u_\varepsilon(0) = x, \quad \varepsilon > 0;$$

2) the **auxiliary boundary conditions method**, where the regularized solution is a solution of the boundary value problem

$$\widetilde{u}'_\varepsilon(t) = A\widetilde{u}_\varepsilon(t), \quad 0 < t < T + \tau, \quad \tau > 0,$$
$$\widetilde{u}_\varepsilon(0) + \varepsilon\widetilde{u}_\varepsilon(T + \tau) = x, \quad \varepsilon > 0;$$

3) **Carasso's method** of reducing an ill-posed problem to a well-posed Dirichlet problem.

In Section 3.2, we obtain error estimates for these regularization methods and compare them. We discuss connections between 'differential' methods and variational methods. Furthermore, we establish the connection between regularization methods for the ill-posed Cauchy problem and the C-regularized semigroup method.

Chapter 0

Illustration and Motivation

In this chapter we use very basic model examples to illustrate some of the main ideas, notions, and connections from the discussion above.

0.1 Heat equation

For simplicity, let $\Omega = (0,1)$. In general, Ω can be an open set from $\mathbb{R}^n$. Consider the Cauchy-Dirichlet problem on $X = L^2(\Omega)$:

$$\frac{\partial u(x,t)}{\partial t} - \frac{\partial^2 u(x,t)}{\partial x^2} = 0, \quad t \in [0,T], \; x \in \Omega, \qquad (0.1.1)$$
$$u(0,t) = u(1,t) = 0,$$
$$u(x,0) = u^0(x).$$

Note that $L^2(\Omega)$ is a separable Hilbert space. Consider operator $A = \frac{d^2}{dx^2}$ in $L^2(\Omega)$ with the domain

$$D(A) = H^2(\Omega) \cap H_0^1(\Omega),$$

where $H^2(\Omega)$ and $H_0^1(\Omega)$ are the following classical Sobolev spaces

$$H_0^1(\Omega) = \left\{ u \in L^2(\Omega) \; \Big| \; \frac{\partial u}{\partial x} \in L^2(\Omega), \; u(0,t) = u(1,t) = 0 \right\},$$
$$H^2(\Omega) = \left\{ u \in L^2(\Omega) \; \Big| \; \frac{\partial^2 u}{\partial x^2} \in L^2(\Omega) \right\}.$$

The operator $-A$ has a self-adjoint compact inverse and therefore its spectrum consists of discrete eigenvalues. Eigenfunctions and eigenvalues of $-A$

1

can be obtained by solving

$$\frac{d^2 e_k}{dx^2} = -\mu_k e_k, \quad e_k(0) = e_k(1) = 0, \ k \in \mathbb{N},$$

which gives

$$\mu_k = k^2 \pi^2 > 0, \ e_k = \sqrt{2} \sin k\pi x, \ k \in \mathbb{N}.$$

Since $\{e_k\}_{k=1}^{\infty}$ forms an orthonormal basis in $L^2(\Omega)$, any $f \in L^2(\Omega)$ can be written in the form

$$f = \sum_{k=1}^{\infty} f_k e_k, \quad \text{where} \ \ f_k = (f, e_k)_{L^2(\Omega)},$$

and we have

$$\|f\|_{L^2(\Omega)} = \left(\sum_{k=1}^{\infty} |f_k|^2 \right)^{1/2}.$$

Assuming that $u^0 \in L^2(\Omega)$ and $u(t) \in L^2(\Omega)$ for each $t \in [0, T]$, we write

$$u(t) = \sum_{k=1}^{\infty} u_k(t) e_k, \quad \text{and} \ \ u^0 = \sum_{k=1}^{\infty} u_k^0 e_k.$$

Then the equation in (0.1.1) becomes

$$\sum_{k=1}^{\infty} \left(\frac{du_k}{dt} + \mu_k u_k \right) e_k = 0.$$

By uniqueness of expansions in $L^2(\Omega)$, we deduce

$$\frac{du_k}{dt} + \mu_k u_k = 0, \qquad k \in \mathbb{N},$$

$$u_k(0) = u_k^0,$$

and the unique solution of this problem is

$$u_k(t) = u_k^0 e^{-\mu_k t}.$$

Now we show that the unique solution of (0.1.1) is given by

$$u(t) = \sum_{k=1}^{\infty} u_k(t) e_k = \sum_{k=1}^{\infty} e^{-\mu_k t} u_k^0 e_k. \tag{0.1.2}$$

Since we assumed that $u^0 \in L^2(\Omega)$, we have

$$\sum_{k=1}^{\infty} |u_k(t)|^2 = \sum_{k=1}^{\infty} |u_k^0|^2 e^{-2\mu_k t} \le \sum_{k=1}^{\infty} |u_k^0|^2 < \infty, \tag{0.1.3}$$

that is for any $t \in [0, T]$, $u(t) \in L^2(\Omega)$. Moreover we have some regularity with respect to t, namely $u \in C\{[0, T], L^2(\Omega)\}$, the Banach space of continuous functions over $[0, T]$ with values in $L^2(\Omega)$ equipped with the norm

$$\|f\|_{C\{[0,T], L^2(\Omega)\}} = \sup_{t \in [0,T]} \left(\int_0^1 |f(x,t)|^2 dx \right)^{1/2}.$$

To verify this we note that the estimate

$$\sup_{t \in [0,T]} \left(\sum_{k=m}^{m+p} |u_k(t)|^2 \right)^{1/2} \leq \left(\sum_{k=m}^{m+p} |u_k^0|^2 \right)^{1/2}, \quad \forall m \geq 1, \; p > 0,$$

implies that the sequence of partial sums of the series $\sum_{k=1}^{\infty} u_k(t)$ is a Cauchy sequence in $C\{[0, T], L^2(\Omega)\}$. Now we have to find out in what sense the Cauchy problem (0.1.1) is satisfied. First, since

$$|u_k(t) - u_k^0|^2 \leq |u_k^0|^2 \left(1 - e^{-\mu_k t} \right)^2,$$

we can write

$$\sum_{k=1}^{\infty} |u_k(t) - u_k^0|^2 \leq \left(1 - e^{-\mu_N t} \right)^2 \sum_{k=1}^{N} |u_k^0|^2 + \sum_{k=N+1}^{\infty} |u_k^0|^2,$$

where N is large enough. This implies that

$$\lim_{t \to 0} (u(t) - u^0) = 0 \quad \text{in} \quad L^2(\Omega),$$

and therefore $u(0) = u^0$ in $L^2(\Omega)$.

Secondly, we demonstrate that $u \in L^2\{[0, T], H_0^1(\Omega)\}$, which clearly implies that the boundary condition $u(0, t) = u(1, t) = 0$ is satisfied almost everywhere in $t \in [0, T]$. It is well-known that $v \in H_0^1(\Omega)$ if and only if

$$\sum_{k=1}^{\infty} \mu_k |v_k|^2 < \infty, \quad \text{where} \quad v_k = (v, e_k).$$

Since

$$|u_k(t)|^2 \leq |u_k^0|^2 e^{-2\mu_k t}, \quad k \in \mathbb{N},$$

then

$$\int_0^T |u_k(t)|^2 dt \leq \frac{1}{\mu_k} |u_k^0|^2,$$

which implies

$$\int_0^T \left(\sum_{k=m}^{m+p} \mu_k |u_k(t)|^2 \right) dt \leq \sum_{k=m}^{m+p} |u_k^0|^2, \quad \forall m \geq 1, \; p > 0,$$

and therefore

$$\int_0^T \Big(\sum_{k=1}^\infty \mu_k |u_k(t)|^2 \Big) dt < \infty,$$

that is $u \in L^2\{[0, T], H_0^1(\Omega)\}$ for any $u^0 \in L^2(\Omega)$.

Note that if $u^0 \in H_0^1(\Omega)$, then the series $\sum_{k=1}^\infty \mu_k |u_k^0|^2$ is convergent, and since

$$\sup_{t \in [0,T]} \Big(\sum_{k=m}^{m+p} \mu_k |u_k(t)|^2 \Big)^{1/2} \leq \Big(\sum_{k=m}^{m+p} \mu_k |u_k^0|^2 \Big)^{1/2}, \quad \forall m \geq 1, \ p > 0,$$

we have $u \in C\{[0, T], H_0^1(\Omega)\}$.

Finally, we discuss in what sense the equation in (0.1.1) is satisfied. We set

$$\mathbf{a}(u, v) = \int_\Omega \frac{du}{dx} \frac{dv}{dx} \, dx, \quad u, v \in H_0^1(\Omega),$$

$$U_m(t) = \sum_{k=1}^m u_k(t) e_k, \quad U_m'(t) = \sum_{k=1}^m \frac{du_k}{dt} e_k.$$

Then, since

$$\frac{du_k}{dt} + \mu_k u_k = 0, \qquad k \in \mathbb{N},$$

we have

$$\big(U_m'(t), e_k\big) + \mathbf{a}\big(U_m(t), e_k\big) = 0, \quad 1 \leq k \leq m. \tag{0.1.4}$$

Now fix $k < m$ and let $\phi \in \mathcal{D}(0, T)$ (the space of Schwartz test functions). Multiplying (0.1.4) by ϕ, and integrating by parts, we obtain

$$-\int_0^T \big(U_m(t), e_k\big) \phi'(t) dt + \int_0^T \mathbf{a}\big(U_m(t), e_k\big) \phi(t) dt = 0.$$

As $U_m \xrightarrow{m \to \infty} U$ uniformly over $[0, T]$, we have that the function $t \to \big(U_m(t), e_k\big) \phi'(t)$ converges to the function $t \to \big(u(t), e_k\big) \phi'(t)$, and therefore

$$\lim_{m \to \infty} \int_0^T \big(U_m(t), e_k\big) \phi'(t) dt = \int_0^T \big(u(t), e_k\big) \phi'(t) dt.$$

Since $u \in L^2\{[0, T], H_0^1(\Omega)\}$ and

$$\left| \int_0^T \mathbf{a}\big(U_m(t) - u(t), e_k\big) \phi(t) dt \right| \leq K \int_0^T \|U_m(t) - u(t)\|^2 dt,$$

we have

$$\lim_{m \to \infty} \int_0^T \mathbf{a}\big(U_m(t), e_k\big) \phi(t) dt = \int_0^T \mathbf{a}\big(u(t), e_k\big) \phi(t) dt.$$

Thus, for any e_k, we have

$$-\int_0^T (u(t), e_k)\phi'(t)dt + \int_0^T \mathbf{a}(u(t), e_k)\phi(t)dt = 0,$$

and since the system $\{e_k\}_{k=1}^\infty$ is dense in $H_0^1(\Omega)$, we obtain

$$-\int_0^T (u(t), v)\phi'(t)dt + \int_0^T \mathbf{a}(u(t), v)\phi(t)dt = 0, \qquad (0.1.5)$$

for all $v \in H_0^1(\Omega)$ and $\phi \in \mathcal{D}(0, T)$.

So we conclude that for any $u^0 \in L^2(\Omega)$, the function u, defined by formula (0.1.2), satisfies the equation in (0.1.1) in the following weak sense

$$\frac{d}{dt}(u(\cdot), v) + \mathbf{a}(u(\cdot), v) = 0, \quad \text{for all } v \in H_0^1(\Omega), \qquad (0.1.6)$$

in the sense of the distributions from $\mathcal{D}'(0, T)$, where the distribution $(u(\cdot), v) \in \mathcal{D}'(0, T)$ is defined by

$$\left\langle (u(\cdot), v), \phi \right\rangle := \int_0^T (u(t), v)\phi(t)dt, \quad \text{for any } \phi \in \mathcal{D}(0, T),$$

and similarly

$$\left\langle \mathbf{a}(u(\cdot), v), \phi \right\rangle := \int_0^T \mathbf{a}(u(t), v)\phi(t)dt, \quad \text{for any } \phi \in \mathcal{D}(0, T).$$

Also, we can rewrite the variational equation (0.1.6) in terms of distributions both in temporal and spatial variables. Since $\mathcal{D}(\Omega)$ is dense in $H_0^1(\Omega)$, we can choose $v \in \mathcal{D}(\Omega)$, and (0.1.6) is equivalent to

$$-\int_{\Omega_T} u(x, t)v(x)\phi'(t)d(x, t) - \int_{\Omega_T} u(x, t)v''(x)\phi(t)d(x, t) = 0, \qquad (0.1.7)$$

where $\Omega_T = \Omega \times (0, T)$ and $d(x, t) = dx\, dt$. Set $\psi = v \otimes \phi := v(x)\phi(t)$, then $\psi \in \mathcal{D}(\Omega_T)$ and (0.1.7) becomes

$$-\int_{\Omega_T} u\frac{\partial \psi}{\partial t}d(x, t) - \int_{\Omega_T} u\frac{\partial^2 \psi}{\partial x^2}d(x, t) = 0, \qquad (0.1.8)$$

for all $\psi \in \mathcal{D}(\Omega_T)$ with $v \in \mathcal{D}(\Omega), \phi \in \mathcal{D}(0, T)$. As the set of linear combinations of functions $v \otimes \phi$ with $v \in \mathcal{D}(\Omega), \phi \in \mathcal{D}(0, T)$, is dense in $\mathcal{D}(\Omega_T)$, we have that (0.1.8) holds for all $\psi \in \mathcal{D}(\Omega_T)$:

$$\left\langle \frac{\partial u}{\partial t} - \frac{\partial^2 u}{\partial x^2}, \psi \right\rangle = 0, \quad \forall \psi \in \mathcal{D}(\Omega_T), \qquad (0.1.9)$$

that is the equation in (0.1.1) is satisfied in the sense of distributions in Ω_T. Thus, we have demonstrated that the variational formula (0.1.6) implies the distributional formula (0.1.9). The converse is not true in general, but it is true in this example since we know that u had values in $H_0^1(\Omega)$.

Now if we assume some regularity for u^0, say $u^0 \in H_0^1(\Omega)$ (or $D(A)$), then

$$\frac{\partial u}{\partial t} \in L^2\{[0,T], L^2(\Omega)\} \quad \text{and} \quad Au \in L^2(\Omega).$$

Therefore, in this case the equation in (0.1.1)

$$\frac{\partial u}{\partial t} = Au$$

is satisfied in $L^2\{[0,T], L^2(\Omega)\}$ and we say that u is a strong solution of the Cauchy problem (0.1.1).

Next, we rephrase this discussion using the semigroup terminology. Firstly, we rewrite (0.1.1) in the following abstract form on $X = L^2(\Omega)$:

$$u'(t) = Au(t), \quad t \in [0,T], \ u(0) = u^0, \qquad (0.1.10)$$

where, as before

$$A = \frac{d^2}{dx^2} \quad \text{with} \quad D(A) = H^2(\Omega) \cap H_0^1(\Omega),$$

is an unbounded linear operator on $L^2(\Omega)$. Note that $v \in D(A)$ if and only if $\sum_{k=1}^{\infty} \mu_k^2 |v_k|^2 < \infty$, where $v_k = (v, e_k)$. For each $t \geq 0$, define a linear operator on $L^2(\Omega)$ by

$$U(t)v := \sum_{k=1}^{\infty} e^{-\mu_k t} v_k e_k, \quad v \in L^2(\Omega).$$

We note that, due to (0.1.3), $U(t)$ is a bounded operator for each $t \geq 0$. Furthermore, since we have

$$\|U(t)v\|_{L^2(\Omega)}^2 = \sum_{k=1}^{\infty} \left| e^{-\mu_k t} v_k \right|^2 \leq \sum_{k=1}^{\infty} |v_k|^2 = \|v\|^2,$$

then $\|U(t)v\|_{L^2(\Omega)} \leq 1$ for each $t \geq 0$. Now we show that the semigroup property

$$U(t)U(s) = U(t+s)$$

is satisfied for all $t, s \geq 0$. Let $v \in X$. Then, using the self-adjointness of $U(t)$, we obtain

$$U(t)U(s)v \ = \ U(t)\big(U(s)v\big) = \sum_{k=1}^{\infty} e^{-\mu_k t} \big(U(s)v, e_k\big) e_k$$

$$= \sum_{k=1}^{\infty} e^{-\mu_k t} \big(v, U(s) e_k \big) e_k$$

$$= \sum_{k=1}^{\infty} e^{-\mu_k t} \big(v, e^{-\mu_k s} e_k \big) e_k = \sum_{k=1}^{\infty} e^{-\mu_k (t+s)} \big(v, e_k \big) e_k$$

$$= U(t+s)v.$$

Thus, U is a C_0-semigroup of contractions. Now we consider

$$u(t) := U(t)u^0, \quad t \geq 0, u^0 \in X.$$

The estimate (0.1.3) implies that u is continuous in $t \geq 0$ and $u(0) = u^0$ in X. Let $u^0 \in D(A)$. Then

$$\sum_{k=1}^{\infty} |\mu_k e^{-\mu_k t} u_k^0|^2 \leq \sum_{k=1}^{\infty} |\mu_k u_k^0|^2 < \infty$$

implies that

$$u'(t) = -\sum_{k=1}^{\infty} \mu_k e^{-\mu_k t} u_k^0 e_k$$

exists for all $t \geq 0$ and is continuous in $t \geq 0$. Furthermore,

$$\sum_{k=1}^{\infty} |\mu_k \big(U(t)u^0, e_k \big)|^2 = \sum_{k=1}^{\infty} |\mu_k e^{-\mu_k t} u_k^0|^2 < \infty$$

for all $t \geq 0$, hence $u(t) \in D(A)$ and

$$-Au(t) = \sum_{k=1}^{\infty} \mu_k \big(U(t)u^0, e_k \big) e_k = \sum_{k=1}^{\infty} \mu_k e^{-\mu_k t} u_k^0 e_k = -u'(t).$$

Therefore, if $u^0 \in D(A)$, then the function $u(\cdot) = U(\cdot)u^0$ is a (strong) solution of the Cauchy problem (0.1.10). Also it is not difficult to show that

$$A = \lim_{h \to 0+} \frac{U(h) - I}{h},$$

which means that the operator A is the generator of the semigroup U.

Now again let $u^0 \in D(A)$ and consider a (Schwartz) distribution $\mathcal{U}$ from $\mathcal{D}'$ defined by

$$\mathcal{U}(\phi) := \langle u(\cdot)H(\cdot), \phi \rangle = \int_0^{\infty} u(t)\phi(t)dt, \quad \phi \in \mathcal{D} \equiv \mathcal{D}(-\infty, T),$$

where u is defined by (0.1.2) and

$$H(t) = \begin{cases} 1, & t \geq 0 \\ 0, & t < 0 \end{cases}$$

is the Heaviside function. Then $\mathcal{U}$ satisfies the equation in (0.1.1) in the follwing sense

$$\mathcal{U}' - A\mathcal{U} = \delta \otimes u^0,$$

where δ is the Dirac distribution (delta function), or

$$\langle \mathcal{U}', \phi \rangle - \langle A\mathcal{U}, \phi \rangle = \langle \delta \otimes u^0, \phi \rangle, \quad \phi \in \mathcal{D},$$

The operator-valued distribution S defined by

$$
\begin{aligned}
S(\phi)u^0 &\equiv \langle S, \phi \rangle u^0 := \langle \mathcal{U}, \phi \rangle \\
&= \int_0^T u(t)\phi(t)dt = \int_0^T \phi(t) \sum_{k=1}^{\infty} e^{-\mu_k t} u_k^0 e_k dt
\end{aligned}
$$

is a distribution semigroup.

0.2 The reversed Cauchy problem for the Heat equation

In the same setting as above, we consider the ill-posed Cauchy problem

$$
\begin{aligned}
&\frac{\partial u(x,t)}{\partial t} + \frac{\partial^2 u(x,t)}{\partial x^2} = 0, \quad t \in [0,T], \; x \in \Omega, \\
&u(0,t) = u(1,t) = 0, \\
&u(x,0) = u^0(x),
\end{aligned}
$$

which can be written in the abstract form

$$u'(t) = Bu(t), \quad t \in [0,T], \; u(0) = u^0 \qquad (0.2.11)$$

with $B = -A$. Formally, we can write the solution of (0.2.11) at $t = T$ as

$$u(T) = U(T)u^0 = \sum_{k=1}^{\infty} e^{\mu_k T} u_k^0 e_k,$$

where operators $U(t)$, $t \geq 0$, are in general unbounded, and therefore they do not form a C_0-semigroup. A solution to problem (0.2.11) can be found using the quasi-reversibility method, which consists of solving the following well-posed Cauchy problem

$$
\begin{aligned}
&u_\varepsilon'(t) = (B - \varepsilon B^2)u_\varepsilon(t), \quad t \in [0,T], \; \varepsilon > 0, \qquad (0.2.12) \\
&u_\varepsilon(0) = u^0,
\end{aligned}
$$

and then showing that $u_\varepsilon(T) \longrightarrow u(T)$ in some sense. The unique solution of (0.2.12) is given by

$$u_\varepsilon(T) = U_\varepsilon(T)u^0 = \sum_{k=1}^{\infty} e^{(\mu_k - \varepsilon\mu_k^2)T} u_k^0 e_k,$$

and here U_ε is a C-regularized semigroup with the operator C defined by

$$Cv = \sum_{k=1}^{\infty} e^{-\varepsilon\mu_k^2 T} v_k e_k, \quad v \in X.$$

Alternatively, one can use the auxiliary boundary conditions method, which consists of solving the well-posed boundary value problem

$$u_\varepsilon'(t) = Bu_\varepsilon(t), \quad t \in [0, T+\tau], \ \varepsilon > 0, \ \tau > 0,$$
$$u_\varepsilon(0) + \varepsilon u_\varepsilon(T+\tau) = u^0,$$

The solution to this problem is given by

$$u_\varepsilon(T) = U_\varepsilon(T)u^0 = \sum_{k=1}^{\infty} \frac{e^{\mu_k T}}{1 + \varepsilon e^{\mu_k(T+\tau)}} u_k^0 e_k,$$

where again U_ε is a C-regularized semigroup with the operator C defined by

$$Cv = \sum_{k=1}^{\infty} \frac{1}{1 + \varepsilon e^{\mu_k(T+\tau)}} v_k e_k, \quad v \in X.$$

0.3 Wave equation

Consider the problem

$$\frac{\partial^2 u(x,t)}{\partial t^2} - \frac{\partial^2 u(x,t)}{\partial x^2} = 0, \quad t \in [0,T], \ x \in \Omega = (0,1),$$

$$\text{(0.3.13)}$$

$$u(0,t) = u(1,t) = 0,$$
$$u(x,0) = u^0(x), \quad \frac{\partial u}{\partial t}(x,0) = u^1(x),$$

in $L^2(\Omega)$. Let

$$u^0 = \sum_{k=1}^{\infty} u_k^0 e_k, \quad u^1 = \sum_{k=1}^{\infty} u_k^1 e_k,$$

where $\{e_k\}_{k=1}^\infty$ is an orthonormal basis in $L^2(\Omega)$ consisting of eigenvectors $e_k = \sqrt{2}\sin k\pi x$, $k \in \mathbb{N}$, of the operator $A = \frac{d^2}{dx^2}$ with the domain

$$D(A) = H^2(\Omega) \cap H_0^1(\Omega) \subset L^2(\Omega).$$

We are looking for a $L^2(\Omega)$ solution of (0.3.13) by setting

$$u(t) = \sum_{k=1}^\infty u_k(t)e_k.$$

The unique solution of each problem

$$\frac{d^2 u_k}{dt^2} + \mu_k u_k = 0, \qquad k \in \mathbb{N}, \tag{0.3.14}$$
$$u_k(0) = u_k^0, \ u_k'(0) = u_k^1,$$

is

$$u_k(t) = u_k^0 \cos\left(\sqrt{\mu_k}t\right) + u_k^1 \frac{\sin\left(\sqrt{\mu_k}t\right)}{\sqrt{\mu_k}}, \qquad k \in \mathbb{N}. \tag{0.3.15}$$

Thus, we can formally write

$$
\begin{aligned}
u(t) &= \sum_{k=1}^\infty u_k(t)e_k \tag{0.3.16}\\
&= \sum_{k=1}^\infty \cos\left(\sqrt{\mu_k}t\right) u_k^0 e_k + \sum_{k=1}^\infty \frac{\sin\left(\sqrt{\mu_k}t\right)}{\sqrt{\mu_k}} u_k^1 e_k.
\end{aligned}
$$

Now we investigate the convergence of the series in (0.3.16).

Case 1. $u^0 \in L^2(\Omega)$, $u^1 \in H^{-1}(\Omega)$.

Here $H^{-1}(\Omega) = \mathcal{L}(H_0^1(\Omega), \mathbb{R})$ is the dual space of $H_0^1(\Omega)$, and we have the following continuous injections

$$H_0^1(\Omega) \hookrightarrow L^2(\Omega) \hookrightarrow H^{-1}(\Omega).$$

Recall that $u^0 \in L^2(\Omega)$ and $u^1 \in H^{-1}(\Omega)$ if and only if

$$\sum_{k=1}^\infty |u_k^0|^2 < \infty \quad \text{and} \quad \sum_{k=1}^\infty \frac{|u_k^1|^2}{\mu_k} < \infty,$$

respectively. Under these assumptions, the sequence

$$U_m(\cdot) = \sum_{k=1}^m u_k(\cdot)e_k$$

is a Cauchy sequence in $C\{[0,T], L^2(\Omega)\}$ since

$$\sup_{t\in[0,T]} |u_k(t)|^2 \le 2\left(|u_k^0|^2 + \frac{|u_k^1|^2}{\mu_k}\right), \qquad k \in \mathbb{N}.$$

Further, since

$$u_k'(t) = -\sqrt{\mu_k}u_k^0 \sin\left(\sqrt{\mu_k}t\right) + u_k^1 \cos\left(\sqrt{\mu_k}t\right),$$

we have

$$\frac{1}{\mu_k}\sup_{t\in[0,T]} |u_k'(t)|^2 \le 2\left(|u_k^0|^2 + \frac{|u_k^1|^2}{\mu_k}\right), \qquad k \in \mathbb{N},$$

therefore, for all $m \in \mathbb{N}$ and all $p \ge 0$

$$\sup_{t\in[0,T]} \left\|U_{m+p}'(t) - U_m'(t)\right\|_{H^{-1}(\Omega)}^2 = \sup_{t\in[0,T]} \sum_{k=m+1}^{m+p} \frac{|u_k'(t)|^2}{\mu_k}$$

$$\le 2 \sum_{k=m+1}^{m+p} \left(|u_k^0|^2 + \frac{|u_k^1|^2}{\mu_k}\right),$$

and

$$U_m'(\cdot) = \sum_{k=1}^{m} u_k'(\cdot)e_k$$

is a Cauchy sequence in $C\{[0,T], H^{-1}(\Omega)\}$. Thus we have shown that U_m and U_m' converge to

$$u \in C\{[0,T], L^2(\Omega)\} \quad \text{and} \quad \frac{\partial u}{\partial t} \in C\{[0,T], H^{-1}(\Omega)\},$$

respectively. Also taking into account

$$U_m(0) = \sum_{k=1}^{m} u_k^0 e_k, \qquad U_m'(0) = \sum_{k=1}^{m} u_k^1 e_k,$$

we have

$$u(0) = u^0 \quad \text{and} \quad \frac{\partial u}{\partial t}(0) = u^1.$$

So, if $u^0 \in L^2(\Omega)$, $u^1 \in H^{-1}(\Omega)$, then the function u defined by (0.3.16), can satisfy the equation in (0.3.13) in some 'very' weak sense, which we will discuss after the next case.

Case 2. $u^0 \in H_0^1(\Omega)$, $u^1 \in L^2(\Omega)$.

In this case we have

$$\sum_{k=1}^{\infty} \mu_k |u_k^0|^2 < \infty, \qquad \sum_{k=1}^{\infty} |u_k^1|^2 < \infty,$$

and

$$\sup_{t\in[0,T]} \mu_k |u_k(t)|^2 \leq 2\left(\mu_k |u_k^0|^2 + |u_k^1|^2\right), \qquad k \in \mathbb{N},$$

$$\sup_{t\in[0,T]} |u_k'(t)|^2 \leq 2\left(\mu_k |u_k^0|^2 + |u_k^1|^2\right), \qquad k \in \mathbb{N},$$

which implies that

$$u \in C\{[0,T], H_0^1(\Omega)\} \quad \text{and} \quad \frac{\partial u}{\partial t} \in C\{[0,T], L^2(\Omega)\},$$

with $u(0) = u^0$, $\frac{\partial u}{\partial t}(0) = u^1$. Note that

$$\frac{\partial^2 u}{\partial t^2} \in C\{[0,T], H^{-1}(\Omega)\},$$

as

$$\frac{1}{\mu_k} \sup_{t\in[0,T]} |u_k''(t)|^2 \leq \mu_k \sup_{t\in[0,T]} |u_k(t)|^2$$

$$(0.3.17)$$

$$\leq 2\left(\mu_k |u_k^0|^2 + |u_k^1|^2\right), \qquad k \in \mathbb{N}.$$

In a fashion similar to the Heat equation (see (0.1.4)), from (0.3.14) we write

$$\left(U_m''(t), e_k\right) + \mathbf{a}\left(U_m(t), e_k\right) = 0, \quad 1 \leq k \leq m.$$

Now let $\phi \in \mathcal{D}(0,T)$, then for any fixed $k < m$, we have

$$-\int_0^T \left(U_m'(t), e_k\right)\phi'(t)dt + \int_0^T \mathbf{a}\left(U_m(t), e_k\right)\phi(t)dt = 0.$$

The uniform convergence of U_m to u in $C\{[0,T], H_0^1(\Omega)\}$ implies that

$$\lim_{m\to\infty} \int_0^T \mathbf{a}\left(U_m(t), e_k\right)\phi(t)dt = \int_0^T \mathbf{a}\left(u(t), e_k\right)\phi(t)dt,$$

and uniform convergence of U_m' to $\frac{\partial u}{\partial t}$ in $C\{[0,T], L^2(\Omega)\}$ gives

$$\lim_{m\to\infty} \int_0^T \left(U_m'(t), e_k\right)\phi'(t)dt = \int_0^T \left(\frac{\partial u}{\partial t}(t), e_k\right)\phi'(t)dt.$$

We conclude that u satisfies the equation in (0.3.13) in the following sense

$$-\int_0^T \left(\frac{\partial u}{\partial t}(t), v\right)\phi'(t)dt + \int_0^T \mathbf{a}(u(t), v)\phi(t)dt = 0,$$

for all $v \in H_0^1(\Omega)$ and $\phi \in \mathcal{D}(0, T)$. In other words, the equation

$$\frac{d}{dt}\left(\frac{\partial u}{\partial t}(\cdot), v\right) + \mathbf{a}(u(\cdot), v) = 0 \qquad (0.3.18)$$

is satisfied in $\mathcal{D}'(0, T)$ for all $v \in H_0^1(\Omega)$.

Now we return to the **Case 1.** From (0.3.14) we obtain

$$\int_0^T \left(U_m(t), e_k\right)\phi''(t)dt - \int_0^T \left(U_m(t), Ae_k\right)\phi(t)dt = 0,$$

for all fixed $k < m$ and for all $\phi \in \mathcal{D}(0, T)$, therefore

$$\int_0^T (u(t), e_k)\phi''(t)dt - \int_0^T (u(t), Ae_k)\phi(t)dt = 0,$$

for any $k \in \mathbb{N}$ and for all $\phi \in \mathcal{D}(0, T)$. Since $\{e_k\}_{k=1}^\infty$ forms a Hilbert basis in $D(A)$, we obtain

$$\int_0^T (u(t), v)\phi''(t)dt - \int_0^T (u(t), Av)\phi(t)dt = 0,$$

for all $v \in D(A)$, $\phi \in \mathcal{D}(0, T)$, or in other words, the equation

$$\frac{d^2}{dt^2}(u(\cdot), v) = (u(\cdot), Av) \qquad (0.3.19)$$

is satisfied in $\mathcal{D}'(0, T)$ for all $v \in D(A)$.

Case 3. $u^0 \in D(A)$, $u^1 \in H_0^1(\Omega)$.

Here we have

$$\sum_{k=1}^\infty \mu_k^2|u_k^0|^2 < \infty, \qquad \sum_{k=1}^\infty \mu_k|u_k^1|^2 < \infty.$$

Therefore (0.3.17) implies the uniform convergence of U_m'' to $\frac{\partial^2 u}{\partial t^2}$, and of AU_m to Au, in the space $C\{[0, T], L^2(\Omega)\}$. Thus, the function u defined by (0.3.16), satisfies the equation

$$\frac{\partial^2 u}{\partial t^2} = Au(t)$$

in $L^2(\Omega)$.

Now we describe these three cases using the semigroup terminology. Problem (0.3.13) can be written in the abstract form

$$u''(t) = Au(t), \quad t \in [0, T], \tag{0.3.20}$$
$$u(0) = u^0, \ u'(0) = u^1,$$

where

$$A = \frac{d^2}{dx^2} \quad \text{with} \ \ D(A) = H^2(\Omega) \cap H_0^1(\Omega) \subset L^2(\Omega) \equiv X$$

is an unbounded linear operator on $L^2(\Omega)$. Taking into account the discussion of **Case 1**, we define the following bounded linear operators on X:

$$\mathcal{C}(t)v := \sum_{k=1}^{\infty} \cos\left(\sqrt{\mu_k} t\right) v_k e_k, \quad \mathcal{S}(t)v := \sum_{k=1}^{\infty} \frac{\sin\left(\sqrt{\mu_k} t\right)}{\sqrt{\mu_k}} v_k e_k,$$

which are the *cos*- and *sin*-operator-functions corresponding to our problem. The unique solution of (0.3.13) is given by

$$u(t) = \mathcal{C}(t)u^0 + \mathcal{S}(t)u^1,$$

where the equation in (0.3.13) is satisfied as described in **Cases 1–3** according to the choice of the initial data.

Now we write the problem (0.3.20) in the form of the first order equation in a product space

$$w'(t) = \Phi w(t), \quad t \in [0, T], \quad w(0) = \begin{pmatrix} u^0 \\ u^1 \end{pmatrix}, \tag{0.3.21}$$

where

$$w(t) = \begin{pmatrix} u(t) \\ u'(t) \end{pmatrix} \in L^2(\Omega) \times L^2(\Omega),$$

$$\Phi = \begin{pmatrix} 0 & I \\ A & 0 \end{pmatrix}, \quad D(\Phi) = D(A) \times L^2(\Omega).$$

The solution operator of (0.3.21) can be formally written as

$$w(t) = U(t) \begin{pmatrix} u^0 \\ u^1 \end{pmatrix} := \begin{pmatrix} \mathcal{C}(t) & \mathcal{S}(t) \\ \mathcal{C}'(t) & \mathcal{S}'(t) \end{pmatrix} \begin{pmatrix} u^0 \\ u^1 \end{pmatrix}$$

$$= \begin{pmatrix} \mathcal{C}(t)u^0 + \mathcal{S}(t)u^1 \\ \mathcal{C}'(t)u^0 + \mathcal{S}'(t)u^1 \end{pmatrix}, \quad t \geq 0, \ u^0, u^1 \in L^2(\Omega),$$

and noting that for any $v \in L^2(\Omega)$, $\mathcal{S}'(t)v = \mathcal{C}(t)v$, we write U in the form

$$U(t) = \begin{pmatrix} \mathcal{C}(t) & \mathcal{S}(t) \\ \mathcal{C}'(t) & \mathcal{C}(t) \end{pmatrix}, \qquad t \geq 0.$$

From the definition of $\mathcal{C}$ it is clear that $\mathcal{C}$ is not necessarily differentiable in t on $L^2(\Omega)$, implying that the operators $U(t)$ are in general unbounded on $L^2(\Omega) \times L^2(\Omega)$, and therefore they do not form a C_0-semigroup on this space.

Let us consider a smaller space $H_0^1(\Omega) \times L^2(\Omega)$, and let

$$\Psi = \begin{pmatrix} 0 & I \\ A & 0 \end{pmatrix}, \qquad \text{with} \qquad D(\Psi) = D(A) \times H_0^1(\Omega).$$

From the calculations in **Case 2** we deduce that the operators $U(t)$ are bounded on $H_0^1(\Omega) \times L^2(\Omega)$, and they form a C_0-semigroup generated by Ψ. Further, if $u^0 \in D(A)$, $u^1 \in H_0^1(\Omega)$, then U gives the unique strong solution of (0.3.21), and if $u^0 \in H_0^1(\Omega)$, $u^1 \in L^2(\Omega)$, then U gives the unique solution of (0.3.21) in the variational sense.

On the initial space $L^2(\Omega) \times L^2(\Omega)$ we consider the integral of the operator-function $U(\cdot)$:

$$V(t) = \begin{pmatrix} \mathcal{S}(t) & \int_0^t \mathcal{S}(s)ds \\ \mathcal{C}(t) - I & \mathcal{S}(t) \end{pmatrix}, \qquad t \geq 0.$$

The operators $V(t)$ are bounded and strongly continuous on $L^2(\Omega) \times L^2(\Omega)$. Moreover, they form an integrated semigroup generated by Φ, which gives the unique solution of the integrated problem (0.3.21):

$$v(t) - t \begin{pmatrix} u^0 \\ u^1 \end{pmatrix} = \int_0^t \Phi v(s)ds, \qquad t \in [0, T], \qquad (0.3.22)$$

$$v = \begin{pmatrix} v_1 \\ v_2 \end{pmatrix}, \qquad u^0, u^1 \in L^2(\Omega),$$

which is equivalent to

$$v_1(t) - tu^0 = \int_0^t v_2(s)ds,$$

$$v_2(t) - tu^1 = \int_0^t Av_1(s)ds.$$

This solution is equivalent to the solution obtained in **Case 1** in the following sense: a function $v(\cdot)$ is a solution of (0.3.22) if and only if $u(\cdot) = v_2(\cdot) + u^0$ is a solution of (0.3.19) in $\mathcal{D}'(0,T)$ for all $v \in D(A)$.

Note that in this case $v_1(t) = \int_0^t u(s)ds$. Also if we take $\begin{pmatrix} u^0 \\ u^1 \end{pmatrix} \in D(\Phi^2)$, then $V(\cdot)\begin{pmatrix} u^0 \\ u^1 \end{pmatrix}$ is differentiable in t, and $V'(\cdot)\begin{pmatrix} u^0 \\ u^1 \end{pmatrix}$ is the unique solution of (0.3.21).

Finally, we note here that the operator Φ is the generator of the distribution semigroup $\mathcal{V}$ defined by

$$\mathcal{V}(\phi)v = -\int_0^\infty \phi'(t)V(t)v\,dt, \qquad \phi \in \mathcal{D}, \ v \in L^2(\Omega) \times L^2(\Omega),$$

which can be regarded as the distributional derivative of the integrated semigroup V.

Chapter 1

Semigroup Methods

1.1 C_0-semigroups

We begin with a brief summary of very basic notions and results from the theory of C_0-semigroups.

1.1.1 Definitions and main properties

Let X be a Banach space.

Definition 1.1.1 *A one-parameter family of bounded linear operators $\{U(t), \ t \geq 0\}$ is called a C_0-semigroup (or semigroup of class C_0) if the following conditions hold*

(U1) $U(t + h) = U(t)U(h), \ \ t, h \geq 0;$

(U2) $U(0) = I;$

(U3) $U(t)$ *is strongly continuous with respect to $t \geq 0$.*

Definition 1.1.2 *The operator defined by*

$$U'(0)x := \lim_{h \to 0} h^{-1}(U(h) - I)x,$$

$$D(U'(0)) = \left\{ x \in X \ \Big| \ \exists \lim_{h \to 0} \frac{U(h) - I}{h} x \right\},$$

is called the (infinitesimal) generator of the semigroup U.

The properties of C_0-semigroups are established in the following proposition.

Proposition 1.1.1 *Let $\{U(t),\ t \geq 0\}$ be a C_0-semigroup and A be its generator. Then*

1) $U(t)U(h) = U(h)U(t)$ *for all* $t, h \geq 0$;

2) *U is exponentially bounded:*
$\exists K > 0,\ \omega \in \mathbb{R} : \forall t \geq 0,$

$$\|U(t)\| \leq Ke^{\omega t}; \tag{1.1.1}$$

3) $\overline{D(A)} = X$ *and operator A is closed;*

4) $U'(t)x = U(t)Ax = AU(t)x$ *for all* $x \in D(A)$;

5) $\forall \lambda \in \mathbb{C} :\ \mathrm{Re}\,\lambda > \omega,\ \ \exists(\lambda I - A)^{-1} =: R_A(\lambda)$ *and*

$$R_A(\lambda)x = \int_0^\infty e^{-\lambda t}U(t)x\,dt,\quad x \in X. \tag{1.1.2}$$

Proof 1) The commutativity of $U(t)$ and $U(h)$, $t, h \geq 0$, follows from property (U1).

2) An arbitrary $t > 0$ can be written in the form $t = n + \tau$, where $n \in \mathbb{N},\ 0 \leq \tau < 1$. Then we have $U(t) = U^n(1)U(\tau)$, and

$$\|U(t)\| \leq \|U(1)\|^n K = Ke^{n\ln\|U(1)\|},$$

where $K = \sup_{0 \leq \tau \leq 1}\|U(\tau)\| < \infty$.

If $\ln\|U(1)\| \leq 0$, then $\|U(t)\| \leq K$, that is (1.1.1) holds for $\omega = 0$. If $\ln\|U(1)\| > 0$, then

$$\|U(t)\| \leq Ke^{(n+\tau)\ln\|U(1)\|)} = Ke^{t\ln\|U(1)\|} = Ke^{\omega t},$$

where $\omega = \ln\|U(1)\|$.

3) To prove that $\overline{D(A)} = X$, we consider the set

$$\mathcal{U} := \left\{ v_{a,b} = \int_a^b U(\tau)u\,d\tau,\ x \in X,\ b > a > 0 \right\}.$$

We show that $\mathcal{U} \subset D(A)$:

$$\begin{aligned}
h^{-1}\Big[U(h) - I\Big]v_{a,b} &= h^{-1}\int_a^b [U(h+\tau) - U(\tau)]x\,d\tau \\
&= h^{-1}\left[\int_{a-h}^{b-h} U(t)x\,dt - \int_a^b U(\tau)x\,d\tau\right] \\
&= h^{-1}\left[\int_{a-h}^a U(\tau)x\,d\tau - \int_{b-h}^b U(\tau)x\,d\tau\right] \\
&\to \Big[U(a) - U(b)\Big]x \text{ as } h \to 0.
\end{aligned}$$

Now we will show that $\overline{\mathcal{U}} = X$. Suppose that there exists $f \in X^*$ such that $f(\mathcal{U}) = 0, f \not\equiv 0$. Then

$$\forall b > a > 0, \qquad f(v_{a,b}) = \int_a^b f(U(\tau)x)d\tau = 0.$$

Hence

$$\forall x \in X, \quad f(U(\tau)x) = 0, \quad \tau > 0,$$

and $f(U(\tau)x) \rightarrow_{\tau \to 0} f(x) = 0$, that is $f \equiv 0$. This contradiction proves the equality $\overline{\mathcal{U}} = X$, therefore $\overline{D(A)} = X$.

To prove the closedness of the generator A, we take a sequence $\{x_n\} \subset D(A)$ such that $x_n \to x$ and $Ax_n \to v$. We show that Ax exists and is equal to v :

$$h^{-1} \int_0^h U(\tau)U'(0)x_n d\tau = h^{-1} \int_0^h \frac{dU(\tau)x_n}{d\tau}d\tau = h^{-1}\left[U(h) - I\right]x_n.$$

Let $n \to \infty$, then we have

$$h^{-1} \int_0^h U(\tau)v d\tau = h^{-1}\left[U(h) - I\right]x.$$

Let $h \to 0$, then the left-hand side converges to v, that is $x \in D(U'(0))$ and $v = U'(0)x = Ax$.

4) For $x \in D(A)$ it follows from Definition 1.1.2 that

$$\begin{aligned}
U(t)Ax &= U(t) \lim_{h \to 0} h^{-1}\left[U(h) - I\right]x \\
&= \lim_{h \to 0} h^{-1}\left[U(h) - I\right]U(t)x = AU(t)x, \\
U'(t)x &= \lim_{h \to 0} h^{-1}\left[U(t + h) - U(t)\right]x \\
&= \lim_{h \to 0} h^{-1}\left[U(t)U(h) - U(t)\right]x \\
&= U(t) \lim_{h \to 0} h^{-1}\left[U(h) - I\right]x = U(t)Ax = AU(t)x.
\end{aligned}$$

5) In view of estimate (1.1.1) the integral in the right-hand side of (1.1.2):

$$\int_0^\infty e^{-\lambda t}U(t)x dt,$$

exists for all $x \in X$ and $\lambda \in \mathbb{C}$ with $\text{Re}\,\lambda > \omega$. Integrating by parts and taking into account the fact that A is closed, we have for $x \in D(A)$

$$\begin{aligned}
\int_0^\infty e^{-\lambda t}U(t)Ax dt &= A \int_0^\infty e^{-\lambda t}U(t)x dt = \int_0^\infty e^{-\lambda t}U'(t)x dt \\
&= -x + \lambda \int_0^\infty e^{-\lambda t}U(t)x dt.
\end{aligned}$$

After extending the equality by continuity to $X = \overline{D(A)}$ we obtain

$$(\lambda I - A) \int_0^\infty e^{-\lambda t} U(t) x \, dt = x, \quad x \in X. \tag{1.1.3}$$

On the other hand,

$$\int_0^\infty e^{-\lambda t} U(t)(\lambda I - A) x \, dt = x, \quad x \in D(A). \tag{1.1.4}$$

Equalities (1.1.3), (1.1.4) imply the existence of a bounded on X operator $R_A(\lambda) = (\lambda I - A)^{-1}$ and relation (1.1.2). $\square$

Proposition 1.1.2 *Let U be a strongly continuous operator-function such that*

$$\exists K > 0, \ \omega \in \mathbb{R}: \quad \|U(t)\| \leq K e^{\omega t}, \quad t \geq 0.$$

Let

$$R(\lambda) := \int_0^\infty e^{-\lambda t} U(t) \, dt, \qquad \operatorname{Re} \lambda > \omega.$$

Then $R(\lambda)$ satisfies the resolvent identity

$$(\mu - \lambda) R(\lambda) R(\mu) = R(\lambda) - R(\mu), \quad \operatorname{Re} \lambda, \ \operatorname{Re} \mu > \omega, \tag{1.1.5}$$

if and only if U satisfies (U1).

Proof Let $\operatorname{Re} \lambda, \operatorname{Re} \mu > \omega$. The result follows from the uniqueness theorem for Laplace transforms given the following observations

$$R(\mu) R(\lambda) = \int_0^\infty e^{-\mu t} \int_0^\infty e^{-\lambda t} U(s) U(t) \, ds \, dt,$$

and

$$\frac{R(\lambda) - R(\mu)}{\mu - \lambda}$$

$$= \int_0^\infty e^{(\lambda - \mu)t} R(\lambda) \, dt - \int_0^\infty (\mu - \lambda)^{-1} e^{(\lambda - \mu)t} e^{-\lambda t} U(t) \, dt$$

$$= \int_0^\infty e^{(\lambda - \mu)t} \int_0^\infty e^{-\lambda s} U(s) \, ds \, dt - \int_0^\infty e^{(\lambda - \mu)t} \int_0^t e^{-\lambda s} U(s) \, ds \, dt$$

$$= \int_0^\infty e^{(\lambda - \mu)t} \int_t^\infty e^{-\lambda s} U(s) \, ds \, dt$$

$$= \int_0^\infty e^{-\mu t} \int_t^\infty e^{-\lambda(s-t)} U(s) \, ds \, dt$$

$$= \int_0^\infty e^{-\mu t} \int_0^\infty e^{-\lambda s} U(s + t) \, ds \, dt. \ \square$$

1.1.2 The Cauchy problem

Consider the Cauchy problem

$$u'(t) = Au(t), \quad t \geq 0, \ u(0) = x, \qquad \text{(CP)}$$

where A is a linear operator with $D(A) \subseteq X$.

Definition 1.1.3 *A function*

$$u(\cdot) \in C^1\{[0,\infty), X\} \cap C\{[0,\infty), D(A)\}$$

is called a solution of the Cauchy problem (CP) *if $u(t)$ satisfies the equation for $t \geq 0$ and the initial condition for $t = 0$.*

Definition 1.1.4 *The Cauchy problem* (CP) *is said to be uniformly well-posed on $E \subset X$, (where $\overline{E} = X$) if*

(a) *a solution exists for any $x \in E$;*

(b) *the solution is unique and for any $T > 0$ is uniformly stable for $t \in [0, T]$ with respect to the initial data.*

Remark 1.1.1 According to Hadamard's definition of correctness, Definition 1.1.4 includes the condition of stability of a solution with respect to the initial data. Now we show that the Cauchy problem (CP) has the following special property.

Lemma 1.1.1 *If for any $x \in D(A)$ there exists a unique solution of* (CP) *and the resolvent set $\rho(A)$ is not empty, then this solution is stable with respect to the initial data.*

Proof Let $u(t) = U(t)x$, $t \geq 0$, be a unique solution of the Cauchy problem with initial value $x \in D(A)$, where

$$U(\cdot) : [D(A)] \to C\{[0,T], [D(A)]\}$$

are the solution operators for any $T > 0$. Here $[D(A)]$ is the Banach space

$$\left\{ D(A), \ \|x\|_A = \|x\| + \|Ax\| \right\}.$$

We show that all $U(t)$ are closed. Suppose $x_n \to x$ in $[D(A)]$ and $U(t)x_n = u_n(t) \to v(t)$ in $C\{[0,T], [D(A)]\}$. Then $u'_n(t) = Au_n(t) \to Av(t)$ in X uniformly in t, therefore from

$$u_n(t) = x_n + \int_o^t u'_n(\tau)d\tau$$

we obtain

$$v(t) = x + \int_o^t Av(\tau)d\tau,$$

that is v is a solution of the Cauchy problem with initial value x, continuously differentiable on $[0, T]$. Since a solution is unique, $v(t) = U(t)x$, $t \geq 0$. Thus, operators $U(t)$ are everywhere defined on $[D(A)]$ and closed, therefore by the Banach theorem the operator-function $U(\cdot)$ is continuous and

$$\sup_{t\in[0,T]} \|u(t)\|_A \leq K\|x\|_A, \qquad (1.1.6)$$

which implies that the Cauchy problem is uniformly well-posed in the space $[D(A)]$.

Since the resolvent set $\rho(A)$ of operator A is not empty, consider $\lambda_0 \in \rho(A)$, $x \in D(A)$ and $y = R(\lambda_0)x$, then we have $y \in D(A^2)$ and

$$U(t)x = -U(t)Ay + \lambda_0 U(t)y = -AU(t)y + \lambda_0 U(t)y,$$

therefore

$$\|U(t)x\| \leq \|AU(t)y\| + |\lambda_0|\|U(t)y\| \leq K_0\|U(t)y\|_A.$$

From (1.1.6) we obtain

$$
\begin{aligned}
\|U(t)x\| &\leq K_1\|y\|_A = K_1\Big(\|y\| + \|Ay\|\Big) \\
&= K_1\Big(\|y\| + \|x - \lambda_0 y\|\Big) \\
&\leq K_2\Big(\|x\| + \|y\|\Big) \\
&\leq K_2\Big(\|x\| + \|R(\lambda_0)\|\,\|x\|\Big) \\
&\leq K\|x\|, \qquad t \in [0,T],
\end{aligned}
$$

that is (CP) is uniformly well-posed in the space X. $\square$

Theorem 1.1.1 *Suppose that A is a closed densely defined linear operator on X. Then the following statements are equivalent:*

(I) *the Cauchy problem* (CP) *is uniformly well-posed on $D(A)$;*

(II) *the operator A is the generator of a C_0-semigroup $\{U(t),\ t \geq 0\}$;*

(III) *Miyadera-Feller-Phillips-Hille-Yosida (MFPHY) condition holds for the resolvent of operator A: there exist $K > 0$, $\omega \in \mathbb{R}$ so that*

$$\left\|R_A^{(k)}(\lambda)\right\| \leq \frac{K\,k!}{(\mathrm{Re}\,\lambda - \omega)^{k+1}}, \qquad (1.1.7)$$

for all $\lambda \in \mathbb{C}$ with $\mathrm{Re}\,\lambda > \omega$, and all $k = 0, 1, \dots$.

In this case the solution of (CP) *has the form*

$$u(\cdot) = U(\cdot)x, \qquad x \in D(A).$$

Proof (I) $\implies$ (II). The solution $u(t)$, $t \geq 0$, of the Cauchy problem (CP) exists and is unique for any $x \in D(A)$. Denote it by $U(\cdot)x$. Since for any $T > 0$ the solution is uniformly stable with respect to $t \in [0, T]$, defined in such a way linear operators $U(t)$ are uniformly bounded with respect to $t \in [0, T]$ on $D(A)$. In view of the condition $\overline{D(A)} = X$, this implies that $U(t)$ can be extended by continuity to the whole space with preservation of the norm estimate. We show that the obtained family of linear bounded operators $\{U(t), \ t \geq 0\}$ is a C_0-semigroup.

Since $U(t)x$ satisfies the equation $U'(t)x = AU(t)x$ for all $x \in D(A)$ and $t \geq 0$, we have $U(t)x \in D(A)$ whenever $x \in D(A)$. For $x \in D(A)$ the functions $U(t+h)x$ and $U(t)U(h)x$ are solutions of (CP) with initial condition $U(h)x$. The uniqueness of the solution gives us the equality

$$\forall x \in D(A), \qquad U(t+h)x = U(t)U(h)x, \quad t, h \geq 0,$$

which can be extended to the whole X.

Thus (U1) is proved, and (U2) follows from the initial condition. Since $\|U(t)\|$ is uniformly bounded with respect to $t \in [0, T]$, and $U(t)x$ is continuous with respect to $t \geq 0$ on $D(A)$ ($\overline{D(A)} = X$), then the operator-function $U(\cdot)$ is strongly continuous with respect to $t \geq 0$, so (U3) holds true. Furthermore for $x \in D(A)$ we have

$$\lim_{h \to 0} h^{-1}\Big[U(h) - I\Big]x = U'(0)x = AU(0)x = Ax,$$

that is $D(A) \subseteq D(U'(0)) = D$, and $U'(0) = A$ on $D(A)$. To prove that $D \subset D(A)$, we show that the resolvent

$$R_{U'(0)}(\lambda) = \int_0^\infty e^{-\lambda t} U(t)dt, \quad \mathrm{Re}\,\lambda > \omega,$$

is equal to the resolvent of operator A. For $x \in D(A) \subset D$ we have

$$
\begin{aligned}
(\lambda I - A)R_{U'(0)}(\lambda)x &= (\lambda I - A)\int_0^\infty e^{-\lambda t}U(t)xdt \\
&= -\int_0^\infty e^{-\lambda t}AU(t)xdt + \lambda\int_0^\infty e^{-\lambda t}U(t)xdt \\
&= x.
\end{aligned}
$$

Since A is closed, this equality can be extended to the whole X. The operator $R_{U'(0)}(\lambda)$ maps X on D, therefore $D \subset D(A)$, and A is the generator of the C_0-semigroup $\{U(t), \ t \geq 0\}$.

(II) $\Longrightarrow$ (III). The estimates (1.1.7) follow from (1.1.2) and (1.1.1). In fact,

$$
\begin{aligned}
\|R_A(\lambda)\| &= \left\| \int_0^\infty e^{-\lambda t} U(t) dt \right\| \leq \int_0^\infty |e^{-\lambda t}| \, \|U(t)\| dt \\
&\leq K \int_0^\infty e^{-[\mathrm{Re}\,\lambda - \omega]t} dt = \frac{K}{\mathrm{Re}\,\lambda - \omega},
\end{aligned}
$$

$$
\begin{aligned}
\left\| \frac{d}{d\lambda} R_A(\lambda) \right\| &= \left\| \int_0^\infty -t e^{-\lambda t} U(t) dt \right\| \leq K \int_0^\infty t e^{-[\mathrm{Re}\,\lambda - \omega]t} dt \\
&= \frac{K \, 1!}{(\mathrm{Re}\,\lambda - \omega)^2}
\end{aligned}
$$

and so on

$$
\begin{aligned}
\left\| \frac{d^k}{d\lambda^k} R_A(\lambda) \right\| &= \left\| \int_0^\infty t^k e^{-\lambda t} U(t) dt \right\| \leq K \int_0^\infty t^k e^{-[\mathrm{Re}\,\lambda - \omega]t} dt \\
&= \frac{K \, k!}{(\mathrm{Re}\,\lambda - \omega)^{k+1}}.
\end{aligned}
$$

(III) $\Longrightarrow$ (I). First we construct the solution of (CP) for sufficiently smooth initial data. Noting that $\overline{D(A^3)} = X$, let $x \in D(A^3)$ and $\sigma > \max\{\omega, 0\}$. Define

$$
\hat{u}(t,x) = x + tAx + \frac{t^2}{2} A^2 x + \frac{1}{2\pi i} \int_{\sigma - i\infty}^{\sigma + i\infty} \lambda^{-3} e^{\lambda t} R_A(\lambda) A^3 x \, d\lambda, \quad t \geq 0. \quad (1.1.8)
$$

The integral in (1.1.8) vanishes for $t = 0$, consequently $\hat{u}(0,x) = x$. It is obvious that $\|\hat{u}(t,x)\| =_{t \to \infty} \mathcal{O}(e^{\sigma t})$. The possibility of differentiation under the integral sign implies the existence of a continuous derivative of $\hat{u}(t,x)$, $t \geq 0$. On the other hand, in view of the closedness of operator A, $\hat{u}(t,x) \in D(A)$ and

$$
\begin{aligned}
&\hat{u}'(t,x) - A\hat{u}(t,x) \\
&= -\frac{t^2}{2} A^3 x + \frac{1}{2\pi i} \int_{\sigma - i\infty}^{\sigma + i\infty} \lambda^{-3} e^{\lambda t} (\lambda I - A) R_A(\lambda) A^3 x \, d\lambda = 0,
\end{aligned}
$$

for $t \geq 0$. Hence, $\hat{u}(t,x)$, $t \geq 0$, $x \in D(A^3)$, is a solution of (CP). In the way similar to that in the proof of equality (1.1.2) in Proposition 1.1.1, we have

$$
\int_0^\infty e^{-\lambda t} \hat{u}(t,x) dt = R_A(\lambda) x. \quad (1.1.9)
$$

Using the Widder-Post inversion formula for Laplace transform, we obtain

$$
\hat{u}(t,x) = \lim_{n \to \infty} \frac{(-1)^n}{n!} \left(\frac{n}{t}\right)^{n+1} \frac{d^n}{d\lambda^n} R_A(\lambda) x \Bigg|_{\lambda = \frac{n}{t}}, \quad (1.1.10)
$$

therefore, taking into account inequalities (1.1.7) we have

$$\|\hat{u}(t,x)\| \leq K\|x\| \lim_{n\to\infty} \left(1 - \frac{\omega t}{n}\right)^{-(n+1)} = Ke^{\omega t}\|x\|, \qquad (1.1.11)$$

for $t \geq 0$. Denote

$$\hat{U}(t)x \equiv \hat{u}(t,x), \quad x \in D(A^3), \quad t \geq 0.$$

In view of (1.1.11) we can extend $\hat{U}(\cdot)x$ and equality (1.1.9) by continuity to the whole of X so that the extension $U(\cdot)$ is strongly continuous with respect to $t \geq 0$ and

$$\|U(t)\| \leq Ke^{\omega t}, \quad t \geq 0.$$

To complete the proof we show that any solution $u(\cdot)$ of (CP) can be represented in the form

$$u(\cdot) = U(\cdot)u(0). \qquad (1.1.12)$$

In fact, it follows from the definition of U that for $x \in D(A^3)$ we have the equality $R_A(\lambda)U(t)x = R_A(\lambda)\hat{U}(t)x = \hat{U}(t)R_A(\lambda)x$, $t \geq 0$, which can be extended to whole X. Therefore

$$U(t)Ax = AU(t)x, \quad x \in D(A), \ t \geq 0,$$

function $U(\cdot)x$ is differentiable for $x \in D(A)$, and for any solution $u(\cdot)$ of (CP) we have

$$\frac{d}{ds}U(t-s)u(s) = -AU(t-s)u(s) + U(t-s)Au(s) = 0, \quad 0 \leq s \leq t.$$

Hence

$$U(0)u(t) = U(t)u(0) = u(t), \quad t \geq 0. \ \square$$

Remark 1.1.2 Since we proved in Proposition 1.1.1 that the generator of a C_0-semigroup is densely defined, we could include the assumption of the density of $D(A)$ in statements (I) and (III) only.

Remark 1.1.3 Using the equality

$$\frac{d^k}{d\lambda^k}R_A(\lambda)x = (-1)^k k! R_A^{k+1}(\lambda)x, \quad x \in X, \qquad (1.1.13)$$

we can rewrite the MFPHY condition (1.1.7) in the following form:
$\exists K > 0, \ \omega \in \mathbb{R}$:

$$\|R_A^k(\lambda)\| \leq \frac{K}{(\operatorname{Re}\lambda - \omega)^k}, \qquad (1.1.14)$$

for all $\lambda \in \mathbb{C}$ with $\text{Re}\,\lambda > \omega$, and all $k = 1, 2 \ldots$.

Remark 1.1.4 The proof of the implication (III) $\implies$ (I) (via (II)) is given in Theorem 1.1.1 due to Fattorini [84] and is based on the inversion formula for Laplace transform. Now we outline alternative proofs of the implication (III) $\implies$ (II), which exploit some different ideas used in various generalizations of C_0-semigroups. We refer to these later on.

First, we note that the right-hand side of (1.1.10) (the inversion formula for Laplace transform) can be rewritten in the form

$$\lim_{n \to \infty} \left(\frac{n}{t}\right)^n R_A^n\left(\frac{n}{t}\right) = \lim_{n \to \infty} \left(I - \frac{n}{t}A\right)^{-n},$$

and (1.1.10) is an operator generalization of the formula

$$e^{ta} = \lim_{n \to \infty} \left(1 - \frac{n}{t}a\right)^{-n}$$

for the numerical exponential function.

Using this formula one can construct a semigroup U generated by an operator A as a strong limit of the following functions

$$U_n(t) = \left(I - \frac{n}{t}A\right)^{-n}, \quad t \geq 0, \ n = 1, 2, \ldots .$$

See [130] for one of the possible proofs.

Now we give another important alternative proof. Consider bounded operators

$$A_n = -\lambda_n I + \lambda_n^2 R_A(\lambda_n),$$

where $\lambda_n \in \mathbb{R}$ and $\lambda_n \to \infty$. We show that $\lim_{n \to \infty} A_n x = Ax$ for any $x \in D(A)$. Consider $x \in D(A)$, then

$$A_n x = \lambda_n R_A(\lambda_n) Ax.$$

Noting that $\lim_{n \to \infty} \lambda_n R_A(\lambda_n)x = x$ for any $x \in X$, we obtain

$$A_n x \to Ax, \quad x \in D(A).$$

Since operators A_n are bounded, we can define a semigroup generated by each operator A_n in the form of the following series

$$e^{tA_n} = e^{-\lambda_n t + \lambda_n^2 R_A(\lambda_n)t} = e^{-\lambda_n t} \sum_{k=0}^{\infty} \frac{(\lambda_n^2 t)^k R_A(\lambda_n)^k}{k!},$$

where

$$\|e^{tA_n}\| \leq e^{-\lambda_n t} \sum_{k=0}^{\infty} \frac{\lambda_n^{2k} t^k K}{k!(\lambda_n - \omega)^k} = K e^{-\lambda_n t} e^{\frac{\lambda_n^2 t}{\lambda_n - \omega}}$$

$$= K e^{\frac{t\lambda_n \omega}{\lambda_n - \omega}} \leq K e^{2\omega t}, \quad \lambda_n > 2\omega. \tag{1.1.15}$$

Consider the identity

$$e^{tA_n} - e^{tA_m} = \int_0^t e^{(t-s)A_n}(A_n - A_m)e^{sA_m}ds,$$

then, for $x \in D(A)$, we have

$$\|e^{tA_n}x - e^{tA_m}x\| \leq \int_0^t \|e^{(t-s)A_n}\| \, \|e^{sA_m}\| \, \|(A_n - A_m)x\|ds,$$

therefore, in view of (1.1.15), functions $e^{tA_n}x$ converge uniformly in $t \in [0,T]$. Taking into account density of $D(A)$ and uniform boundness of e^{tA_n} on $[0,T]$, we obtain that $e^{tA_n}x$ converge uniformly in $t \in [0,T]$ for any $x \in X$. We denote

$$U(t)x := \lim_{n\to\infty} e^{tA_n}x,$$

then U is obviously a C_0-semigroup with the estimate $\|U(t)\| \leq Ke^{2\omega t}$, which follows from the inequality (1.1.15) by a passage to the limit as $n \to \infty$.

Let us prove that the operator A is the generator of this semigroup. The identity

$$e^{tA_n}x - x = \int_0^t e^{sA_n}A_n xds, \quad x \in D(A),$$

implies that

$$U(t)x - x = \int_0^t U(s)Axds, \quad x \in D(A),$$

and hence $U'(0)x = Ax$, that is $U'(0) \supset A$, and

$$\left[\lambda I - U'(0)\right]R_A(\lambda) = \left[\lambda I - A\right]R_A(\lambda) = I.$$

In view of Proposition 1.1.1 the operator $U'(0)$ has a resolvent for $\lambda \in \mathbb{C}$ with $\mathrm{Re}\,\lambda > \omega$, therefore $R_A(\lambda) = R_{U'(0)}(\lambda)$ and $U'(0) = A$. $\square$

We also note that this construction of a C_0-semigroup implies that the condition

$\exists K > 0, \; \omega \in \mathbb{R}$:

$$\left\|R_A^{(k)}(\lambda)\right\| \leq \frac{K\,k!}{(\lambda - \omega)^{k+1}},$$

for all $\lambda > \omega$ and all $k = 0, 1, \dots$,

is equivalent to condition (1.1.7) and is also referred to as the MFPHY condition.

Remark 1.1.5 It follows from the equality (1.1.13) that the condition

$$\exists\, \omega \geq 0 : \quad \|R_A(\lambda)\| \leq \frac{1}{\operatorname{Re}\lambda - \omega}, \quad \operatorname{Re}\lambda > \omega$$

or

$$\exists\, \omega \geq 0 : \quad \|R_A(\lambda)\| \leq \frac{1}{\lambda - \omega}, \quad \lambda > \omega,$$

which is called the 'Hille-Yosida condition', is sufficient for condition (1.1.7) to hold with $K = 1$, and therefore for uniform well-posedness of (CP).

In general, the following sufficient condition for the unique solvability of (CP) holds.

Theorem 1.1.2 *Suppose that for a linear operator A the following condition holds*

$$\exists K > 0,\ \omega \geq 0 \ : \ \|R_A(\lambda)\| \leq K(1 + |\lambda|)^\gamma, \quad \operatorname{Re}\lambda > \omega, \qquad (1.1.16)$$

for some $\gamma \geq -1$. Then for any $x \in D(A^{[\gamma]+3})$ the function

$$\hat{J}(t) = \begin{cases} J(t) = \text{v.p.} \ \frac{1}{2\pi i} \int_{\sigma-i\infty}^{\sigma+i\infty} e^{\lambda t} R_A(\lambda) x\, d\lambda, & t > 0,\ \sigma > \omega \\ J(+0), & t = 0 \end{cases}$$

yields the unique solution of (CP).

Proof Any $x \in D(A^{[\gamma]+3})$ we can write in the form

$$x = R_A(\mu)^{[\gamma]+3} x_1,$$

where $x_1 = (\mu I - A)^{[\gamma]+3} x$. Then, using the resolvent identity (1.1.5), we obtain

$$
\begin{aligned}
R_A(\lambda)x &= \frac{\left[R_A(\lambda) - R_A(\mu)\right] R_A(\mu)^{[\gamma]+2}}{\mu - \lambda} x_1 \\
&= \frac{R_A(\lambda)}{(\mu - \lambda)^{[\gamma]+3}} x_1 - \frac{R_A(\mu)^{[\gamma]+3}}{\mu - \lambda} x_1 \\
&\quad - \frac{R_A(\mu)^{[\gamma]+2}}{(\mu - \lambda)^2} x_1 - \ldots - \frac{R_A(\mu)}{(\mu - \lambda)^{[\gamma]+3}} x_1
\end{aligned}
$$

for $\operatorname{Re}\lambda,\ \operatorname{Re}\mu > \omega,\ \lambda \neq \mu$. By Jordan's lemma we have

$$
\begin{aligned}
J(t)x &= \text{v.p.} \ \frac{1}{2\pi i} \int_{\sigma-i\infty}^{\sigma+i\infty} e^{\lambda t} R_A(\lambda) x\, d\lambda \\
&= \text{v.p.} \ \frac{1}{2\pi i} \int_{\sigma-i\infty}^{\sigma+i\infty} e^{\lambda t} \frac{R_A(\lambda)}{(\mu - \lambda)^{[\gamma]+3}} x_1\, d\lambda, \quad t > 0.
\end{aligned}
$$

It follows from (1.1.16) that the latter integral converges absolutely and uniformly in $t \geq 0$. After differentiating under the integral sign we obtain the integral

$$\text{v.p.} \quad \frac{1}{2\pi i} \int_{\sigma - i\infty}^{\sigma + i\infty} e^{\lambda t} \frac{\lambda R_A(\lambda)}{(\mu - \lambda)^{[\gamma]+3}} x_1 d\lambda,$$

which also converges absolutely and uniformly in $t \geq 0$, and therefore is equal to $J'(t)$ for $t > 0$. It is not difficult to check that $J(t)$ satisfies the equation in (CP) for $t > 0$. The existence of $\lim_{t \to 0} J'(t) := \hat{J}'(0)$ and the closedness of A imply that $\hat{J}'(0) = A\hat{J}(0)$. By the Cauchy theorem we have

$$\hat{J}(0) = R_A(\mu)^{[\gamma]+3} x_1 = x,$$

i.e., $\hat{J}(t)$ is a solution of the Cauchy problem (CP).

Now we show the uniqueness of the solution. Suppose that $u(\cdot)$ is a solution of (CP). Since $u(\cdot)$ is continuously differentiable in $t \geq 0$, $\int_0^t e^{-\lambda \tau} u(\tau) d\tau$ can be integrated by parts

$$\int_0^t e^{-\lambda \tau} u(\tau) d\tau = \lambda^{-1} \left[u(0) - u(t)e^{-\lambda t} \right] + \lambda^{-1} \int_0^t e^{-\lambda \tau} A u(\tau) d\tau.$$

Suppose $u(0) = 0$, then taking into account closedness of A, we have

$$(A - \lambda I) \int_0^t e^{-\lambda \tau} u(\tau) d\tau = u(t)e^{-\lambda t},$$

therefore

$$-\int_0^t e^{\lambda(t-\tau)} u(\tau) d\tau = R_A(\lambda) u(t),$$

which, together with the condition (1.1.16), implies that $u(\tau) = 0$ in $[0, t]$, i.e. the solution is unique. $\square$

Remark 1.1.6 The following result guarantees the uniqueness of the solution of (CP) under weaker than (1.1.16) conditions on operator A. It can be proved similar to the uniqueness in the theorem above.

Lyubich's Uniqueness Theorem *Suppose that a linear operator A has a resolvent for all $\lambda > \omega$ and*

$$\limsup_{\lambda \to +\infty} \frac{\ln \|R_A(\lambda)\|}{\lambda} = 0,$$

then the Cauchy problem (CP) has at most one weak solution (that is a solution that is defined and continuous for $t \geq 0$ and continuously differentiable for $t > 0$).

In Section 1.2, using the technique of integrated semigroups, we show that conditions of the type (1.1.16) on the resolvent of A, are necessary and sufficient for the existence of the solution and its stability with respect to small variations of initial data measured in the graph-norm of A^n, where n depends on γ.

Remark 1.1.7 Another important class of operators closely related to a well-posed (CP) is the generators of analytic semigroups. Let A be a densely defined operator, $0 \leq \varphi \leq \frac{\pi}{2}$. Then A is the generator of a semigroup analytic in

$$\left\{ \zeta \in \mathbb{C} \;\middle|\; |\arg \zeta| \leq \varphi, \; \zeta \neq 0 \right\}$$

and strongly continuous in

$$\left\{ \zeta \in \mathbb{C} \;\middle|\; |\arg \zeta| \leq \varphi \right\}$$

if and only if for every $0 < \varphi' < \varphi$ there exists a real number $\alpha = \alpha(\varphi')$ such that

$$\|R_A(\lambda)\| \leq \frac{K}{|\lambda - \alpha|} \tag{1.1.17}$$

for λ from the region

$$\left\{ \lambda \in \mathbb{C} \;\middle|\; |\arg(\lambda - \alpha)| \leq \varphi' + \frac{\pi}{2}, \; \lambda \neq \alpha \right\}$$

with $K = K_{\varphi'}$.

We note that the estimate (1.1.17) need be assumed only in the half-plane $\operatorname{Re}\lambda > \alpha$. (For proofs see [84] Section 4.2)

Remark 1.1.8 One of the remarkable advantages of using semigroup methods for treating differential equations in applications is the presence of the following perturbation result.

Let A (closed densely defined linear operator) be the generator of a C_0-semigroup U and let B be a closed densely defined operator such that $D(B) \supset D(A)$ and

$$\|BU(t)x\| \leq \alpha(t)\|x\|, \quad x \in D(A), \; t > 0,$$

where α is a finite, measurable function from $L^1(0,1)$. Then the operator $A + B$ with domain $D(A + B) = D(A)$ is also the generator of a C_0-semigroup.

In particular, if B is everywhere defined and bounded, then $A + B$ generates a C_0-semigroup U_1 with

$$\|U_1(t)\| \leq Ke^{(\omega + \|B\|)t}, \quad t \geq 0,$$

given that

$$\|U(t)\| \le Ke^{\omega t}, \quad t \ge 0.$$

For proofs see [84] Chapter 5 and [130] Chapter 9, where one can also find perturbation results for m-dissipative operators, essentially self-adjoint operators and operators generating analytic semigroups.

1.1.3 Examples

Example 1.1.1

In Chapter 0 we used the Fourier method to construct various semi-groups related to the Heat and Wave equations. Now we use a very simple first-order equation to illustrate the semigroup methods.

Consider the Cauchy problem

$$\frac{\partial u(x,t)}{\partial t} + \frac{\partial u(x,t)}{\partial x} = 0, \quad t \ge 0, \ x \in \mathbb{R}, \qquad (1.1.18)$$
$$u(x,0) = f(x)$$

on the space $X = L^2(\mathbb{R})$. We can write it in the abstract form

$$u'(t) = Au(t), \quad t \ge 0, \ u(0) = f,$$

where $A = -d/dx$ with the domain

$$D(A) = \left\{ u \in L^2(\mathbb{R}) \ \middle| \ u' \in L^2(\mathbb{R}) \right\},$$

To find the resolvent of A, we solve the equation

$$(\lambda I - A)g = \lambda g + g' = f, \quad g \in D(A), \qquad (1.1.19)$$

assuming that $f \in X$ is given. If $\lambda > 0$ then the solution is

$$g(x) = (R_A(\lambda)f)(x) = \int_{-\infty}^{x} e^{-\lambda(x-s)} f(s)ds, \ x \in \mathbb{R}.$$

Using Fourier transforms, it is not difficult to verify that the Hille-Yosida condition:

$$\|R_A(\lambda)\| \le \frac{1}{\lambda},$$

holds for $\lambda > 0$. Hence (1.1.18) is uniformly well-posed on $D(A)$. Furthermore, A is the generator of a C_0-semigroup defined by

$$(U(t)f)(x) := f(x - t), \quad x \in \mathbb{R}, \ t \ge 0,$$

and for any $f \in D(A)$, the function

$$u(x,t) = (U(t)f)(x), \quad t \geq 0, \ x \in \mathbb{R}$$

is the unique solution of (1.1.18), which is stable with respect to f.

Now we consider the Cauchy problem (1.1.18) on the space $X = L^2[0, \infty)$. In this case,

$$D(A) = \left\{ u \in L^2[0, \infty) \ \middle| \ u' \in L^2[0, \infty), \ u(0) = 0 \right\},$$

and

$$(R_A(\lambda)f)(x) = \int_0^x e^{-\lambda(x-s)} f(s)ds, \quad \lambda > 0, \ x \in [0, \infty).$$

The C_0-semigroup generated by A is defined by

$$(U(t)f)(x) = \begin{cases} f(x-t), & x \geq t \\ 0, & 0 \leq x \leq t \end{cases}.$$

Finally, if $X = L^2(-\infty, 0]$ and

$$D(A) = \left\{ u \in L^2(-\infty, 0] \ \middle| \ u' \in L^2(-\infty, 0], \ u(0) = 0 \right\},$$

then all $\lambda > 0$ do not belong to the resolvent set of A. In this case, (1.1.18) is solvable only for $f \equiv 0$.

Note that these results agree with the well-known geometric interpretation: the solution is constant on the characteristics $x = t + C$, $C \in \mathbb{R}$.

Example 1.1.2 (A class of operators generating C_0-semigroups)

Let

$$X = \left\{ L^p(\mathbb{R}) \times L^p(\mathbb{R}), \ \|u\| = \|u_1\|_{L^p} + \|u_2\|_{L^p} \right\}, \quad \text{where} \quad u = \begin{pmatrix} u_1 \\ u_2 \end{pmatrix}.$$

Consider the operator A defined by

$$Au = \begin{pmatrix} -g & -f \\ 0 & -g \end{pmatrix} u$$

with

$$D(A) = \left\{ \begin{pmatrix} u_1 \\ u_2 \end{pmatrix} \in X \ \middle| \ gu_1 + fu_2 \in L^p(\mathbb{R}), \ gu_2 \in L^p(\mathbb{R}) \right\},$$

where $g(x) = 1 + |x|$, $f(x) = |x|^\gamma$, $\gamma > 0$.

We show that if $\gamma \in (0,1]$, then A generates a C_0-semigroup on X. For the formally defined operator-function

$$U(t) := e^{At} = I + tA + \frac{t^2 A^2}{2!} + \dots$$

we can write

$$U(t)u = e^{-tg} \begin{pmatrix} 1 & -tf \\ 0 & 1 \end{pmatrix} u$$

and

$$
\begin{aligned}
\|U(t)\| &= \max_{x \in R}\left[(1 + t|x|^\gamma)e^{-t(1+|x|)} \right] \\
&= (1 + \gamma^\gamma t^{1-\gamma})e^{-t\left(1+\frac{\gamma}{t}\right)} \underset{t \to 0}{=} \mathcal{O}(t^{1-\gamma}).
\end{aligned}
$$

Let $\gamma \in (0,1]$, then $U(t)$ is bounded as $t \to 0$. Now we show that the operator $(\lambda I - A)^{-1}$, $\lambda > 0$, is the resolvent of A and we verify the MFPHY condition for $R_A(\lambda)$. Since

$$\left[(\lambda I - A)^{-1} \right]^k u = \frac{1}{(\lambda + g)^{2k}} \begin{pmatrix} (\lambda + g)^k & -k(\lambda + g)^{k-1}f \\ 0 & (\lambda + g)^k \end{pmatrix} u$$

for $k = 1, 2, \dots$, then

$$
\begin{aligned}
\|(\lambda I - A)^{-k}u\| \leq \ &\|(\lambda + g)^{-k}u_1\|_{L^p} + \| -k(\lambda + g)^{-(k+1)}fu_2\|_{L^p} \\
&+ \|(\lambda + g)^{-k}u_2\|_{L^p}.
\end{aligned}
$$

If $\lambda > 0$ then we have the following estimates

$$
\begin{aligned}
\|(\lambda + g)^{-k}u_i\|_{L^p} &= \left(\int_0^\infty |\lambda|^{-kp}\left|1 + \frac{g}{\lambda}\right|^{-kp}|u_i|^p dx \right)^{1/p} \\
&\leq \frac{1}{\lambda^k}\|u_i\|_{L^p}, \quad i = 1, 2,
\end{aligned}
$$

and

$$
\begin{aligned}
\|k(\lambda + g)^{-k+1}fu_2\|_{L^p} &= \left(\int_{-\infty}^\infty \left| \frac{kfu_2}{(\lambda + g)^{k+1}} \right|^p dx \right)^{1/p} \\
&\leq \frac{1}{|\lambda|^k}\left(\int_{-\infty}^\infty \frac{k^p}{|1 + \frac{g}{\lambda}|^{kp}}\left|\frac{f}{\lambda + g}\right|^p |u_2|^p dx \right)^{1/p} \\
&\leq \frac{1}{|\lambda|^k}\left(\int_{-\infty}^\infty \frac{|f|^p |u_2|^p}{|g|^p}dx \right)^{1/p} \leq \frac{1}{\lambda^k}\|u_2\|_{L^p}.
\end{aligned}
$$

Therefore $(\lambda I - A)^{-1} = R_A(\lambda)$ and

$$\|R_A^k(\lambda)u\| \leq \frac{1}{|\lambda|^k}\|u_1\|_{L^p} + \frac{K+1}{|\lambda|^k}\|u_2\|_{L^p} \leq \frac{K_1}{\lambda^k}\|u\|$$

for any $\gamma \in (0,1]$ and $\lambda > 0$. Thus, if $\gamma \in (0,1]$, then A is the generator of a C_0-semigroup U. In general, we have

$$(\lambda I - A)^{-1}u = \frac{1}{(\lambda+g)^2}\begin{pmatrix} (\lambda+g) & -f \\ 0 & (\lambda+g) \end{pmatrix}u.$$

If $1 < \gamma \leq 2$, then the operator $(\lambda I - A)^{-1}$, $\lambda > 0$, is bounded and therefore

$$\|(\lambda I - A)^{-1}\| = \|R_A(\lambda)\| \leq K, \quad \lambda > 0,$$

but the resolvent $R_A(\lambda)$ does not satisfy the MFPHY condition. If $\gamma > 2$, then the operator $(\lambda I - A)^{-1}$ is unbounded.

1.2 Integrated semigroups

In Section 1.1 it was proved that the Cauchy problem

$$u'(t) = Au(t), \quad t \geq 0, \ u(0) = x, \tag{CP}$$

with A being the generator of a C_0-semigroup, is uniformly well-posed. Now we begin to consider families of bounded operators connected with problems (CP) that are not uniformly well-posed. The first of these families are n-times integrated semigroups. They are related to (CP) well-posed on $D(A^{n+1})$. If A generates a C_0-semigroup U, then the n-times integrated semigroup generated by A is the n-th order primitive of U. In contrast to C_0-semigroups, integrated semigroups may be not exponentially bounded, may be locally defined, and their generators may be not densely defined. In this section we consider nondegenerate exponentially bounded and local integrated semigroups, and their connection with the Cauchy problem.

1.2.1 Exponentially bounded integrated semigroups

Definition 1.2.1 *Let $n \in \mathbb{N}$. A one-parameter family of bounded linear operators $\{V(t), \ t \geq 0\}$ is called an exponentially bounded n-times integrated semigroup if the following conditions hold*

(V1) $V(t)V(s) =$

$$\frac{1}{(n-1)!}\int_0^s \left[(s-r)^{n-1}V(t+r) - (t+s-r)^{n-1}V(r)\right]dr,$$

for $s, t \geq 0$;

(V2) $V(t)$ *is strongly continuous with respect to* $t \geq 0$;

(V3) $\exists K > 0, \omega \in \mathbb{R}:$ $\|V(t)\| \leq Ke^{\omega t}, t \geq 0$.

Semigroup V is called nondegenerate if

(V4) *if* $V(t)x = 0$ *for all* $t \geq 0$, *then* $x = 0$.

A C_0-semigroup is called 0-times integrated semigroup.

It follows from (V1), (V4) that $V(0) = 0$.

Suppose U is a C_0-semigroup. Consider the integrals

$$
\begin{aligned}
V_1(t) &= \int_0^t U(s)\,ds, \quad \dots \\
V_n(t) &= \int_0^t V_{n-1}(s)ds = \int_0^t \frac{(t-s)^{n-1}}{(n-1)!} U(s)ds, \quad t \geq 0.
\end{aligned}
$$

For the generator A we have

$$
\begin{aligned}
\lambda^{-1} R_A(\lambda) &= \int_0^\infty e^{-\lambda t} V_1(t)dt, \quad \dots \\
\lambda^{-n} R_A(\lambda) &= \int_0^\infty e^{-\lambda t} V_n(t)dt, \quad \operatorname{Re}\lambda > \omega
\end{aligned}
$$

or

$$
R_A(\lambda) = \int_0^\infty \lambda^n e^{-\lambda t} V_n(t)dt.
$$

We show that the operator $R_A(\lambda)$, which is defined by this equality, satisfies the resolvent identity if and only if V_n satisfies (V1).

Proposition 1.2.1 *Let $n \in \mathbb{N}$ and V be a strongly continuous operator-function such that*

$$
\exists K > 0, \omega \in \mathbb{R}: \quad \|V(t)\| \leq Ke^{\omega t}, \quad t \geq 0.
$$

Let

$$
R(\lambda) := \int_0^\infty \lambda^n e^{-\lambda t} V(t)dt, \qquad \operatorname{Re}\lambda > \omega.
$$

Then $R(\lambda)$ satisfies the resolvent identity

$$
(\mu - \lambda)R(\lambda)R(\mu) = R(\lambda) - R(\mu)
$$

for $\lambda, \mu \in \mathbb{C}$ with $\operatorname{Re}\lambda, \operatorname{Re}\mu > \omega$, if and only if V satisfies (V1).

Proof Let $\operatorname{Re}\mu > \operatorname{Re}\lambda > \omega$. From the resolvent identity we have

$$\lambda^{-n}R(\lambda)\mu^{-n}R(\mu) \;=\; \frac{R(\lambda) - R(\mu)}{(\lambda - \mu)\lambda^n\mu^n} \tag{1.2.1}$$

$$\equiv \frac{\lambda^{-n}R(\lambda) - \mu^{-n}R(\mu)}{(\lambda - \mu)\mu^n} + \frac{R(\mu)(\mu^{-n} - \lambda^{-n})}{(\lambda - \mu)\mu^n}.$$

The result follows from the uniqueness theorem for Laplace transforms and from the following observations:

$$\lambda^{-n}R(\lambda)\mu^{-n}R(\mu) = \int_0^\infty e^{-\lambda t}\int_0^\infty e^{-\mu s}V(t)V(s)dsdt$$

and

$$\frac{\lambda^{-n}R(\lambda) - \mu^{-n}R(\mu)}{\lambda - \mu}$$

$$= -\int_0^\infty e^{(\mu-\lambda)t}\mu^{-n}R(\mu)dt + \int_0^\infty (\mu - \lambda)^{-1}e^{-\lambda t}V(t)dt$$

$$= \int_0^\infty e^{(\mu-\lambda)t}\int_0^\infty e^{-\mu s}V(s)dsdt - \int_0^\infty e^{(\mu-\lambda)t}\int_0^t e^{-\mu s}V(s)dsdt$$

$$= \int_0^\infty e^{(\mu-\lambda)t}\int_t^\infty e^{-\mu s}V(s)dsdt$$

$$= \int_0^\infty e^{-\lambda t}\int_t^\infty e^{-\mu(s-t)}V(s)dsdt$$

$$= \int_0^\infty e^{-\lambda t}\int_0^\infty e^{-\mu s}V(t + s)dsdt\ .$$

Integrating n times by parts we obtain the first term on the right-hand side of (1.2.1):

$$\frac{\lambda^{-n}R(\lambda) - \mu^{-n}R(\mu)}{(\lambda - \mu)\mu^n}$$

$$= -\int_0^\infty e^{-\lambda t}\int_0^\infty e^{-\mu s}\int_0^s \frac{1}{(n-1)!}(s-r)^{n-1}V(r+t)drdsdt\ .$$

The second term:

$$\frac{R(\mu)(\mu^{-n} - \lambda^{-n})}{(\lambda - \mu)\mu^n}$$

$$= \left(\sum_{k=0}^{n-1}\frac{1}{\lambda^{k+1}\mu^{n-k}}\right)\frac{R(\mu)}{\mu^n} = \sum_{k=0}^{n-1}\lambda^{-(k+1)}\int_0^\infty \mu^{k-n}e^{-\mu s}V(s)ds$$

$$= \sum_{k=0}^{n-1}\lambda^{-(k+1)}\int_0^\infty e^{-\mu s}\int_0^s \frac{(s-r)^{n-k-1}}{(n-k-1)!}V(r)drds$$

$$= \sum_{k=0}^{n-1} \int_0^\infty \frac{e^{-\lambda t} t^k}{k!} dt \int_0^\infty e^{-\mu s} \int_0^s \frac{(s-r)^{n-k-1}}{(n-k-1)!} V(r) dr ds$$

$$= \int_0^\infty e^{-\lambda t} \int_0^\infty e^{-\mu s} \int_0^s \frac{1}{(n-1)!} \times$$

$$\left[\sum_{k=0}^{n-1} \binom{n-1}{k} (s-r)^{n-k-1} t^k \right] V(r) dr ds$$

$$= \int_0^\infty e^{-\lambda t} \int_0^\infty e^{-\mu s} \frac{1}{(n-1)!} \int_0^s (t+s-r)^{n-1} V(r) dr ds dt . \quad \Box$$

If the semigroup $\{V(t), \ t \geq 0\}$ is nondegenerate, then the operator $R(\lambda)$ is invertible. It follows from the resolvent identity that $\lambda I - R(\lambda)^{-1}$ is independent of λ, that is there exists a unique operator A such that $R(\lambda)^{-1} = (\lambda I - A)$ for $\mathrm{Re}\,\lambda > \omega$. This operator $A := \lambda I - R(\lambda)^{-1}$ is called the generator of n-times integrated semigroup $\{V(t), \ t \geq 0\}$.

It follows from Proposition 1.1.2 that for a C_0-semigroup this definition of generator coincides with the generally accepted definition by means of the semigroup's derivative at zero.

Now we clarify the connection between integrated semigroups and the Cauchy problem (CP).

Proposition 1.2.2 *Let A be the generator of n-times integrated semigroup $\{V(t), \ t \geq 0\}$ (where $n \in \mathbb{N} \cup \{0\}$). Then*

1) *for $x \in D(A)$, $t \geq 0$*

$$V(t)x \in D(A), \quad AV(t)x = V(t)Ax,$$

and

$$V(t)x = \frac{t^n}{n!}x + \int_0^t V(s)Ax \ ds; \tag{1.2.2}$$

2) *for $x \in \overline{D(A)}$, $t \geq 0$*

$$\int_0^t V(s)x ds \in D(A),$$

and

$$A \int_0^t V(s)x ds = V(t)x - \frac{t^n}{n!}x; \tag{1.2.3}$$

3) *for $x \in D(A^n)$, $n \in \mathbb{N}$*

$$V^{(n)}(t)x = V(t)A^n x + \sum_{k=0}^{n-1} \frac{t^k}{k!} A^k x; \tag{1.2.4}$$

4) *for* $x \in D(A^{n+1})$

$$\frac{d}{dt}V^{(n)}(t)x = AV^{(n)}(t)x = V^{(n)}Ax. \qquad (1.2.5)$$

Proof Let $\mu \in \rho(A)$, then for $\operatorname{Re}\lambda > \omega$, $x \in X$ we have

$$
\begin{aligned}
\int_0^\infty e^{-\lambda t}V(t)R_A(\mu)x\,dt &= \lambda^{-n}R_A(\lambda)R_A(\mu)x = \lambda^{-n}R_A(\mu)R_A(\lambda)x \\
&= \int_0^\infty e^{-\lambda t}R_A(\mu)V(t)x\,dt.
\end{aligned}
$$

Hence $R_A(\mu)V(t) = V(t)R_A(\mu)$, that implies $AV(t) = V(t)A$, $x \in D(A)$.
Let $x \in D(A)$, then for $\operatorname{Re}\lambda > \omega$ we have

$$
\begin{aligned}
\frac{\lambda^{n+1}}{n!}&\int_0^\infty e^{-\lambda t}t^n x\,dt \\
&= x = \lambda R_A(\lambda) - R_A(\lambda)Ax \\
&= \int_0^\infty \lambda^{n+1}e^{-\lambda t}V(t)x\,dt - \int_0^\infty \lambda^n e^{-\lambda t}V(t)Ax\,dt \\
&= \int_0^\infty \lambda^{n+1}e^{-\lambda t}V(t)x\,dt - \int_0^\infty \lambda^{n+1}e^{-\lambda t}\int_0^t V(s)Ax\,ds\,dt,
\end{aligned}
$$

from where (1.2.2) follows by the uniqueness theorem for Laplace transforms. Since A is closed, (1.2.3) follows from (1.2.2); differentiation of (1.2.2) gives (1.2.4); (1.2.5) follows from (1.2.4). $\square$

Theorem 1.2.1 (Arendt-Widder) *Let* $a \geq 0$ *and* $r : (a,\infty) \to X$ *be an infinitely differentiable function. For* $K > 0$, $\omega \in (-\infty, a\,]$ *the following statements are equivalent:*

(I)

$$\|r^{(k)}(\lambda)\| \leq \frac{K\,k!}{(\lambda - \omega)^{k+1}}, \quad \lambda > a, \; k = 0,1,\dots;$$

(II) *there exists a function* $V : [0,\infty) \to X$ *satisfying* $V(0) = 0$ *and*

$$\limsup_{\substack{\delta \to 0 \\ h \leq \delta}} h^{-1}\|V(t+h) - V(t)\| \leq Ke^{\omega t}, \; t \geq 0, \qquad (1.2.6)$$

such that

$$r(\lambda) = \int_0^\infty \lambda e^{-\lambda t}V(t)\,dt, \; \lambda > a.$$

Moreover, $r(\lambda)$ *has an analytic extension to*

$$\left\{ \lambda \in \mathbb{C} \;\middle|\; \operatorname{Re}\lambda > \omega \right\}.$$

The Arendt-Widder theorem and Proposition 1.2.1 imply

Theorem 1.2.2 *Let* $n \in \{0\} \cup \mathbb{N}$, $\omega \in \mathbb{R}$, $K > 0$. *A linear operator A is the generator of an $(n+1)$-times integrated semigroup V satisfying condition (1.2.6) if and only if there exists $a \geq \max\{\omega, 0\}$ such that $(a, \infty) \subset \rho(A)$ and*

$$\left\| \left[\frac{R_A(\lambda)}{\lambda^n} \right]^{(k)} \right\| \leq \frac{K\,k!}{(\lambda - \omega)^{k+1}} \qquad (1.2.7)$$

for all $\lambda > a$, and $k = 0, 1, \ldots$. In this case

$$R_A(\lambda) = \int_0^\infty \lambda^{n+1} e^{-\lambda t} V(t)\,dt, \qquad \lambda > a. \; \square$$

Hence, for $n = 0$ we have the equivalence of existence of an integrated semigroup and MFPHY-type condition. At first sight, this equivalence is weaker than in Theorem 1.1.1, but the fact is that integrated semigroups, in contrast to C_0-semigroups, may have not densely defined generators. If $\overline{D(A)} = X$, then the following extension of Theorem 1.1.1 holds.

Theorem 1.2.3 *Let A be a densely defined linear operator with $(a, \infty) \subset \rho(A)$, where $a \geq 0$, $K > 0$, $\omega \in (-\infty, a]$. Then condition (1.2.7) is equivalent to the statement: A is the generator of an n-times integrated semigroup $\{V(t),\ t \geq 0\}$, such that*

$$\|V(t)\| \leq K e^{\omega t}.$$

Proof Suppose that condition (1.2.7) is fulfilled, then A is the generator of an $(n+1)$-times integrated semigroup $\{V_{n+1}(t),\ t \geq 0\}$. In view of (1.2.6) for V_{n+1}, the set F_1 of all $x \in X$ such that $V_{n+1}(\cdot)x \in C^1\{(0, \infty), X\}$ is closed. By Proposition 1.2.2, $D(A) \subset F_1$, hence $\overline{D(A)} = X \subset F_1$ and $F_1 = X$. Then the relation $V(t)x = V'_{n+1}(t)x$, $x \in X$, defines an n-times integrated semigroup $\{V(t),\ t \geq 0\}$ generated by A. $\square$

1.2.2 (n, ω)-well-posedness of the Cauchy problem

We consider the Cauchy problem

$$u'(t) = Au(t), \quad t \geq 0,\ u(0) = x, \qquad \text{(CP)}$$

where A is a linear closed densely defined operator on a Banach space X. We denote by $[D(A^n)]$ the Banach space

$$\Big\{ D(A^n),\ \ \|x\|_{A^n} = \|x\| + \|Ax\| + \ldots + \|A^n x\| \Big\}.$$

Definition 1.2.2 *Let $n \in \mathbb{N}$. The Cauchy problem (CP) is said to be (n, ω)-well-posed on E if for any $x \in E \subseteq D(A^{n+1})$*

(a) *there exists a unique solution*

$$u(\cdot) \in C\{[0, \infty], D(A)\} \cap C^1\{[0, \infty], X\};$$

(b) $\exists K > 0, \ \omega \in \mathbb{R}: \ \|u(t)\| \ \leq K e^{\omega t}\|x\|_{A^n}.$

If $E = D(A^{n+1})$, then we say that the problem (CP) is (n, ω)-well-posed.

Theorem 1.2.4 *Let A be a densely defined linear operator on X with nonempty resolvent set. Then the following statements are equivalent:*

(I) *A is the generator of an n-times integrated semigroup $\{V(t), \ t \geq 0\}$;*

(II) *the Cauchy problem (CP) is (n, ω)-well-posed.*

Proof (I) $\Longrightarrow$ (II). Let $x \in D(A^{n+1})$, consider the function $V(\cdot)x$. By Proposition 1.2.2, it is $(n+1)$-times continuously differentiable, $V^{(n)}(t)x \in D(A)$, and

$$V^{(n)}(t)x = V(t)A^n x + \sum_{k=0}^{n-1} \frac{t^k}{k!} A^k x, \quad \frac{d}{dt} V^{(n)}(t)x = AV^{(n)}(t)x.$$

Let $u(t) := V^{(n)}(t)x$, then $u(0) = x$ and $u(t) \in D(A)$ for $t \geq 0$. Furthermore, $\|u(t)\| \leq K e^{\omega t}\|x\|_{A^n}$, and $u'(t) = Au(t)$.

We now show that $u(\cdot)$ is unique. Let $v(\cdot)$ be a solution of (CP), then for $\lambda \in \rho(A)$, $R_A^n(\lambda)v(\cdot)$ is the solution of (CP) with the initial value $R^n(\lambda)x \in D(A^{n+1})$. As in the proof of uniqueness in Theorem 1.1.1, we have

$$\frac{d}{ds}V^n(t-s)R_A^n(\lambda)v(s)$$
$$= -AV^n(t-s)R_A^n(\lambda)v(s) + V^n(t-s)AR_A^n(\lambda)v(s) = 0,$$

for $0 \leq s \leq t$. Hence

$$V^n(0)R_A^n(\lambda)v(t) = R_A^n(\lambda)V^n(0)v(t) = R_A^n(\lambda)V^n(t)v(0),$$

and $v(\cdot) = V^n(\cdot)x$.

(II) $\Longrightarrow$ (I). Let $x \in D(A^{n+1})$, then there exists a unique solution $u(\cdot)$ of (CP) such that $\|u(t)\| \leq K e^{\omega t}\|x\|_{A^n}$. For $\mu \in \rho(A)$ the function $w(\cdot) := R_A(\mu)u(\cdot)$ is the solution of the Cauchy problem with the initial value $R_A(\mu)x$ and with the estimate $\|w(t)\| \leq K_1 e^{\omega t}\|x\|_{A^{n-1}}$. Let

$$u_1(t) = \int_0^t u(s)ds,$$

then

$$u_1(t) = w(t) - \mu \int_0^t w(s)ds - R_A(\mu)x.$$

Therefore, $\|u_1(t)\| \leq K_2 e^{\omega t} \|x\|_{A^{n-1}}$. By induction we obtain an n-times integrated solution

$$u_n(t) = \int_0^t \frac{1}{(n-1)!}(t-s)^{n-1} u(s)ds, \ t \geq 0,$$

which is exponentially bounded. For $t \geq 0$ we define an operator-function $V(t) : D(A^{n+1}) \to X$ by the relation $V(t)x = u_n(t)$, where $u_n(t)$ is the unique n-times integrated solution of (CP) with the initial value $x \in D(A^{n+1})$. Since $\overline{D(A)} = \overline{D(A^{n+1})} = X$, we can extend V to the whole X. The operator-function $V(t)$, $t \geq 0$, is exponentially bounded and for $x \in D(A^{n+1})$, $V(t)x$ is continuous in t, therefore $V(\cdot)$ is strongly continuous. In addition, it is not difficult to show that the operator $R(\mu) = \int_0^\infty \mu^n e^{-\mu t} V(t)dt$ coincides with the resolvent $R_A(\mu)$ of operator A. Hence $\{V(t), \ t \geq 0\}$ is an n-times integrated semigroup with the generator A. $\square$

Remark 1.2.1 If we do not assume that $\overline{D(A)} = X$, then after constructing $u_1(t), ..., u_n(t)$ in the proof of the implication (II) $\Longrightarrow$ (I), we can extend $u_1(t)$ to $D(A^n)$, ... , $u_n(t)$ to $D(A)$, and as a result, construct an $(n+1)$-times integrated semigroup $V(t)x := u_{n+1}(t)$ on X.

1.2.3 Local integrated semigroups

Definition 1.2.3 *Let* $n \in \mathbb{N}$, $T \in (0, \infty)$. *A one-parameter family of bounded linear operators* $\{V(t), \ 0 \leq t < T\}$ *is called a* local n-times integrated semigroup *in* X *if* (V1) *in Definition 1.2.1 is fulfilled for* $t, s \in [0, T)$ *such that* $s + t \in [0, T)$, *and* (V2) *holds for* $t \in [0, T)$.

If $\{V(t), \ t \geq 0\}$ is an exponentially bounded n-times integrated semigroup then its generator A is defined from the equality

$$(\lambda I - A)^{-1} x = \int_0^\infty \lambda^n e^{-\lambda t} V(t)x \, dt, \qquad x \in X, \ \lambda > \omega. \qquad (1.2.8)$$

For a local n-times integrated semigroup $\{V(t), \ t \in [0, T)\}$ this integral may not exist. For this reason the infinitesimal generator A_0 of a local semigroup $V(t)$ is defined in the following way

$$A_0 x := \lim_{t \to 0} t^{-1} \left[V^{(n)}(t)x - x \right]$$

$$D(A_0) = \left\{ x \in \bigcup_{0 < \delta < T} C^n(\delta) \ \middle| \ \lim_{t \to 0} t^{-1} \left[V^{(n)}(t)x - x \right] \text{ exists} \right\},$$

where

$$C^n(\delta) = \Big\{ x \in X \text{ such that}$$

$$V(t)x : [0, \delta) \to X \text{ is } n\text{-times continuously differentiable} \Big\}.$$

It is proved in [138] that A_0 is closable and we call $\overline{A_0}$ the complete infinitesimal generator or the generator of a local n-times integrated semigroup $\{V(t),\ t \in [0, T)\}$. For a densely defined operator, this definition of the complete infinitesimal generator of an exponentially bounded semigroup is equivalent to the definition of a generator via formula (1.2.8).

We list below the main properties of local integrated semigroups which are similar to those in Proposition 1.2.2.

Proposition 1.2.3 *Let $n \in \mathbb{N}$ and let A be the generator of a local n-times integrated semigroup $\{V(t),\ t \in [0, T)\}$, then*

1) *for $x \in D(A),\ t \in [0, T)$*

$$V(t)x \in D(A) \quad and \quad AV(t)x = V(t)Ax;$$

2) *for $x \in D(A),\ t \in [0, T)$*

$$V(t)x = \frac{t^n}{n!}x + \int_0^t V(s)Ax\,ds; \tag{1.2.9}$$

3) *if $\overline{D(A)} = X$, then for $x \in X,\ t \in [0, T)$*

$$\int_0^t V(s)x\,ds \in D(A), \qquad A \int_0^t V(s)x\,ds = V(t)x - \frac{t^n}{n!}x;$$

4) *$\overline{D(A)} = X$ if and only if $\overline{C^n(T)} = X$.*

Definition 1.2.4 *The local Cauchy problem*

$$u'(t) = Au(t), \quad t \in [0, T),\ u(0) = x, \tag{CP}$$

is said to be n-well-posed if for any $x \in D(A^{n+1})$ there exists a unique solution satisfying

$$\sup_{t \in [0, \tau] \subset [0, T)} \|u(t, x)\| \leq K_\tau \|x\|_{A^n} \tag{1.2.10}$$

for some constant K_τ.

The following result is similar to Lemma 1.1.1.

Lemma 1.2.1 *If for any* $x \in D(A^{n+1})$ *there exists a unique solution of local* (CP) *and* $\rho(A) \neq \emptyset$, *then this solution satisfies (1.2.10).*

Proof Let us consider the solution operator

$$S : [D(A^{n+1})] \to C\{[0, \tau], [D(A)]\}$$

which is defined everywhere on $[D(A^{n+1})]$ by

$$Sx = u, \quad x \in [D(A^{n+1})],$$

and is closed. Let $x_j \to x$ and $u_j(t) \to y$, then we have

$$u_j'(t) = Au_j(t) \to Ay(t) \quad \text{and} \quad u_j(t) - x_j \to y(t) - x = \int_0^t Ay(t)dt,$$

hence $y'(t) = Ay(t)$, $y(0) = x$. By the Banach theorem, the operator S is continuous, that is

$$\sup_{t \in [0,\tau]} \|u(t)\|_A \leq K_\tau \|x\|_{A^{n+1}}.$$

Furthermore, as in the proof of Lemma 1.1.1, we have

$$\sup_{t \in [0,\tau]} \|u(t)\| \leq K_\tau \|x\|_{A^n}.$$

Thus, this solution is stable with respect to the initial data in the corresponding n-norm. $\square$

Proposition 1.2.4 *If A is the generator of a local n-times integrated semigroup* $\{V(t), \ t \in [0, T)\}$, *then local* (CP) *is n-well-posed.*

Proof Let $k \in \mathbb{N}$ with $1 \leq k \leq n$. Then, from (1.2.9) we have

$$V^{(k)}(t) = \sum_{i=1}^{k} \frac{t^{n-i}}{(n-i)!} A^{k-i} x + V(t)A^k x$$

for $x \in D(A^k)$ and $0 \leq t < T$. Now for $x \in D(A^{n+1})$ we define $u(t) := V^n(t)x$, $t \in [0, T)$. Using (1.2.9) one can easily show that $u(\cdot)$ is a solution of local (CP). In Theorem 1.2.4 it is proved that every solution of this problem has the form $V^n(\cdot)x$. $\square$

Theorem 1.2.5 *Let A be the generator of a local n-times integrated semigroup* $\{V(t), \ t \in [0, T)\}$ *and* $\overline{D(A)} = X$. *Then there exists region*

$$\Lambda = \left\{ \lambda \in \mathbb{C} \ \Big| \ \operatorname{Re} \lambda > \frac{n}{\tau} \log(1 + |\lambda|) + \frac{1}{\tau} \log \frac{C}{\gamma} \right\} \subset \rho(A),$$

$$\tau \in (0,T), \ C > 0, \ 0 < \gamma < 1,$$

such that

$$\exists K > 0 : \forall \lambda \in \Lambda, \quad \|R_A(\lambda)\| \le \frac{K\,|\lambda|^n}{\log(1+|\lambda|)}.$$

Proof If A is the generator of a local n-times integrated semigroup $\{V(t), 0 \le t < T\}$, then the function

$$R(\lambda, \tau) = \int_0^\tau \lambda^n e^{-\lambda t} V(t) dt$$

is defined for any $\tau \in (0,T)$. Taking into account (1.2.9), for $x \in D(A)$ we have

$$R(\lambda, \tau)Ax$$

$$= \lambda^n e^{-\lambda \tau} \int_0^\tau V(s)Ax\,ds \ + \ \int_0^\tau \lambda^{n+1} e^{-\lambda t} \int_0^t V(s)Ax\,ds\,dt$$

$$= \lambda^n e^{-\lambda \tau} \left(V(\tau)x - \frac{\tau^n}{n!}x \right) \ + \ \int_0^\tau \lambda^{n+1} e^{-\lambda t} \left(V(t)x - \frac{t^n}{n!}x \right) dt,$$

$$R(\lambda, \tau)(\lambda I - A)x$$

$$= \int_0^\tau \lambda^{n+1} e^{-\lambda t} V(t) x\,dt - \lambda^n e^{-\lambda \tau} \left(V(\tau)x - \frac{\tau^n}{n!}x \right)$$

$$- \int_0^\tau \lambda^{n+1} e^{-\lambda t} \left(V(t)x - \frac{t^n}{n!}x \right) dt$$

$$= -\lambda^n e^{-\lambda \tau} V(\tau)x + \lambda^n e^{-\lambda \tau} \frac{\tau^n}{n!}x \ + \ \int_0^\tau \lambda^{n+1} e^{-\lambda t} \frac{t^n}{n!}x\,dt.$$

Since

$$\int_0^\tau \lambda^{n+1} e^{-\lambda t} \frac{t^n}{n!}x\,dt \ = \ -\lambda^n e^{-\lambda \tau} \frac{\tau^n}{n!}x \ + \ \int_0^\tau \lambda^n e^{-\lambda t} \frac{t^{n-1}}{(n-1)!}x\,dt$$

$$= \ \ldots = -\sum_{k=0}^n \frac{(\lambda \tau)^k}{k!} e^{-\lambda \tau} x + x,$$

then

$$R(\lambda, \tau)(\lambda I - A)x = (\lambda I - A)R(\lambda, \tau)x = (I - G(\lambda))x, \quad x \in D(A),$$

where

$$G(\lambda)x = \lambda^n e^{-\lambda \tau} V(\tau)x + \sum_{k=0}^{n-1} \frac{(\lambda \tau)^k}{k!} e^{-\lambda \tau} x,$$

and

$$\|G(\lambda)\| \le C(1 + |\lambda|)^n e^{-\tau \operatorname{Re} \lambda}, \quad C = C(\tau, n).$$

We also have that $G(\lambda)$ commutes with $R(\lambda, \tau)$ on X and with A on $D(A)$. Using this estimate for $\|G(\lambda)\|$, we can find a region $\Lambda \subset \mathbb{C}$ such that $\|G(\lambda)\| < 1$ for any $\lambda \in \Lambda$. After taking logarithms of the inequality

$$C(1 + |\lambda|)^n e^{-\tau \operatorname{Re} \lambda} < \gamma < 1,$$

we obtain that the estimates

$$\|G(\lambda)\| < \gamma, \quad \|(I - G(\lambda))^{-1}\| < \frac{1}{1 - \gamma}$$

hold in the region

$$\Lambda = \left\{ \lambda \in \mathbb{C} \;\middle|\; \operatorname{Re} \lambda > \frac{n}{\tau} \log(1 + |\lambda|) + \frac{1}{\tau} \log \frac{C}{\gamma} \right\}.$$

Therefore, there exists the operator $(\lambda I - A)^{-1}$, which is bounded on $\overline{D(A)} = X$, and

$$\exists K > 0 : \ \forall \lambda \in \Lambda, \qquad \|(\lambda I - A)^{-1}\| \le \frac{K\,|\lambda|^n}{\log(1 + |\lambda|)}.$$

Since $\overline{D(A)} = X$, we have $(\lambda I - A)^{-1} = R_A(\lambda)$. $\square$

Remark 1.2.2 The converse result that polynomial estimates on the resolvent of operator A in some region Λ imply that A is the generator of some local integrated semigroup is proved later in Theorem 2.1.5.

The following necessary and sufficient condition in terms of estimates on powers of the resolvent is the generalization of (1.1.14).

Theorem 1.2.6 *If for an operator A*

$$\exists \omega : \ \forall \tau \in (0, T),$$

$$\sup \left\{ \|\lambda^k R_A^k(\lambda) x\| \;\middle|\; 0 \le \frac{k}{\lambda} \le \tau, \ \lambda > \omega, \ k = 0, 1, \dots \right\}$$

$$\le C_\tau \|x\|_{A^n} \tag{1.2.11}$$

for some constant C_τ, then A is the generator of a local n-times integrated semigroup $\{V(t), \ t \in [0, T)\}$. Conversely, if A is the generator of a local n-times integrated semigroup $\{V(t), \ t \in [0, T)\}$ and $\overline{D(A)} = X$ or $\rho(A) \ne \emptyset$, then (1.2.11) holds.

Proof If A is the generator of a local n-times integrated semigroup $\{V(t),$ $t \in [0, T)\}$, then by Proposition 1.2.4 for any $x \in D(A^{n+1})$ there exists a unique solution of the local Cauchy problem (CP):

$$u(\cdot) = V^{(n)}(\cdot)x =: U(\cdot)x$$

such that

$$\|U(t)x\| \le K_\tau \|x\|_{A^n}, \quad 0 \le t \le \tau < T. \tag{1.2.12}$$

For $U(t)x$ we have

$$A \int_0^\tau e^{-\lambda t} U(t)x \, dt = e^{-\lambda \tau} U(\tau)x - x + \lambda \int_0^\tau e^{-\lambda t} U(t)x \, dt.$$

Hence by Theorem 1.2.5,

$$R_A(\lambda)x = \int_0^\tau e^{-\lambda t} U(t)x \, dt + e^{-\lambda \tau} R_A(\lambda)U(\tau)x, \quad \lambda \in \rho(A).$$

By the Cauchy's integral formula

$$
\begin{aligned}
\lambda^k \left[R_A(\lambda)\right]^k x &= \frac{(-1)^{k-1}\lambda^k}{(k-1)!} R_A^{(k-1)}(\lambda)x \\
&= \frac{\lambda^k}{(k-1)!} \int_0^\tau e^{-\lambda t} t^{k-1} U(t)x \, dt \\
&\quad + \frac{1}{2\pi i} \int_\gamma e^{-\xi \tau} \left(1 - \frac{\xi}{\lambda}\right)^{-k} R_A(\xi)U(\tau)x \, d\xi \\
&\equiv I_1 + I_2,
\end{aligned}
$$

for some contour γ, which is described in [226]. The following estimates are obtained in [226] for these integrals

$$\|I_1\| \le C_1 \|x\|_{A^n}, \quad x \in D(A^n), \ \lambda > 0,$$

$$\exists \omega \ : \ \|I_2\| \le C_2 \|x\|_{A^n}, \quad x \in D(A^n), \ \frac{k}{\lambda} \in [0, \tau], \ \lambda > \omega.$$

This and (1.2.12) imply (1.2.11).

Conversely, it is proved in [213] that if (1.2.11) is fulfilled for operator A, then

$$U(t)x := \lim_{k \to \infty} \left(I - \frac{t}{k}A\right)^{-k} x, \quad t \in [0, T),$$

is defined for $x \in D(A^n)$. For $x \in D(A^{n+1})$, $u(\cdot) = U(\cdot)x$ is the unique solution of the local Cauchy problem with the stability property (1.2.12). In this case (see [255]) A is the generator of a local n-times integrated semigroup $\{V(t), 0 \le t < T\}$. $\square$

Remark 1.2.3 We showed that if A is the generator of an n-times integrated semigroup $\{V(t),\ 0 \le t < T\}$, $T \le \infty$ (defined as the closure of operator $V^{(n+1)}(0)$), then we have (1.2.9):

$$
V(t)x - \frac{t^n}{n!}x
$$

$$
= \int_0^t V(s)Ax\,ds = \int_0^t AV(s)x\,ds, \quad 0 \le t < T, \quad x \in D(A),
$$

and if A is densely defined, then

$$
V(t)x - \frac{t^n}{n!}x = A \int_0^t V(s)x\,ds, \quad x \in X. \tag{1.2.13}
$$

Now we prove that if A and a family of bounded linear operators $\{V(t),\ 0 \le t < T\}$ satisfy (1.2.9), (1.2.13), then V satisfies the 'semigroup relation' (V1) from the definition of an n-times integrated semigroup. Hence, an n-times integrated semigroup with a densely defined generator A may be equivalently defined as follows:

Definition 1.2.5 *A family of bounded linear operators* $\{V(t),\ 0 \le t < T\}$ *is called a local n-times integrated semigroup generated by operator A, if for $x \in D(A)$, $AV(t)x = V(t)Ax$, and*

$$
A \int_0^t V(s)x\,ds = V(t)x - \frac{t^n}{n!}x, \quad x \in X.
$$

The operator A is called the generator of V.

If A is the generator of V then

$$
D(A) = D := \left\{ x \in X \ \middle|\ \exists y : V(t)x = \frac{t^n}{n!}x + \int_0^t V(s)y\,ds,\ t \in [0, T) \right\},
$$

where $y = Ax$. In fact, let $x \in D(A)$, then $x \in D$ and $y = Ax$. Conversely, if $x \in D$, then for $\lambda \in \rho(A)$

$$
V(t)R_A(\lambda)x - \frac{t^n}{n!}R_A(\lambda)x = \int_0^t V(s)R_A(\lambda)y\,ds = \int_0^t V(s)AR_A(\lambda)x.
$$

Hence $x \in D(A)$ and $y = Ax$. Note that it was proved in Theorem 1.2.5, that for such a family $\rho(A) \ne \emptyset$.

Lemma 1.2.2 *Let $\tau > 0$ and A be the generator of a local n-times integrated semigroup $\{V(t),\ 0 \le t < T\}$ in the sense of Definition 1.2.5. Then for any continuous function $H : [0, \tau) \to \mathbb{C}$ and for any $x \in X$*

$$
A \int_0^t (H * V)(s)x\,ds = (H * V)(t)x - (H * F)(t)x, \quad 0 \le t < \tau.
$$

Moreover, if $H \in C^1\{[0,\tau), \mathbb{C}\}$, then for any $x \in X$ and $0 \le t < \tau$

$$A(H * V)(t)x = (H' * V)(t)x - (H' * F)(t)x + H(0)\Big[V(t)x - F(t)x\Big],$$

where

$$(H * V)(t)x := \int_0^t H(s)V(t-s)x\,ds, \quad F(t) := \frac{t^n}{n!}.$$

Proof Since operator A is closed, we have

$$\int_0^t (H * V)(s)x\,ds \in D(A), \quad x \in X, \ 0 \le t < \tau.$$

From Definition 1.2.5 we obtain

$$A \int_0^t (H * V)(s)x\,ds = A \int_0^t \left(\int_0^s H(r)V(s-r)x\,dr \right) ds$$

$$= \int_0^t H(r) \left(A \int_r^t V(s-r)x\,ds \right) dr = \int_0^t H(r) \left(A \int_0^{t-r} V(s)x\,ds \right) dr$$

$$= \int_0^t H(r) \left[V(t-r)x - F(t-r)x \right] dr = (H * V)(t)x - (H * F)(t)x,$$

$$x \in X.$$

Since

$$(H * V)(t)x = \int_0^t H(s)V(t-s)x\,ds = \int_0^t H(t-s)V(s)x\,ds, \quad x \in X,$$

then the second statement of the lemma follows from the first one after integration by parts. $\square$

Lemma 1.2.3 *Let $\tau > 0$ and A be the generator of a local n-times integrated semigroup $\{V(t), \ 0 \le t < T\}$ in the sense of Definition 1.2.5. Then for any $x \in X$ and $0 \le t, s < \tau$*

$$V(t)V(s)x$$

$$= \int_s^{t+s} \frac{(t+s-r)^{n-1}}{(n-1)!} V(r)x\,dr - \int_0^t \frac{(t+s-r)^{n-1}}{(n-1)!} V(r)x\,dr.$$

Proof Let $h > 0$, define

$$E_h(t) := E(t+h), \quad \text{where} \quad E(t) = \frac{t^{n-1}}{(n-1)!}.$$

For $0 \le s < \tau$ and fixed x, define

$$w(t) := \int_s^{t+s} E(t+s-r)V(r)x\,dr - \int_0^t E(t+s-r)V(r)x\,dr,$$

where $x \in X$, $0 \le t < \tau$. We have

$$
\begin{aligned}
w(t) &= \int_0^{t+s} E(t+s-r)V(r)x\,dr \\
&\quad - \int_0^t E(t+s-r)V(r)x\,dr - \int_0^s E(t+s-r)V(r)x\,dr \\
&= (E*V)(t+s)x - (E_s*V)(t)x - (E_t*V)(s)x, \quad x \in X.
\end{aligned}
$$

Using Lemma 1.2.2, we obtain

$$
\begin{aligned}
A \int_0^t & w(r)dr \\
&= A\int_0^t (E*V)(s+r)x\,dr - A\int_0^t (E_s*V)(r)x\,dr \\
&\quad - A\int_0^t (E_r*V)(s)x\,dr \\
&= A\int_0^{t+s} (E*V)(r)x\,dr - A\int_0^s (E*V)(r)x\,dr \\
&\quad - A\int_0^t (E_s*V)(r)x\,dr - A\Big[(F_t*V)(s) - (F*V)(s)\Big] \\
&= (E*V)(t+s)x - (E*F)(t+s)x - (E*V)(s)x + (E*F)(s)x \\
&\quad - (E_s*V)(t)x + (E_s*F)(t)x - (E_t*V)(s)x + (E_t*F)(s)x \\
&\quad - F(t)V(s)x + F(t)F(s)x + (E*V)(s)x - (E*F)(s)x \\
&= (E*V)(t+s)x - (E_s*V)(t)x - (E_t*V)(s)x - F(t)V(s)x \\
&\quad + \Big[F(t)F(s) - (E*F)(t+s) + (E_s*F)(t) + (E_t*F)(s)\Big]x \\
&= w(t) - F(t)V(s)x, \quad x \in X,
\end{aligned}
$$

where

$$
F(t) = \int_0^t E(s)ds = \frac{t^n}{n!}.
$$

Here we used the following relation

$$
\begin{aligned}
F(t)F(s)x \\
&= (E*F)(t+s)x - (E_s*F)(t)x - (E_t*V)(s)x \\
&= \int_s^{t+s} E(t+s-r)F(r)x\,dr - \int_0^t E(t+s-r)F(r)x\,dr
\end{aligned}
$$

for $x \in X$, which can be easily proved using integration by parts. Therefore $v(\cdot) = w(\cdot)$ is a solution of the problem

$$v'(t) = Av(t) + F(t)V(s)x, \quad 0 \le t < \tau , \qquad (1.2.14)$$

$$v(0) = 0 \, , v \in C\Big\{[0, \tau), \, D(A) \cap C^1\{[0, \tau), \, X\}\Big\}.$$

Since $V(t)V(s)x$, $t, s \in [0, \tau)$, is also a solution of this problem, then by uniqueness we obtain that $w(t) = V(t)V(s)x$, $t, s \in [0, \tau)$. $\square$

1.2.4   Examples

Example 1.2.1 (An integrated semigroup with the generator that is non-densely defined)

Consider the operator $A = -d/dx$ on $X = C[0, \infty)$ with

$$D(A) = \Big\{u \in C[0, \infty) \, \Big| \, u' \in C[0, \infty), \, u(0) = 0\Big\}.$$

Since $\overline{D(A)} \ne X$ then A cannot be a generator of a C_0-semigroup (compare this with Example 1.1.1). In this case A is the generator of the 1-times integrated semigroup V defined by

$$(V(t)f)(x) = \begin{cases} \int_x^{x-t} f(s)\, ds, & x \ge t \\[2mm] \int_x^{o} f(s)\, ds, & 0 \le x \le t \end{cases}.$$

Note that A is the generator of a C_0-semigroup in the space $C_0[0, \infty)$ of continuous on $[0, \infty)$ functions vanishing at infinity.

Example 1.2.2 (A class of operators generating integrated semigroups)

As in Example 1.1.2 let

$$X = \Big\{L^p(\mathbb{R}) \times L^p(\mathbb{R}), \quad \|u\| = \|u_1\|_{L^p} + \|u_2\|_{L^p}\Big\}, \quad \text{where} \quad u = \begin{pmatrix} u_1 \\ u_2 \end{pmatrix}.$$

Consider the operator A defined by

$$Au = \begin{pmatrix} -g & -f \\ 0 & -g \end{pmatrix} u$$

with

$$D(A) = \left\{ \begin{pmatrix} u_1 \\ u_2 \end{pmatrix} \in X \,\middle|\, gu_1 + fu_2 \in L^p(\mathbb{R}), \, gu_2 \in L^p(\mathbb{R}) \right\},$$

where $g(x) = 1+|x|$, $f(x) = |x|^\gamma$, $\gamma > 0$. For the formally defined operator-function

$$V(t) = \int_0^t e^{As}ds, \quad t \geq 0,$$

we can write

$$V(t)u \;=\; \frac{1}{h}\left(\begin{array}{cc} 1 - e^{-ht} & tfe^{-ht} + (e^{-ht} - 1)f/h \\ 0 & 1 - e^{-ht} \end{array} \right) u, \quad u \in X.$$

If $\gamma \in (1,2]$, then the family of bounded linear operators $\{V(t),\ t \geq 0\}$ satisfies conditions (V1) – (V4) for a 1-time integrated semigroup. The operator A is the generator of this semigroup since

$$\begin{aligned} \lambda I - R(\lambda)^{-1} &= \lambda I - \left(\int_0^\infty \lambda e^{-\lambda t} V(t)dt \right)^{-1} \\ &= \lambda I - \left(\begin{array}{cc} \frac{1}{\lambda+h} & -\frac{f}{(\lambda+h)^2} \\ 0 & \frac{1}{\lambda+h} \end{array} \right)^{-1} = A. \end{aligned}$$

Furthermore, if $\gamma \leq 2$, then the operator-functions $V_k(t)$, $t \geq 0$, defined by

$$V_k(t)u = \int_0^t V_{k-1}(s)uds, \quad u \in X,\ k \geq 2,\ V_1 = V,$$

form k-times integrated semigroups on X. In particular,

$$\begin{aligned} V_2(t)u &= \int_0^t V(s)uds \\ &= \frac{1}{h}\left(\begin{array}{cc} t - \frac{1-e^{-ht}}{h} & -tfe^{-ht} + \frac{(1-e^{-ht})f}{h} + \frac{(1-e^{-ht})f}{h^2} - \frac{tf}{h} \\ 0 & t - \frac{1-e^{-ht}}{h} \end{array} \right) u, \end{aligned}$$

$u \in X$,

defines a 2-times integrated semigroup V_2 on X. We also note that if $\gamma > 2$, then (see Example 1.1.2) all $\lambda > 0$ do not belong to the resolvent set of A, and therefore for any n, the operator A cannot generate an n-times integrated semigroup on X.

Example 1.2.3

The differential operator

$$A = \sum_{|\alpha| \leq k} a_\alpha D^\alpha,$$

where

$$D^\alpha = \left(\frac{\partial}{\partial x_1}\right)^{\alpha_1} \cdots \left(\frac{\partial}{\partial x_n}\right)^{\alpha_n}, \quad |\alpha| = \sum_{j=1}^n \alpha_j, \ \max|\alpha| > n/2,$$

and its symbol

$$p(x) = \sum_{|\alpha| \leq k} a_\alpha (ix)^\alpha$$

is an elliptic polynomial with $\operatorname{Re} p(x) < \infty$, is the generator of $(\llbracket n/2 \rrbracket + 2)$-times integrated semigroup in the spaces $C_0(\mathbb{R}^n)$, $L^p(\mathbb{R}^n)$, $1 \leq p \leq \infty$.

Example 1.2.4 (An integrated semigroup associated with the second order Cauchy problem)

This example generalizes the Cauchy problem for Wave equation which was discussed in Chapter 0. Consider the second order Cauchy problem

$$u''(t) = Bu(t), \qquad t \geq 0, \ u(0) = x, \ u'(0) = y \qquad (1.2.15)$$

in a Banach space X, with linear operator B generating cosine and sine operator-functions $\mathcal{C}(\cdot)$ and $\mathcal{S}(\cdot)$. In Section 1.7 we study the properties of $\mathcal{M}, \mathcal{N}$-functions, which in particular give the properties of $\mathcal{C}, \mathcal{S}$-functions. The unique solution of (1.2.15) has the form

$$u(t) = \mathcal{C}(t)x + \mathcal{S}(t)y.$$

The problem (1.2.15) can be reduced to the Cauchy problem for the first order system

$$w'(t) = \Phi w(t), \qquad w(0) = \begin{pmatrix} x \\ y \end{pmatrix},$$

where

$$\Phi = \begin{pmatrix} 0 & I \\ B & 0 \end{pmatrix}, \quad w(t) = \begin{pmatrix} u(t) \\ u'(t) \end{pmatrix}.$$

Using cosine and sine operator-functions, we can rewrite w in the following form

$$w(t) = \begin{pmatrix} \mathcal{C}(t)x + \mathcal{S}(t)y \\ \mathcal{C}'(t)x + \mathcal{C}(t)y \end{pmatrix} \equiv U(t) \begin{pmatrix} x \\ y \end{pmatrix},$$

where operator $U(t)$ is not defined everywhere on $X \times X$ for all $t \geq 0$, because function $\mathcal{C}(\cdot)$ is not necessarily differentiable on X.

The operator Φ is the generator of the integrated semigroup

$$V(t) = \begin{pmatrix} \mathcal{S}(t) & \int_0^t \mathcal{S}(\tau)d\tau \\ \mathcal{C}(t) - I & \mathcal{S}(t) \end{pmatrix}.$$

Conditions (V1) – (V4) are satisfied due to the properties of $\mathcal{C}, \mathcal{S}$-functions.

Example 1.2.5 (An integrated semigroup that is not exponentially bounded)

Let $X = l_2$ and

$$Ax := \{a_m x_m\}_{m=1}^{\infty}, \qquad a_m = m + i\{e^{2m^2} - m^2\}^{1/2}, \quad \alpha > 1,$$

then the operators

$$U(t)x = \{e^{a_m t} x_m\}_{m=1}^{\infty}$$

form an unbounded semigroup. The operators

$$V(t)x = \int_0^t U(s)x\,ds = \left\{\frac{e^{a_m t}}{a_m} x_m\right\} \equiv \{b_m x_m\}$$

are bounded for any $t \geq 0$ since

$$\|V(t)\| = \sup_m |b_m| = \sup_m \{e^{mt - m^2}\} = e^{t^2/4},$$

and form an integrated semigroup that is not exponentially bounded.

Example 1.2.6 (Local n-times integrated semigroup)

Let $X = l_2$, $T > 0$,

$$Ax := \{a_m x_m\}_{m=1}^{\infty}, \qquad a_m = \frac{m}{T} + i\left\{\frac{e^{2m}}{m^2} - \frac{m^2}{T^2}\right\}^{1/2},$$

$$D(A) = \left\{x \in l_2 \;\middle|\; Ax \in l_2\right\}.$$

For operator A we have

$$\sigma(A) = \left\{\lambda \in \mathbb{C} \;\middle|\; \lambda = a_m, \; m \in \mathbb{N}\right\}$$

and since $\operatorname{Re} a_m = \frac{m}{T}$, then for any $\omega \in \mathbb{R}$ there exists $\lambda \in \sigma(A)$ such that $\operatorname{Re} \lambda > \omega$. The operators $T(t)x := \{e^{a_m t} x_m\}_{m=1}^{\infty}$ form an unbounded semigroup. Integrating $e^{a_m t}$ we obtain the factor me^{-m}, which on the n-th step makes the product bounded for $t < nT$, and, as a result, we obtain a local n-times integrated semigroup $V_n(t)$:

$$
\begin{aligned}
V_n(t)x &= \left\{\int_0^{\infty} \frac{(t-s)^{n-1}}{(n-1)!} e^{a_m s} x_m\,ds\right\}_{m=1}^{\infty} \\
&= \left\{\left[a_m^{-n} e^{a_m t} - \sum_{p=1}^{n} (a_m)^{-p} \frac{t^{n-p}}{(n-p)!}\right] x_m\right\}.
\end{aligned}
$$

Since $|e^{a_m t}| = e^{\frac{mt}{T}}$ and $|a_m| = \frac{e^m}{m}$, it follows that

$$m^n e^{m(\frac{t}{T}-n)} \quad - \quad \sum_{p=1}^{n} m^p e^{-pm} \frac{t^{n-p}}{(n-p)!}$$

$$\leq \quad \left| \int_0^t \frac{(t-s)^{n-1}}{(n-1)!} e^{a_m s} ds \right|$$

$$\leq \quad m^n e^{m(\frac{t}{T}-n)} + \sum_{p=1}^{n} m^p e^{-pm} \frac{t^{n-p}}{(n-p)!},$$

therefore

$$\|V_n(t)\| = \sup_{m \in N} \left\{ \left| \int_0^t \frac{(t-s)^{n-1}}{(n-1)!} e^{a_m s} ds \right| \right\}$$

is bounded if and only if $0 \leq t < nT$.

1.3 κ-convoluted semigroups

In the previous section we showed that an n-times integrated semigroup with the densely defined generator A can be defined as a family of bounded linear operators $\{V(t),\ t \in [0, T)\}$ satisfying

$$V(t)x - \frac{t^n}{n!}x \quad = \quad \int_0^t V(s)Axds = \int_0^t AV(s)xds,$$

$$0 \leq t < T, \quad x \in D(A),$$

and

$$V(t)x - \frac{t^n}{n!}x = A \int_0^t V(s)xds, \quad x \in X.$$

A similar approach is used in this section for κ-convoluted semigroups.

1.3.1   Generators of κ-convoluted semigroups

Definition 1.3.1 *Let A be a closed operator and $\kappa(\cdot)$ be a continuous function on $[0, T)$, $T \leq \infty$. If there exists a strongly continuous operator-family $\{S_\kappa(t),\ t \in [0, T)\}$ such that $S_\kappa(t)Ax = AS_\kappa(t)x$ for $x \in D(A)$, $t \in [0, T)$, and for all $x \in X$*

$$\int_0^t S_\kappa(s)xds \in D(A)$$

and

$$S_\kappa(t)x = A \int_0^t S_\kappa(s)xds + \Theta(t)x, \quad 0 \leq t < T, \qquad (1.3.1)$$

where $\Theta(t) = \int_0^t \kappa(s)ds$, *then* S_κ *is called a κ-convoluted semigroup generated by* A, *and* A *is called the generator of* S_κ.

In the particular case $\Theta(t) = t^n/n!$, the κ-convoluted semigroup S_κ is an n-times integrated semigroup. If A is the generator of a C_0-semigroup U, then

$$S_\kappa(t)x = \int_0^t \frac{(t-s)^{n-1}}{(n-1)!} U(s)x ds$$

is an n-times integrated semigroup and a κ-convoluted semigroup. This is the reason that S_κ is called a κ-convoluted semigroup, but not a Θ-convoluted semigroup.

The corresponding Cauchy problem

$$v'(t) = Av(t) + \Theta(t)x, \quad 0 \le t < T, \quad v(0) = 0 \tag{1.3.2}$$

is called the Θ-convoluted Cauchy problem. If there exists a solution of the Cauchy problem

$$u'(t) = Au(t), \quad 0 \le t < T, \quad u(0) = x, \tag{1.3.3}$$

then, as usual for a nonhomogeneous equation, we have $v = \Theta * u$.

The following results connect the existence of a κ-convoluted semigroup and the well-posedness of a Θ-convoluted Cauchy problem with the behaviour of the resolvent $R_A(\lambda)$. In contrast to the case of an exponentially bounded and local n-times integrated semigroup, where the resolvent has polynomial estimates in a half-plane or in a logarithmic region, respectively, in the 'convoluted' case the resolvent exists in some smaller region and is allowed to increase exponentially:

$$\|R_A(\lambda)\| \le K e^{M(\lambda)}.$$

Here, a real-valued function $M(\lambda)$, $\lambda \in \mathbb{C}$, grows not faster than λ and is defined by the order of decreasing of $\widetilde{\kappa}(\lambda)$, the Laplace transform of $\kappa(t)$.

Theorem 1.3.1 *Let* $\kappa(\cdot)$ *be an exponentially bounded function such that*

$$|\widetilde{\kappa}(\lambda)| \underset{|\lambda| \to \infty}{=} \mathcal{O}\left(e^{-M(\lambda)}\right).$$

Suppose that A *is the generator of a κ-convoluted semigroup* $\{S_\kappa(t), 0 \le t < T\}$, *then there exists a region*

$$\Lambda = \left\{\lambda \in \mathbb{C} \mid \operatorname{Re}\lambda > \frac{M(\lambda)}{T} + \beta\right\}$$

such that $\|R_A(\lambda)\| \le K e^{M(\lambda)}$ *for any* $\lambda \in \Lambda$ *and some* $K > 0$.

Proof Let $|\kappa(t)| \leq K e^{\omega t}$. Consider the operator

$$R(\lambda, t) := \tilde{\kappa}^{-1} \int_0^t e^{-\lambda s} S_\kappa(s)\, ds,$$

where $\tilde{\kappa}^{-1} = 1/\tilde{\kappa}(\lambda)$. As in the case of integrated semigroups (Theorem 1.2.6) we now show that $R(\lambda, t)$ is 'nearly' the resolvent of A:

$$R_A(\lambda) = R(\lambda, t)\Big[I - B_t(\lambda)\Big]^{-1}, \quad \lambda \in \Lambda,$$

where $B_t(\lambda)$ are linear operators such that $\|B_t(\lambda)\| \leq \delta < 1$ for $t \in [0, T)$ and $\lambda \in \Lambda$. Since $\int_0^t e^{-\lambda s} S_\kappa(s)\, ds x \in D(A)$ for any $x \in X$, we can apply operator $(\lambda I - A)$ to $R(\lambda, t)x$

$$(\lambda I - A) R(\lambda, t) x =$$
$$\lambda \tilde{\kappa}^{-1}\left[\int_0^t e^{-\lambda s} S_\kappa(s)\, ds - A \int_0^t e^{-\lambda s} S_\kappa(s)\, ds\right] x, \quad x \in X.$$

Using the equation for the κ-convoluted semigroup for the first term in the right-hand side and integrating by parts the second term, we obtain

$$(\lambda I - A) R(\lambda, t)$$
$$= \tilde{\kappa}^{-1}\left(\lambda \int_0^t e^{-\lambda s}\Big[A \int_0^s S_\kappa(\tau)\, d\tau + \Theta(s)I\Big]\, ds\right.$$
$$\left. - A\lambda \int_0^t e^{-\lambda s} \int_0^s S_\kappa(\tau)\, d\tau\, ds - A e^{-\lambda t} \int_0^t S_\kappa(\tau)\, d\tau\right)$$
$$= \tilde{\kappa}^{-1}\left[\lambda \int_0^t e^{-\lambda s}\Theta(s)I\, ds - e^{-\lambda t} A \int_0^t S_\kappa(s)\, ds\right]$$
$$= \tilde{\kappa}^{-1}\left[\int_0^t e^{-\lambda s}\kappa(s)I\, ds - e^{-\lambda t} S_\kappa(t)\right]$$
$$= I - \tilde{\kappa}^{-1}\left[e^{-\lambda t} S_\kappa(t) + \int_t^\infty e^{-\lambda s}\kappa(s)I\, ds\right]$$
$$=: I - B_t(\lambda).$$

We estimate norms of operators $B_t(\lambda)$ for $\operatorname{Re}\lambda \geq \omega_1 > \omega$:

$$\|B_t(\lambda)\| \leq |\tilde{\kappa}^{-1}|\left(e^{-\operatorname{Re}\lambda t}\|S_\kappa(t)\| + \int_t^\infty e^{-\operatorname{Re}\lambda s}|\kappa(s)|\, ds\right)$$
$$\leq K|\tilde{\kappa}^{-1}|\left(e^{-\operatorname{Re}\lambda t} + \int_t^\infty e^{\omega s - \operatorname{Re}\lambda s}\, ds\right)$$
$$\leq K|\tilde{\kappa}^{-1}|e^{(\omega - \operatorname{Re}\lambda)t}\left(1 + \frac{1}{\omega_1 - \omega}\right) \leq K e^{M(\lambda) - (\operatorname{Re}\lambda - \omega)t}.$$

Let $Ke^{M(\lambda)-(\operatorname{Re}\lambda-\omega)t} \leq \delta < 1$, then

$$M(\lambda) - (\operatorname{Re}\lambda - \omega)t - \ln\frac{\delta}{K} \leq 0.$$

For $t = T$ we obtain

$$\operatorname{Re}\lambda \geq \frac{M(\lambda) - \ln\frac{\delta}{K}}{T} + \omega =: \frac{M(\lambda)}{T} + \beta,$$

and for these λ

$$\|B_t(\lambda)\| \leq \delta.$$

Since for $x \in D(A)$ we have

$$(\lambda I - A)R(\lambda, t)x = R(\lambda, t)(\lambda I - A)x = (I - B_t(\lambda))x,$$

then the operator $(\lambda I - A)^{-1} = R_A(\lambda)$ exists in the region

$$\overline{\Lambda} = \left\{\lambda \in \mathbb{C} \;\middle|\; \operatorname{Re}\lambda \geq \frac{M(\lambda)}{T} + \beta\right\},$$

and for these λ

$$\|R(\lambda)\| \leq |\widetilde{\kappa}^{-1}|\left|\int_0^t e^{-\lambda s}S_\kappa(s)\,ds\right|\frac{1}{1-\delta} \leq Ke^{M(\lambda)}.\;\Box$$

Remark 1.3.1 If, instead of the condition

$$|\widetilde{\kappa}(\lambda)|\underset{|\lambda|\to\infty}{=} \mathcal{O}\left(e^{-M(\lambda)}\right)$$

we take

$$|\widetilde{\kappa}(\lambda)|\underset{|\lambda|\to\infty}{=} \mathcal{O}\left(e^{-M(\alpha\lambda)}\right)$$

for some α, then for any

$$\lambda \in \Lambda_{T,\alpha,\beta} := \left\{\lambda \in \mathbb{C} \;\middle|\; \operatorname{Re}\lambda > \frac{M(\alpha\lambda)}{T} + \beta\right\}$$

we have

$$\|R_A(\lambda)\| \leq Ke^{M(\alpha\lambda)}.\;\Box$$

Remark 1.3.2 Note that the equality (1.3.1) is considered on X, not on $D(A)$. This is the reason that, in contrast to Theorem 1.2.5 for integrated semigroups, here we do not assume that $\overline{D(A)} = X$.

Theorem 1.3.2 *Suppose*

$$\forall \lambda \in \Lambda_{\gamma,\alpha,\beta}, \quad \|R_A(\lambda)\| \leq K e^{M(\alpha\lambda)} \tag{1.3.4}$$

for some $K, \gamma, \alpha, \beta > 0$, and

$$|\widetilde{\kappa}(\lambda)| \underset{|\lambda| \to \infty}{=} \mathcal{O}\left(e^{-M(\varsigma\lambda)}\right) \tag{1.3.5}$$

with some $\varsigma > 0$ such that $(\varsigma/\alpha - 1) > 0$. Then A is the generator of a κ-convoluted semigroup on $[0, T_1) = [0, \gamma(\frac{\varsigma}{\alpha} - 1))$.

Proof Let $R_A(\lambda)$ satisfy (1.3.4) and $\kappa(\cdot)$ satisfy (1.3.5) with some $\varsigma > 0$ such that $(\varsigma/\alpha - 1) > 0$. Consider the inverse Laplace transform of $\widetilde{\kappa}(\lambda) R_A(\lambda)$

$$S^1_\kappa(t) := \int_\Gamma e^{\lambda t} \widetilde{\kappa}(\lambda) R_A(\lambda) \, d\lambda,$$

where $\Gamma = \partial \Lambda_{\gamma,\alpha,\beta}$ is the boundary of $\Lambda_{\gamma,\alpha,\beta}$. Then

$$\|S^1_\kappa(t)\| \leq \int_\Gamma K e^{\operatorname{Re} \lambda t + M(\alpha\lambda) - M(\varsigma\lambda)} \, |d\lambda|.$$

Since

$$e^{\operatorname{Re} \lambda t + M(\alpha\lambda) - M(\varsigma\lambda)} = e^{(t + \gamma - \gamma\frac{\varsigma}{\alpha})\operatorname{Re} \lambda}$$

on Γ, the operators $S^1_\kappa(t)$ are defined for

$$t + \gamma - \gamma\frac{\varsigma}{\alpha} < 0 \quad \text{or} \quad t < \gamma\left(\frac{\varsigma}{\alpha} - 1\right) =: T_1.$$

Now we show that operators $S^1_\kappa(t)$ satisfy the equation (1.3.1) for the κ-convoluted semigroup:

$$S^1_\kappa(t)x = A \int_0^t S^1_\kappa(s)x \, ds + \Theta(t)x, \quad x \in X.$$

We have

$$A \int_0^t S^1_\kappa(s)x \, ds$$

$$= (A \pm \lambda I) \int_0^t \int_\Gamma e^{\lambda s} R_A(\lambda) \widetilde{\kappa}(\lambda) x \, d\lambda ds$$

$$= \lambda \int_0^t \int_\Gamma e^{\lambda s} R_A(\lambda) \widetilde{\kappa}(\lambda) x d\lambda ds - \int_0^t \int_\Gamma e^{\lambda s} \widetilde{\kappa}(\lambda) x \, d\lambda ds$$

$$= \int_\Gamma R_A(\lambda) \widetilde{\kappa}(\lambda) \lambda \int_0^t e^{\lambda s} x \, ds d\lambda - \int_0^t \kappa(s)x \, ds$$

$$= S^1_\kappa(t)x - \int_\Gamma R_A(\lambda) \widetilde{\kappa}(\lambda) x \, d\lambda - \Theta(t)x.$$

Since on Γ we have

$$\|R_A(\lambda)\tilde{\kappa}(\lambda)\| \leq K e^{M(\alpha\lambda) - M(\varsigma\lambda)} = K e^{\gamma(1 - \frac{\varsigma}{\alpha})\mathrm{Re}\,\lambda},$$

and $(1 - \frac{\varsigma}{\alpha}) < 0$, then the integral $\int_\Gamma R_A(\lambda)\tilde{\kappa}(\lambda)x\,d\lambda$ is equal to zero. Hence for any $x \in X$, we have

$$A \int_0^t S_\kappa^1(s)x\,ds = S_\kappa^1(t)x - \Theta(t)x. \quad \square$$

Combining these results we obtain the necessary and sufficient conditions for A to be the generator of a κ-convoluted semigroup. They are also sufficient for well-posedness of the Θ-convoluted Cauchy problem (1.3.2).

1.3.2 Θ-convoluted Cauchy problem

Theorem 1.3.3 *Let* $M(\lambda)$, $\lambda \in \mathbb{C}$, *be a real-valued function growing at infinity not faster than* λ^p, $p < 1$. *Then the following conditions are equivalent:*

(K) *A is the generator of a κ-convoluted semigroup on $[0, T)$ for some $T > 0$, with κ such that*

$$\left|\tilde{\kappa}(\lambda)\right|_{|\lambda| \to \infty} = \mathcal{O}\left(e^{-M(\varsigma\lambda)}\right) \quad \text{for some } \varsigma > 0;$$

(R)

$$\exists \gamma, \alpha, \beta > 0 : \quad \forall \lambda \in \Lambda_{\gamma,\alpha,\beta}, \qquad \|R_A(\lambda)\| \leq K e^{M(\alpha\lambda)};$$

(K1) *for any $T > 0$, A is the generator of a κ-convoluted semigroup on $[0, T)$ with κ such that*

$$\left|\tilde{\kappa}(\lambda)\right|_{|\lambda| \to \infty} = \mathcal{O}\left(e^{-M(\varsigma\lambda)}\right) \quad \text{for some } \varsigma > 0;$$

(R1)

$$\forall \gamma > 0 \ \exists \alpha, \beta > 0 : \quad \forall \lambda \in \Lambda_{\gamma,\alpha,\beta}, \ \|R_A(\lambda)\| \leq K e^{M(\alpha\lambda)}.$$

In this case the Θ-convoluted Cauchy problem has the unique solution $v(t) = \int_0^t S_\kappa(s)x\,ds$, $t \in [0, T)$.

Proof From Theorem 1.3.1 and Remark 1.3.1 we have (K) $\implies$ (R) with $\gamma = T$, $\alpha = \varsigma$ and (K1) $\implies$ (R1) with $\gamma = T$, $\alpha = \varsigma$. From Theorem 1.3.2 we have (R) $\implies$ (K1) with ς defined from the equality $T = \gamma(\frac{\varsigma}{\alpha} - 1)$, and (R1) $\implies$ (K) with $T = \gamma$, $\varsigma = \alpha$. $\square$

It follows from the definition of a κ-convoluted semigroup that $v(t) = \int_0^t S_\kappa(s)x\,ds$, $x \in X$, $t \in [0, T)$, is a solution of (1.3.2). By the Lyubich's theorem (Remark 1.1.6), the uniqueness of this solution follows from the condition on $M(\lambda)$ and the estimate (1.3.4):

$$\lim_{\lambda \to \infty} \frac{\ln \|R_A(\lambda)\|}{|\lambda|} \leq \lim_{\lambda \to \infty} \frac{|\lambda^p|}{|\lambda|} = 0.$$

Using the uniqueness of a solution of the Θ-convoluted Cauchy problem we can obtain for a κ-convoluted semigroup its 'semigroup' relation:

$$
\begin{aligned}
S_\kappa(t)S_\kappa(s) &= S_\kappa(s)S_\kappa(t) & (1.3.6)\\
&= \int_s^{t+s} \kappa(t+s-r)S_\kappa(r)dr - \int_s^t \kappa(t+s-r)S_\kappa(r)dr.
\end{aligned}
$$

Note that (V1) from Definition 1.2.1 is the particular case of (1.3.6) with $\kappa(t) = t^{n-1}/(n-1)!$, and (1.3.6) was proved in Lemma 1.2.3 .

Theorem 1.3.4 *Let $\{S_\kappa(t),\ 0 \leq t < T\}$ be a κ-convoluted semigroup, then for any $0 \leq t < T$ the equality (1.3.6) holds.*

Proof See the proof of Lemma 1.2.3 with $E(t) = \kappa(t)$. $\square$

In the Section 2.3, devoted to Cauchy problems in the spaces of (abstract) ultradistributions, we show that the well-posedness of the Cauchy problem in a space of ultradistributions is equivalent to the condition (R), where $M(\lambda)$ is the function associated with a sequence M_n that defines the space of ultradistributions. Hence, any of the conditions (K), (K1), (R1) with such function $M(\lambda)$ are equivalent to well-posedness of the Cauchy problem in the space of ultradistributions.

1.4 C-regularized semigroups

We continue the study of families of bounded operators connected with Cauchy problems that are not uniformly well-posed. C-regularized semigroups are related to problems that are C-well-posed (see Definition 1.4.4) on $CD(A)$ for some bounded invertible operator C.

1.4.1 Generators of C-regularized semigroups

Suppose C is an injective, bounded operator on a Banach space X.

Definition 1.4.1 *A one-parameter family of bounded linear operators $\{S(t),\ t \geq 0\}$ is called an exponentially bounded C-regularized semigroup if*

(C1) $S(t+h)C = S(t)S(h), \qquad t, h \geq 0,\ S(0) = C;$

(C2) $S(t)$ *is strongly continuous with respect to* $t \geq 0;$

(C3) $\exists K > 0, \; \omega \in \mathbb{R} : \quad \|S(t)\| \le K e^{\omega t}, \qquad t \ge 0.$

It follows from (C1) at $t = 0$ that $S(h)$ commutes with C for all $h \ge 0$:

(C4) $S(h)C = CS(h), \qquad h \ge 0.$

If $C = I$, then the C-regularized semigroup is a C_0-semigroup.
   Define the operator

$$L(\lambda)x = \int_0^\infty e^{-\lambda t} S(t)x\,dt, \qquad x \in X. \tag{1.4.1}$$

If $C = I$, then $L(\lambda) = R_A(\lambda)$, where A is the generator of the corresponding C_0-semigroup.
   Property (C3) implies that $L(\lambda)$ is a bounded operator for all $\lambda \in \mathbb{C}$ with $\operatorname{Re}\lambda > \omega$. We show that this operator, called the C-resolvent, is invertible and satisfies the 'regularized' resolvent identity.

Proposition 1.4.1 *Let* $\{S(t), \; t \; \ge \; 0\}$ *be an exponentially bounded C-regularized semigroup. Then for all $\lambda, \mu \in \mathbb{C}$ with $\operatorname{Re}\lambda, \operatorname{Re}\mu > \omega$*

$$(\lambda - \mu)L(\lambda)L(\mu) = L(\mu)C - L(\lambda)C, \tag{1.4.2}$$

and there exists a closed operator Z defined by

$$Zx := \left(\lambda I - L(\lambda)^{-1}C\right)x, \tag{1.4.3}$$

$$D(Z) = \left\{x \in X \;\middle|\; Cx \in \operatorname{ran} L(\lambda)\right\}.$$

Proof Let $x \in X$ and consider $\lambda, \mu \in \mathbb{C}$ with $\operatorname{Re}\lambda, \operatorname{Re}\mu > \omega$. Then

$$
\begin{aligned}
& L(\lambda)L(\mu)x \\
&= \int_0^\infty \int_0^\infty e^{-(\lambda s + \mu t)} S(s + t) Cx\,dt\,ds \\
&= \int_0^\infty \int_0^\tau e^{-(\lambda(\tau - t) + \mu t)} S(\tau) Cx\,dt\,d\tau \\
&= (\mu - \lambda)^{-1}\left[\int_0^\infty e^{-\lambda\tau} S(\tau) Cx\,d\tau - \int_0^\infty e^{-\mu\tau} S(\tau) Cx\,d\tau\right] \\
&= (\mu - \lambda)^{-1}\left[L(\lambda)Cx - L(\mu)Cx\right],
\end{aligned}
$$

i.e. (1.4.2) holds. We now show that $L(\lambda)$ is injective for $\operatorname{Re}\lambda > \omega$. It follows from (1.4.2) that if $L(\lambda) = 0$, then $CL(\mu)x = 0$ and $L(\mu)x = 0$, that is $\ker L(\lambda) = \ker L(\mu)$. From the definition of $L(\lambda)$ and the uniqueness theorem for Laplace transform we obtain that if $L(\lambda)x = 0$, then $S(0)x =$

$Cx = 0$ and $x = 0$. Hence the operator $L(\lambda)$ is invertible for $\operatorname{Re} \lambda > \omega$. Let $Cx \in \operatorname{ran} L(\lambda)$. From (1.4.2) we have

$$L(\mu)\,[\lambda L(\lambda) - C\,]\,x = L(\lambda)\,[\mu L(\mu) - C\,]\,x, \qquad \operatorname{Re}\lambda, \operatorname{Re}\mu > \omega,$$

and

$$L(\lambda)^{-1}\,[\lambda L(\lambda) - C\,]\,x = L(\mu)^{-1}\,[\mu L(\mu) - C\,]\,x.$$

Therefore, the operator Z defined by

$$Zx := L(\lambda)^{-1}\,[\lambda L(\lambda) - C\,]\,x = [\lambda I - L(\lambda)^{-1}C\,]x$$

is independent of λ and is closed. Then we have

$$(\lambda I - Z)^{-1} = C^{-1}L(\lambda), \qquad\qquad\qquad (1.4.4)$$

$$D\left((\lambda I - Z)^{-1}\right) = \left\{x \in X \;\middle|\; L(\lambda)x \in \operatorname{ran} C\right\}$$

and

$$L(\lambda)x = C(\lambda I - Z)^{-1}x, \qquad x \in D\left((\lambda I - Z)^{-1}\right). \;\square$$

Definition 1.4.2 *The operator Z defined by (1.4.3) is called the generator of a C-regularized semigroup $\{S(t),\ t \geq 0\}$.*

Along with the generator Z we introduce the infinitesimal generators A and G by

$$Ax := C^{-1} \lim_{t \to 0} t^{-1}\,[S(t)x - Cx],$$

$$D(A) = \left\{x \in X \;\middle|\; \exists \lim_{t \to 0} t^{-1}\,[S(t)x - Cx] \in \operatorname{ran} C\right\}$$

and

$$Gx := \lim_{t \to 0} t^{-1}\left[C^{-1}(S(t)x - Cx)\right], \qquad\qquad (1.4.5)$$

$$D(G) = \left\{x \in \operatorname{ran} C \;\middle|\; \exists \lim_{t \to 0} t^{-1}\left[C^{-1}(S(t)x - Cx)\right]\right\}.$$

Proposition 1.4.2 *Let $\{S(t),\ t \geq 0\}$ be a C-regularized semigroup. Then for operators G, A, Z the following relation holds*

$$G \subset A = Z. \qquad\qquad\qquad (1.4.6)$$

If $\overline{\operatorname{ran} C} = X$, *then* $\overline{D(G)} = X$.

Proof Let $x \in D(G)$, then

$$C \lim_{t \to 0} \frac{C^{-1}S(t)x - Cx}{t} = \lim_{t \to 0} \frac{S(t)x - Cx}{t} \in \operatorname{ran} C$$

and

$$Ax = C^{-1} \lim_{t \to 0} \frac{S(t)x - Cx}{t} = \lim_{t \to 0} \frac{C^{-1}S(t)x - x}{t} = Gx.$$

Hence $D(G) \subset D(A)$ and $G \subset A$. In order to show $A \subset Z$, let $x \in D(A)$. Then

$$
\begin{aligned}
\frac{dS(t)Cx}{dt} &= \lim_{h \to 0} \frac{S(t)(S(h) - C)x}{h} \\
&= S(t) \lim_{h \to 0} \frac{S(h)x - Cx}{h} = S(t)CAx,
\end{aligned}
$$

and

$$
\begin{aligned}
CL(\lambda)(\lambda I - A)x &= L(\lambda)C(\lambda I - A)x \\
&= \lambda L(\lambda)Cx - \int_0^\infty e^{-\lambda t} S(t)CAx \, dt \\
&= \lambda L(\lambda)Cx - \int_0^\infty e^{-\lambda t} \frac{dS(t)Cx}{dt} dt = C^2 x.
\end{aligned}
$$

Hence $L(\lambda)(\lambda I - A)x = Cx$, i.e. $Cx \in \operatorname{ran} L$ and

$$Zx = \left[\lambda I - L^{-1}(\lambda)C \right] x = Ax.$$

Now let $x \in D(Z)$, then there exists y such that $Cx = L(\lambda)y$. We consider

$$
\begin{aligned}
\frac{S(h)x - Cx}{h} & \\
&= \frac{S(h)C^{-1}L(\lambda)y - L(\lambda)y}{h} \\
&= \frac{1}{h} \left[C^{-1}S(h) \int_0^\infty e^{-\lambda t} S(t)y \, dt - \int_0^\infty e^{-\lambda t} S(t)y \, dt \right] \\
&= \frac{1}{h} \left[\int_0^\infty e^{-\lambda t} S(t + h)y \, dt - \int_0^\infty e^{-\lambda t} S(t)y \, dt \right] \\
&= \frac{e^{\lambda h} - 1}{h} \int_h^\infty e^{-\lambda t} S(t)y \, dt - \frac{1}{h} \int_0^h e^{-\lambda t} S(t)y \, dt \\
&\longrightarrow_{h \to 0} \lambda L(\lambda)y - Cy = C(\lambda x - y) \in \operatorname{ran} C.
\end{aligned}
$$

Hence $x \in D(A)$, $D(A) = D(Z)$ and $Ax = Zx$.

Now suppose $\overline{\operatorname{ran} C} = X$. Taking into account that C is injective, we obtain that $\operatorname{ran} C^2$ is also dense in X. Therefore, in order to prove $\overline{D(G)} = X$, we show that

$$\forall x \in \operatorname{ran} C^2, \ \forall \varepsilon > 0, \ \exists x_\varepsilon \in D(G) : \ \|x - x_\varepsilon\| < \varepsilon.$$

It follows from properties of C-regularized semigroups that the operators

$$U(t) := C^{-1}S(t), \quad t \geq 0$$

with

$$D(U(t)) = \left\{ x \in X \mid S(t)x \in \operatorname{ran} C \right\},$$

form a C_0-semigroup on $\operatorname{ran} C^2$. Put $x = C^2 y$ and

$$x(t) := \int_0^t U(\tau)x d\tau = C \int_0^t S(\tau)y d\tau \in \operatorname{ran} C.$$

We have $\|x(t) - x\| \to_{t \to 0} 0$. Let us show that $x(t) \in D(G)$. Since

$$U(h)x(t) \quad = \quad U(h)C \int_0^t S(\tau)y d\tau = \int_0^t S(h + \tau)Cy d\tau,$$

we obtain

$$h^{-1} \left[U(h)x(t) - x(t) \right]$$

$$= \quad h^{-1} \int_0^t S(h + \tau)Cy d\tau - h^{-1} \int_0^t S(\tau)Cy d\tau$$

$$= \quad h^{-1} \int_h^{t+h} S(\tau)Cy d\tau - h^{-1} \int_0^t S(\tau)Cy d\tau$$

$$= \quad h^{-1} \int_t^{t+h} S(\tau)Cy d\tau - h^{-1} \int_0^h S(\tau)Cy d\tau$$

$$\longrightarrow_{h \to 0} \quad S(t)Cy - C^2 y.$$

Hence, $x(t) \in D(G)$, $Gx(t) = S(t)Cy - C^2 y$, and $\overline{D(G)} = X$. $\square$

Definition 1.4.3 *The closure of operator G defined by (1.4.5) is called the complete infinitesimal generator of the C-regularized semigroup $\{S(t), t \geq 0\}$.*

For the complete infinitesimal generator we have the inclusion $\overline{G} \subset Z = A$.

Proposition 1.4.3 *Let $\{S(t), \, t \geq 0\}$ be a C-regularized semigroup. If $\rho(\overline{G}) \neq \emptyset$, then $\overline{G} = A = Z$.*

Proof Let us show the equality

$$Ax = C^{-1}GCx = C^{-1}\overline{G}Cx, \qquad x \in D(A), \tag{1.4.7}$$

which together with the assumption $\rho(\overline{G}) \neq \emptyset$ implies that $\overline{G} = A$. Take $x \in D(A)$, then $Cx \in D(G)$:

$$GCx \quad = \quad \lim_{h \to 0} h^{-1} \left[C^{-1}S(h)Cx - Cx \right]$$

$$= \quad \lim_{h \to 0} h^{-1} \left[S(h)x - Cx \right] = CAx.$$

Therefore $\overline{G}Cx = CAx$ for all $x \in D(A)$. Applying C^{-1} to both sides of this equality, we obtain (1.4.7).

Let $\lambda \in \rho(\overline{G})$ and $x \in D(A)$, put $y = (\lambda I - A)x$, $z = (\lambda I - \overline{G})^{-1}y$. Then $z \in D(\overline{G})$ and $(\lambda I - \overline{G})z = (\lambda I - A)z = y = (\lambda I - A)x$. In view of invertibility of $(\lambda I - \overline{G})$, the operator

$$(\lambda I - A) = C^{-1}(\lambda I - \overline{G})C$$

is also invertible. Hence $z = x \in D(\overline{G})$ and $D(A) \subset D(\overline{G})$, that is $A = \overline{G}$. $\square$

The following theorem makes clear the connection between a generator of a C-regularized semigroup and the Cauchy problem (CP).

Theorem 1.4.1 *Let Z be the generator of an exponentially bounded C-regularized semigroup $\{S(t),\ t \geq 0\}$. Then for $x \in X$*

$$S(t)x - Cx = Z \int_0^t S(\tau)x d\tau; \qquad (1.4.8)$$

for $x \in D(Z)$

$$S(t)x - Cx = \int_0^t S(\tau)Zx d\tau, \qquad (1.4.9)$$

$$\frac{dS(t)}{dt}x = S(t)Zx = ZS(t)x, \qquad S(0)x = Cx; \qquad (1.4.10)$$

for $x \in CD(Z)$ we have

$$\frac{dU(t)}{dt}x = ZU(t)x, \qquad U(0)x = x, \qquad (1.4.11)$$

where $U(t)x := C^{-1}S(t)x$.

Proof Keeping in mind that $A = Z$, we prove that if $x \in X$, then $\int_0^t S(\tau)x d\tau \in D(A)$ and (1.4.8) holds true. Let $x \in X$, then

$$\frac{S(h)\int_0^t S(\tau)x d\tau - C\int_0^t S(\tau)x d\tau}{h}$$

$$= \frac{\int_0^t S(\tau + h)Cx d\tau - C\int_0^t S(\tau)x d\tau}{h}$$

$$= \frac{C\int_0^{t+h} S(\tau)x d\tau - \int_0^h S(\tau)Cx d\tau}{h}$$

$$\xrightarrow[h\to 0]{} C[S(t) - C]x.$$

Thus

$$A \int_0^t S(\tau)x d\tau = S(t)x - Cx.$$

Since the operator $A = Z$ is closed, we have (1.4.9):

$$\int_0^t S(\tau)Ax d\tau = \int_0^t S(\tau)Zx d\tau = S(t)x - Cx$$

which implies (1.4.10) – (1.4.11). $\square$

Now we can summarize all properties of the generator of a C-regularized semigroup.

Theorem 1.4.2 *Let Z be the generator of an exponentially bounded C-regularized semigroup $\{S(t),\ t \geq 0\}$, then Z is closed. If, in addition, $\overline{\operatorname{ran} C} = X$, then Z is densely defined. For all $\lambda \in \mathbb{C}$ with $\operatorname{Re}\lambda > \omega$ the following conditions hold*

(Z1) $(\lambda I - Z)^{-1}Cx = C(\lambda I - Z)^{-1}x$ *for all* $x \in \operatorname{ran}(\lambda I - Z)$;

(Z2) $\exists K > 0$,

$$\|(\lambda I - Z)^{-m}C\| \leq \frac{K}{(\operatorname{Re}\lambda - \omega)^m}, \qquad m = 1, 2, \ldots.$$

Proof In Proposition 1.4.1 we proved that the operator

$$(\lambda I - Z)^{-1} = C^{-1}L(\lambda)$$

exists on

$$\operatorname{ran}(\lambda I - Z) = \Big\{x \in X \ \Big|\ L(\lambda)x \in \operatorname{ran} C\Big\}.$$

Hence $C(\lambda I - Z)^{-1}x = L(\lambda)x$ for $x \in \operatorname{ran}(\lambda I - Z)$. On the other hand, for all $x \in X$ we have $L(\lambda)Cx = CL(\lambda)x$, therefore (Z1) holds. Let us show that $\operatorname{ran} C \subset D((\lambda I - Z)^{-m})$, $m = 1, 2, \ldots$. If $x \in X$, then $L(\lambda)Cx = CL(\lambda)x \in \operatorname{ran} C$ and the following operators are defined

$$
\begin{aligned}
(\lambda I - Z)^{-1}Cx &= C^{-1}L(\lambda)Cx = C^{-1}\int_0^\infty e^{-\lambda t_1} S(t_1)Cx dt_1 \\
&= \int_0^\infty e^{-\lambda t_1} C^{-1}S(t_1)Cx dt_1 = \int_0^\infty e^{-\lambda t_1} S(t_1)x dt_1, \\
(\lambda I - Z)^{-2}Cx &= \int_0^\infty C^{-1}e^{-\lambda t_2}S(t_2)\int_0^\infty e^{-\lambda t_1}C^{-1}S(t_1)Cx dt_1 dt_2 \\
&= \int_0^\infty \int_0^\infty e^{-\lambda(t_1+t_2)}S(t_1+t_2)x dt_1 dt_2
\end{aligned}
$$

and so on. Hence we obtain the estimates (Z2):

$$\|(\lambda I - Z)^{-m}Cx\| = \|(C^{-1}L(\lambda))\ldots(C^{-1}L(\lambda))Cx\|$$

$$\leq \int_0^\infty \cdots \int_0^\infty e^{-\operatorname{Re}\lambda(t_1+\ldots+t_m)} K e^{\omega(t_1+\ldots+t_m)} \|x\| dt_1 \ldots dt_m$$

$$\leq \frac{K}{(\operatorname{Re}\lambda - \omega)^m} \|x\|. \ \square$$

Property (Z2) generalizes the MFPHY condition to the case of C-regularized semigroups, and Theorem 1.4.2 indicates that the MFPHY-type condition on the C-resolvent is necessary for operator Z to be the generator of a C-regularized semigroup. The next theorem gives sufficient conditions for operator Z to be the generator of a C-regularized semigroup. Theorem 1.4.2 together with Theorem 1.4.3 are the generalization of the MFPHY theorem to the case of C-regularized semigroups with $\overline{\operatorname{ran} C} = X$.

Theorem 1.4.3 *Let Z be a densely defined closed linear operator satisfying (Z1), (Z2) for $\lambda \in \mathbb{C}$ with $\operatorname{Re}\lambda > \omega$. Then there exists an exponentially bounded C-regularized semigroup $\{S(t), \ t \geq 0\}$ with the generator Z.*

This theorem can be proved in a fashion similar to the proof in Remark 1.1.4. First we define operators

$$P_\lambda x = \lambda^2(\lambda I - Z)^{-1} x - \lambda x, \qquad x \in \operatorname{ran}(\lambda I - Z),$$

such that $\lim_{\lambda\to\infty} P_\lambda C x = ZC x$. Next define

$$S_\lambda(t)x := e^{-\lambda t} \sum_{n=1}^\infty \frac{t^n \lambda^{2n}}{n!} (\lambda I - Z)^{-n} C x, \qquad x \in X.$$

These operators are bounded and have all the properties of C-regularized semigroups. Finally, we set

$$S(t)x = \lim_{\lambda\to\infty} S_\lambda(t)x, \qquad x \in X. \ \square$$

1.4.2 C-well-posedness of the Cauchy problem

We consider the Cauchy problem

$$u'(t) = Au(t), \quad t \geq 0, \ u(0) = x, \tag{CP}$$

where A is the generator of an exponentially bounded C-regularized semigroup $\{S(t), \ t \geq 0\}$. It is shown in Theorem 1.4.1 that for $x \in CD(A)$ there exists a solution to this problem: $u(t) = U(t)x = C^{-1}S(t)x, \ t \geq 0$. The semigroup U is not necessarily bounded on X, therefore, in general, the solution $u(\cdot)$ is not stable in X with respect to variation of x. We now show that the well-posedness of the Cauchy problem can be restored in a certain subspace of X with a norm stronger than the norm of X.

Theorem 1.4.4 *Let $\{S(t),\ t \geq 0\}$ be an exponentially bounded C-regularized semigroup. Then the following statements are equivalent:*

(I) *A is the generator of $\{S(t),\ t \geq 0\}$;*

(II) a) *A is a closed operator and it commutes with C:*

$$ACx = CAx, \qquad x \in D(A);$$

b) *there exists a Banach space Σ with the norm $\|\cdot\|_\Sigma$ such that $\operatorname{ran} C \subset \Sigma \subset X$, and the following estimates hold*

$$\exists K_1 > 0 : \forall x \in \Sigma, \quad \|x\| \leq K_1 \|x\|_\Sigma,$$
$$\exists K_2 > 0 : \forall x \in \operatorname{ran} C, \quad \|x\|_\Sigma \leq K_2 \|C^{-1}x\|.$$

Furthermore, the part of the operator A in Σ is the generator of a C_0-semigroup.

Proof (I) $\Longrightarrow$ (II). Let $\{S(t),\ t \geq 0\}$ be an exponentially bounded C-regularized semigroup satisfying (C1) – (C3), then a) holds by virtue of (Z1). Let $b > \omega$, define in X the linear manifold

$$\Sigma = \Big\{ x \in X \ \Big|\ C^{-1}S(t)x \text{ is continuous with respect to } t \geq 0,$$

$$\text{and } \lim_{t \to \infty} e^{-bt}\|C^{-1}S(t)x\| = 0 \Big\}$$

with the norm

$$\|x\|_\Sigma = \sup_{t \geq 0} e^{-bt}\|C^{-1}S(t)x\|.$$

Taking into account that $C^{-1}S(t)S(h)x = S(t+h)x$, we have $\operatorname{ran} S(h) \subset \Sigma$ for $h \geq 0$. If $x \in \operatorname{ran} C = \operatorname{ran} S(0)$, then

$$\|x\|_\Sigma \leq \sup_{t \geq 0} Ke^{(\omega - b)t}\|C^{-1}x\| = K\|C^{-1}x\|.$$

For $x \in \Sigma$ we have

$$\|x\| = \left(e^{-bt}\|C^{-1}S(t)x\|\right)_{t=0} \leq \|x\|_\Sigma.$$

It is not difficult to show that Σ with the norm $\|\cdot\|_\Sigma$ is a Banach space. It is invariant with respect to operators $C^{-1}S(t)$ as

$$C^{-1}S(\tau)C^{-1}S(t)x = C^{-1}S(\tau + t)x, \quad x \in \Sigma.$$

Therefore, the operators $U(t) = C^{-1}S(t)$ map Σ into Σ. Let us show that A_Σ (the part of operator A in Σ) is the generator of a C_0-semigroup $\{U(t),\ t \geq 0\}$ in Σ. We have $U(t+\tau) = U(t)U(\tau)$ and $U(0) = I$. Moreover,

$$
\begin{aligned}
\|U(t)x\|_\Sigma &= \sup_{\tau \geq 0} e^{-b\tau}\|C^{-1}S(\tau + t)x\| \\
&\leq \sup_{\tau \geq 0} e^{-b(\tau + t)}e^{bt}\|C^{-1}S(\tau + t)x\| \\
&= e^{bt}\sup_{\tau \geq t} e^{-bt}\|C^{-1}S(\tau)x\| \\
&\leq e^{bt}\|x\|_\Sigma
\end{aligned}
$$

and

$$
\begin{aligned}
&\|U(h)x - x\|_\Sigma \\
&= \sup_{t \geq 0} e^{-bt}\|C^{-1}S(t+h)x - C^{-1}S(t)x\| \\
&\leq \max\left\{ \sup_{t \in [0,T]} e^{-bt}\|C^{-1}S(t+h)x - C^{-1}S(t)x\|, \right. \\
&\qquad\qquad \left. \sup_{t \geq T} e^{-bt}\|C^{-1}S(t)x\| + e^{bh}\sup_{t \geq T} e^{-b(t+h)}\|C^{-1}S(t+h)x\| \right\} \\
&\equiv \max\{a_T(h),\ c_T(h)\}.
\end{aligned}
$$

By definition of Σ we have

$$
\forall \varepsilon > 0,\ \exists T_0\ :\ \forall h \geq 0, \qquad c_{T_0}(h) < \frac{\varepsilon}{2}.
$$

Also, since the continuous function $C^{-1}S(\cdot)x$ is uniformly continuous on any closed interval, we have

$$
\forall \varepsilon > 0,\ \exists \delta\ :\ \forall h < \delta, \qquad a_{T_0}(h) < \frac{\varepsilon}{2}.
$$

Hence

$$
\|U(h)x - x\|_\Sigma \underset{h \to 0}{\longrightarrow} 0,
$$

i.e., the operators $U(t)$ form a C_0-semigroup on Σ. Let $\widetilde{A}$ be its generator, then for $x \in D(\widetilde{A})$

$$
\begin{aligned}
&\left\| t^{-1}\left[C^{-1}S(t)x - x\right] - \widetilde{A}x \right\| \\
&\leq K_1 \left\| t^{-1}\left[C^{-1}S(t)x - x\right] - \widetilde{A}x \right\|_\Sigma \underset{t \to 0}{\longrightarrow} 0.
\end{aligned}
$$

Therefore $x \in D(A) \cap \Sigma$ and $Ax = \widetilde{A}x \in \Sigma$, that is $x \in D(A_\Sigma)$ and $D(\widetilde{A}) \subset D(A_\Sigma)$. To prove $D(A_\Sigma) \subset D(\widetilde{A})$, we take $x \in D(A)$ such that

$Ax \in \Sigma$, then it follows from (1.4.9) that

$$U(t)x - x = \int_0^t U(\tau)Ax\,d\tau.$$

Hence, taking into account that for any $x \in \Sigma$ the function $U(\cdot)Ax$ is continuous in $\tau > 0$, we obtain

$$\left\| t^{-1}\left[C^{-1}S(t)x - x\right] - Ax \right\|_\Sigma \underset{t \to 0}{\longrightarrow} 0.$$

Thus $x \in D(\widetilde{A})$ and $A_\Sigma = \widetilde{A}$.

(II) $\Longrightarrow$ (I). Let U be a C_0-semigroup generated by the operator A_Σ. It is not difficult to verify that the operator-function $S(\cdot)$ defined by $S(t)x = U(t)Cx$ for all $x \in X$ and $t \geq 0$, satisfies (C1) – (C3). Let Z be the generator of S, then $C^{-1}ZCx = Zx$, $x \in D(Z)$, and

$$(\lambda I - Z)^{-1}Cx = \int_0^\infty e^{-\lambda t}S(t)x\,dt, \qquad x \in X,\ \lambda > \omega.$$

On the other hand, for the C_0-semigroup U we have

$$(\lambda I - A_\Sigma)^{-1}Cx = \int_0^\infty e^{-\lambda t}U(t)Cx\,dt, \qquad x \in X,\ \lambda > \omega,$$

therefore

$$(\lambda I - A_\Sigma)^{-1}Cx = (\lambda I - Z)^{-1}Cx, \qquad x \in X,\ \lambda > \omega. \tag{1.4.12}$$

Hence

$$\begin{aligned}
(\lambda I - Z)^{-1}C(\lambda I - A)x &= (\lambda I - A_\Sigma)^{-1}C(\lambda I - A)x \\
&= (\lambda I - A_\Sigma)^{-1}(\lambda I - A_\Sigma)Cx = Cx
\end{aligned}$$

for $x \in D(A)$. Consequently $Ax = C^{-1}ZCx = Zx$, and $A \subset Z$. In the same way, it follows from (1.4.12) that $Z \subset A$, so $A = Z$. $\square$

By Theorem 1.4.4, we obtain that (CP) with operator A being the generator of a C-regularized semigroup is well-posed in the space Σ. The fact that A_Σ generates a C_0-semigroup in Σ implies that for any $x \in D(A_\Sigma)$ there exists the unique solution of (CP): $u(\cdot) = U(\cdot)x$ such that

$$\sup_{0 \leq t \leq T} \|u(t)\|_\Sigma \leq K\|x\|_\Sigma. \tag{1.4.13}$$

Definition 1.4.4 *The Cauchy problem* (CP) *is said to be C-well-posed if for any $x \in CD(A)$ there exists a unique solution, that is stable with respect to the C^{-1}-graph-norm:*

$$\forall T < \infty,\ \exists K > 0:$$

$$\sup_{0 \leq t \leq T} \|u(t)\| \leq K\|x\|_{C^{-1}} \equiv K\left(\|x\| + \|C^{-1}x\|\right).$$

If A is the generator of a C-regularized semigroup, then for any $x \in CD(A) \subset D(A_\Sigma)$, $u(\cdot) = C^{-1}S(\cdot)x$ is the unique solution of (CP). Taking into account *b)*, we obtain from (1.4.13) that

$$\sup_{0 \leq t \leq T} \|u(t)\| \leq K_1 K K_2 \|C^{-1}x\|.$$

This implies C-well-posedness of (CP) with operator A being the generator of a C-regularized semigroup.

1.4.3 Local C-regularized semigroups

There are operators A (for example, those having arbitrary large positive eigenvalues) that generate C-regularized semigroups, defined only for t from some bounded subset of $\mathbb{R}$. Such semigroups correspond to the local C-well-posedness of the Cauchy problem.

Let $T \in (0, \infty)$ and let $C : X \to X$ be an injective bounded linear operator with dense range: $\operatorname{ran} C = X$.

Definition 1.4.5 *A one-parameter family of bounded linear operators $\{S(t),\ t \in [0, T)\}$ is called a local C-regularized semigroup in X if*

(LC1) $S(t + s)C = S(t)S(s)$ *for* $s, t \geq 0$, $s + t < T$ *and* $S(0) = C$,

(LC2) *for every* $x \in X$, $S(t)x : [0, T) \to X$ *is continuous in t.*

In the case $T = \infty$ a local C-regularized semigroup is a C-regularized semigroup. If there exist $K > 0$ and $\omega \in \mathbb{R}$ such that $\|S(t)\| \leq Ke^{\omega t}$ for $t \geq 0$, then S is an exponentially bounded C-regularized semigroup.

Let S be a local C-regularized semigroup. Define its generator G by

$$Gx := \lim_{h \to 0} h^{-1}(C^{-1}S(h)x - x), \qquad x \in D(G),$$

$$D(G) = \left\{ x \in \operatorname{ran} C \ \middle|\ \lim_{h \to 0} h^{-1}(C^{-1}S(h)x - x) \ \text{exists} \right\}$$

It is proved in [138] that G is a densely defined, closable linear operator. Its closure $\overline{G}$ is called the complete infinitesimal generator of S. The connection of $\overline{G}$ with the local Laplace transform

$$L_\tau(\lambda)x := \int_0^\tau e^{-\lambda t} S(t)x\, dt, \qquad \lambda \in \mathbb{R},\ x \in X, \tag{1.4.14}$$

and with the local Cauchy problem (CP) is described in the following proposition.

Proposition 1.4.4 *Let $\overline{G}$ be the complete infinitesimal generator of a local C-regularized semigroup $\{S(t),\ t \in [0, T)\}$ and let $L_\tau(\lambda)$ be as in (1.4.14). Then $\overline{G}$ is densely defined and the following hold:*

1) *for $x \in D(\overline{G})$, $0 \leq t < T$ we have $S(t)x \in D(G)$,*

$$\overline{G}S(t)x = S(t)\overline{G}x \quad \text{and} \quad \overline{G}x = S'(t)x;$$

2) *for $x \in D(\overline{G})$, $\overline{G}L_\tau(\lambda)x = L_\tau(\lambda)\overline{G}x;$*

3) *for $x \in X$, we have $L_\tau(\lambda)x \in D(\overline{G})$ and*

$$(\lambda I - \overline{G})L_\tau(\lambda)x = Cx - e^{-\tau\lambda}S(\tau)x;$$

4) *for $x \in X$,*

$$L_\tau(\lambda)L_\tau(\mu)x = L_\tau(\mu)L_\tau(\lambda)x$$

and

$$\exists K_\tau: \qquad \left\| \frac{d^{n-1}}{d\lambda^{n-1}}L_\tau(\lambda) \right\| \leq K_\tau \frac{(n-1)!}{\lambda^n} \qquad (1.4.15)$$

for $\lambda > 0$, $n \in \mathbb{N}$.

From properties 2) – 4) we conclude that if $\lambda > 0$ is sufficiently large, then the operator $L_\tau(\lambda)$ is similar to the C-resolvent.

Definition 1.4.6 *Let A be a closed linear operator in X, and let $\tau \in (0, T)$. A family of bounded linear operators $\{L_\tau(\lambda), \ \lambda > \omega\}$ is called an asymptotic C-resolvent of A if*

(I) *for $x \in X$, $L_\tau(\lambda)x$ is an infinitely differentiable function of λ and*

$$L_\tau(\lambda)L_\tau(\mu)x = L_\tau(\mu)L_\tau(\lambda)x;$$

(II) *for $x \in X$, $L_\tau(\lambda)x \in D(A)$ and*

$$(\lambda I - A)L_\tau(\lambda)x = Cx + V_\tau(\lambda)x,$$

where V_τ satisfies the following estimates:
$\exists K_\tau:$

$$\left\| \frac{d^{n-1}}{d\lambda^{n-1}}V_\tau(\lambda)x \right\| \leq K_\tau \tau^{n-1}e^{-\tau\lambda}\|x\|$$

for $\lambda > \omega$, $n \in \mathbb{N}$;

(III) *for $x \in D(A)$, $AL_\tau(\lambda)x = L_\tau(\lambda)Ax$.*

As in Theorems 1.2.5, 1.2.6 for local integrated semigroups, the following theorem gives conditions for an operator to be the complete infinitesimal generator of a C-regularized semigroup.

Theorem 1.4.5 *A closed linear operator A in X is the complete infinitesimal generator of a local C-regularized semigroup $\{S(t),\ t \in [0,T)\}$ if and only if*

(I) $D(A)$ *is dense in X;*

(II) *for every $\tau \in (0,T)$ there exists an asymptotic C-resolvent $L_\tau(\lambda)$ of A such that (1.4.15) is fulfilled for $\frac{n}{\lambda} \in [0,\tau]$, $\lambda > \omega$, $n \in \mathbb{N}$;*

(III) $CD(A)$ *is a core for A, i.e.*

$$\overline{A|_{CD(A)}} = A.$$

Proof The proof of necessity follows from Proposition 1.4.4. The proof of sufficiency is based on the following analogy with the case of C_0-semigroups. If $\overline{G}$ is the complete infinitesimal generator of a semigroup $\{\widetilde{S}(t),\ t > 0\}$, then $\widetilde{S}(t)$ can be represented in the form

$$\widetilde{S}(t)x = \lim_{n \to \infty} \left(I - \frac{t}{n}\,\overline{G} \right)^{-n} Cx, \qquad x \in X.$$

From the equality for the resolvent of operator $\overline{G}$:

$$\frac{d^n R_{\overline{G}}(\lambda)}{d\lambda^n} = (-1)^n\, n!\, R_{\overline{G}}^{n+1}(\lambda)$$

similarly to (1.1.9) we have

$$
\begin{aligned}
\widetilde{S}(t)x &= \lim_{n \to \infty} \left(\frac{n}{t} \right)^n R_{\overline{G}}^n \left(\frac{n}{t} \right) Cx = \\[2mm]
&= \lim_{n \to \infty} \left(\frac{n}{t} \right)^n \frac{(-1)^{n-1}}{(n-1)!} \frac{d^{n-1}}{d\lambda^{n-1}} R_{\overline{G}}(\lambda) Cx \bigg|_{\lambda = \frac{n}{t}}.
\end{aligned}
\tag{1.4.16}
$$

So we can see that a local C-regularized semigroup can be presented in the form (1.4.16) where C-resolvent $R_{\overline{G}}(\frac{n}{t})C$ is replaced by the asymptotic C-resolvent $L_\tau(\frac{n}{t})$.

Let us fix $\tau \in (0,T)$. For an operator A, satisfying the conditions of the theorem and for $n > \omega\tau$ (here constant ω is from the definition of asymptotic C-resolvent), we define the family of bounded linear operators $\{S_{n,\tau}(t),\ 0 \le t \le r\}$ in X:

$$
S_{n,\tau}(t) = \begin{cases} \frac{(-1)^{n-1}}{(n-1)!} \left(\frac{n}{t} \right)^n L_\tau^{(n-1)} \left(\frac{n}{t} \right) & \text{for } 0 < t \le \tau, \\[4mm] C & \text{for } t = 0. \end{cases}
$$

Let $S_\tau(t) = \lim_{n \to \infty} S_{n,\tau}(t)$. It is proved in [133] that the operators $S_\tau(t)$ exist, do not depend on τ, and form a local C-regularized semigroup generated by A. $\square$

Consider the local Cauchy problem

$$u'(t) = Au(t), \quad t \in [0, T), \ u(0) = x \qquad \text{(CP)}$$

Definition 1.4.7 *The local problem* (CP) *is called C-well-posed on $[0, T)$ if for any $x \in CD(A)$ there exists a unique solution $u(\cdot)$ such that*

$$\|u(t)\| \leq K(t)\|C^{-1}x\|,$$

where $K(t)$ is bounded on every compact interval from $[0, T)$.

The following theorem on well-posedness of local (CP) can be proved in a fashion similar to the corresponding theorems on C-well-posedness of (CP).

Theorem 1.4.6 *Suppose A is a densely defined closed linear operator in X satisfying*

a) $\forall x \in D(A), \quad Cx \in D(A) \quad \text{and} \quad ACx = CAx,$

b) $\overline{A|_{CD(A)}} = A.$

Then A is the complete infinitesimal generator of a local C-regularized semigroup $\{S(t), \ t \in [0, T)\}$ if and only if the local (CP) *is C-well-posed on $[0, T)$.*

1.4.4 Integrated semigroups and C-regularized semigroups

The following two theorems clarify the connection of C-regularized semigroups with integrated semigroups.

Theorem 1.4.7 *Let A be a linear operator in X with $\rho(A) \neq \emptyset$. Let $\mu \in \rho(A)$ and $n \geq 0$ be an integer. The following statements are equivalent:*

(I) *A is the generator of $(n+1)$-times integrated semigroup $\{V(t), \ t \geq 0\}$ in X, satisfying*

$$\exists K > 0, \ \omega \in \mathbb{R}: \qquad \|V(t+h) - V(t)\| \leq K h\, e^{\omega(t+h)}$$

for all $t, h \geq 0$;

(II) *A is the generator of a C-regularized semigroup $\{S(t), \ t \geq 0\}$ with operator $C = R_A^{n+1}(\mu)$, satisfying*

$$\exists K > 0, \ \omega \in \mathbb{R}: \qquad \|S(t+h) - S(t)\| \leq K h\, e^{\omega(t+h)}$$

for all $t, h \geq 0$.

In this case, we have

$$V(t)x = (\mu I - A)^{n+1} \int_0^t \int_0^{t_1} \dots \int_0^{t_n} S(t_{n+1})x dt_{n+1} \dots dt_1$$

for all $x \in X$ and $t \geq 0$.

In the case of a densely defined operator A, this theorem and Theorem 1.2.3 lead to the following result.

Theorem 1.4.8 *Let A be a densely defined linear operator in X, $\mu \in \rho(A)$ and $n \in \mathbb{N}$. Then the following statements are equivalent:*

(I) *A is the generator of an n-times integrated exponentially bounded semigroup $\{V(t), \ t \geq 0\}$;*

(II) *A is the generator of an exponentially bounded C-regularized semigroup $\{S(t), \ t \geq 0\}$ with $C = (\mu I - A)^{-n}$.*

Proof We prove the following implications: (R) $\Longrightarrow$ (II) $\Longrightarrow$ (I) $\Longrightarrow$ (R), where

(R) $\exists K > 0$, $a \in \mathbb{R} \ : \ (a, \infty) \subset \rho(A)$,

$$\left\| (\lambda I - A)^{-m} (\mu I - A)^{-n} \right\| \leq \frac{K}{(\lambda - a)^m},$$

for all $\lambda > a$ and $m = 1, 2, \dots$.

The implication (R) $\Longrightarrow$ (II) follows from Theorem 1.4.3.
    Now we prove (II) $\Longrightarrow$ (I). We define the operators $S_k(t)$, $k \geq 0$, by

$$
\begin{aligned}
S_0(t) &= S(t), \\
S_k(t)x &= \int_0^t \dots \int_0^{t_2} S(t_1)x dt_1 \dots dt_k, \qquad x \in X, \ t \geq 0, \ k \geq 1.
\end{aligned}
$$

It is shown in the proof of Theorem 1.4.1 that $\int_0^t S(\tau)x d\tau \in D(A)$ and

$$S(t)x - Cx = A \int_0^t S(\tau)x d\tau, \quad x \in X, \ t \geq 0.$$

Then by induction in k, we have

$$S_k(t)x \in D(A^k), \qquad \int_0^t (\mu I - A)^{k-1} S_{k-1}(\tau)x d\tau \in D(A),$$

and

$$\exists K_k > 0 : \qquad \left\| (\mu I - A)^k S_k(t) \right\| \leq K_k e^{\omega t}.$$

Furthermore, for $x \in X$, $(\mu I - A)^k S_k(t)x$ is continuous in $t \geq 0$.

We define $V(t)x := (\mu I - A)^n S_n(t)x$, $x \in X$. Then V satisfies (V2), (V3) and (V4) (see Definition 1.2.1). Let us prove (V1). It follows from Proposition 1.2.1 that it is sufficient to show that

$$\int_0^\infty \lambda^n e^{-\lambda t} V(t)x\,dt = (\lambda I - A)^{-1}x, \qquad x \in X.$$

Integrating by parts we obtain

$$\begin{aligned}
\int_0^\infty \lambda^n e^{-\lambda t} V(t)x\,dt &= \int_0^\infty \lambda^n e^{-\lambda t}(\mu I - A)^n S_n(t)x\,dt \\
&= (\mu I - A)^n \int_0^\infty e^{-\lambda t} S(t)x\,dt \\
&= (\mu I - A)^n (\lambda I - A)^{-1} Cx = (\lambda I - A)^{-1}x.
\end{aligned}$$

(I) $\Longrightarrow$ (R). By Proposition 1.2.1 for the generator of a nondegenerate n-times integrated semigroup V we have the inclusion $(\omega, \infty) \subset \rho(A)$ and

$$(\lambda I - A)^{-1}x = \int_0^\infty \lambda^n e^{-\lambda t} V(t)x\,dt, \qquad x \in X, \ \lambda > \omega. \qquad (1.4.17)$$

Taking into account (1.2.6):

$$V^{(n)}(t)x = V(t)A^n x + \sum_{k=0}^{n-1} \frac{t^k}{k!} A^k x,$$

we integrate (1.4.17) by parts

$$(\lambda I - A)^{-1}x = \int_0^\infty e^{-\lambda t}\left[\sum_{k=0}^{n-1} \frac{t^k}{k!} A^k + V(t)A^n\right]x\,dt, \qquad x \in D(A^n).$$

Differentiating this equality $(m-1)$-times with respect to λ, we obtain

$$(m-1)!\,\|(\lambda I - A)^{-m}(\mu I - A)^{-n}x\|$$

$$\leq \int_0^\infty t^{m-1} e^{-\lambda t}\left[\sum_{k=0}^{n-1} \frac{t^k}{k!}\|A^k(\mu I - A)^{-n}\|\right.$$

$$\left. +Ke^{\omega t}\|A^n(\mu I - A)^{-n}\|\right]\|x\|\,dt$$

$$\leq (m-1)!\,K_1 \int_0^\infty t^{m-1} e^{-\lambda t}\frac{e^t + e^{\omega t}}{2}\,dt\|x\| \leq \frac{(m-1)!\,K_1}{(\lambda - a)^m}\|x\|$$

for $x \in X$, $\lambda > a$, $m \geq 1$. Here the constant K is from (V3), and

$$a = \max\{1, \omega\},$$

$$K_1 = 2 \max_{k=1,\dots,n-1}\left\{\|A^k(\mu I - A)^{-n}\|,\ K\|A^n(\mu I - A)^{-n}\|\right\}. \qquad \square$$

1.4.5 Examples

Example 1.4.1 The operator

$$A = \begin{pmatrix} 0 & I \\ B & 0 \end{pmatrix} \quad \text{with} \quad D(A) = \left\{ \begin{pmatrix} x \\ y \end{pmatrix} \in X \times X \ \Big| \ x, y \in D(B) \right\},$$

which was considered as the generator of an integrated semigroup in Example 1.2.4, is the generator of an exponentially bounded C-regularized semigroup S defined by

$$S(t) \begin{pmatrix} x \\ y \end{pmatrix} = \begin{pmatrix} \lambda_0 \mathcal{C}(t) + B\mathcal{S}(t) & \mathcal{C}(t) + \lambda_0 \mathcal{S}(t) \\ \lambda_0 \mathcal{C}'(t) + B\mathcal{C}(t) & \mathcal{C}'(t) + \lambda_0 \mathcal{C}(t) \end{pmatrix} R_B(\lambda_0^2) \begin{pmatrix} x \\ y \end{pmatrix}$$

with $C = R_A(\lambda_0)$, $\lambda_0 > \omega$, where ω is defined by the C-function. Here $R_B(\lambda^2) = (\lambda^2 I - B)^{-1}$. Note that $S(\cdot)$ is defined for any $x, y \in X$ since range of $R_B(\lambda_0^2)$ is $D(B)$. The C-resolvent of operator A has the form

$$
\begin{aligned}
R_A(\lambda_0) R_A(\lambda) &= \begin{pmatrix} \lambda_0 & I \\ B & \lambda_0 \end{pmatrix} R_B(\lambda_0^2) \begin{pmatrix} \lambda & I \\ B & \lambda \end{pmatrix} R_B(\lambda^2) \\
&= \begin{pmatrix} B + \lambda\lambda_0 & \lambda + \lambda_0 \\ B(\lambda + \lambda_0) & B + \lambda\lambda_0 \end{pmatrix} \frac{R_B(\lambda^2) - R_B(\lambda_0^2)}{\lambda^2 - \lambda_0^2}.
\end{aligned}
$$

Example 1.4.2 (An exponentially bounded C-regularized semigroup that is not an n-times integrated semigroup)

As in Examples 1.1.2 and 1.2.2, let

$$X = \left\{ L^p(\mathbb{R}) \times L^p(\mathbb{R}), \ \|u\| = \|u_1\|_{L^p} + \|u_2\|_{L^p} \right\}, \quad \text{where} \quad u = \begin{pmatrix} u_1 \\ u_2 \end{pmatrix}.$$

Consider the operator A defined by

$$Au = \begin{pmatrix} -g & -f \\ 0 & -g \end{pmatrix} u$$

with

$$D(A) = \left\{ \begin{pmatrix} u_1 \\ u_2 \end{pmatrix} \in X \ \Big| \ gu_1 + fu_2 \in L^p(\mathbb{R}), \ gu_2 \in L^p(\mathbb{R}) \right\},$$

where $g(x) = 1 + |x|$, $f(x) = |x|^\gamma$, $\gamma > 0$. In Example 1.2.1 we demonstrated that for any $\gamma \in (0, 1]$ the operator A is the generator of a C_0-semigroup U defined by

$$U(t)u = e^{-tg} \begin{pmatrix} 1 & -tf \\ 0 & 1 \end{pmatrix} u, \quad t \geq 0, \ u \in X.$$

Recall that for any $\gamma \geq 0$ we have

$$
\begin{aligned}
\|U(t)\| &= \max_{x \in R}\left[(1 + t|x|^{\gamma})e^{-t(1+|x|)}\right] \\
&= (1 + \gamma^{\gamma}t^{1-\gamma})e^{-t\left(1+\frac{\gamma}{t}\right)} \underset{t \to 0}{=} \mathcal{O}(t^{1-\gamma}).
\end{aligned}
$$

Note that for any $\gamma \in (1,2]$, A is the generator of an integrated semigroup (see Example 1.2.2), therefore by Theorem 1.4.8, A generates a C-regularized semigroup with $C = R_A(\lambda)$.

Now let $\gamma > 2$. We have

$$
(\lambda I - A)^{-1}u = (\lambda + g)^{-2}\begin{pmatrix} (\lambda + g) & -f \\ 0 & (\lambda + g) \end{pmatrix} u, \ \lambda > 0,
$$

therefore, the operator $(\lambda I - A)^{-1}$ is not bounded for $\lambda > 0$. Hence, the resolvent of A does not exist for $\lambda > 0$ and A cannot be a generator of an n-times integrated semigroup for any n. Since $\|U(t)\| = \mathcal{O}(t^{1-\gamma})$ as $t \to 0$, then the semigroup U has a singularity of order $\alpha = \gamma - 1$ at $t = 0$. This semigroup satisfies the following definition of a semigroup of growth α (see [84] and references therein):

(a) $X_0 := \bigcap_{t>0} U(t)[X]$ is dense in X,

(b) U is non-degenerate,

(c) $\|t^{\alpha}U(t)\|$ is bounded as $t \to 0$.

Let $n = [\gamma]$. Define the operator C in X by

$$
Cu = \frac{1}{n!}\int_0^{\infty} t^n e^{-\lambda t}U(t)u\,dt, \quad u \in X, \ \lambda > 0,
$$

then C is injective and $\overline{R(C)} = X$. It is not difficult to verify that the family $\{S(t) := U(t)C, \ t \geq 0\}$ is an exponentially bounded C-regularized semigroup. If $n = 0$, then U has an integrable singularity. In this case, $C = (\lambda I - A)^{-1}$, $\lambda > 0$, is the resolvent of A and A is the generator of an integrated semigroup.

Now let $n \geq 1$. If $(\lambda I - A)^{-1}$ exists and is equal to the resolvent of A, then $C = (-1)^n R_A^{n+1}(\lambda)$ and A is the generator of an $(n + 1)$-times integrated semigroup. Otherwise (which is the case for $\gamma > 2$), A is the generator of a C-regularized semigroup and is not a generator of an integrated semigroup.

Finally, consider the operator C defined by

$$
Cu := (1 + |f|)^{-1}u, \ u \in X,
$$

then we have that C is injective and $\overline{R(C)} = X$. In this case, $\{S(t) = U(t)C,\ t \geq 0\}$ is an exponentially bounded C-regularized semigroup. We note that this semigroup remains exponentially bounded even for functions like $f(x) = e^{(1+|x|)}$ that grow faster than $|x|^{\gamma}$ for any $\gamma > 0$.

Example 1.4.3 (A C-regularized semigroup that is not exponentially bounded)

Consider the operator A defined by

$$Af(z) = zf(z), \qquad D(A) = \left\{ f \in X = L^2(\mathbb{C}) \ \Big| \ zf(z) \in X \right\},$$

which is the generator of the C-regularized semigroup

$$S(t)f := e^{zt - |z|^2} f(z), \qquad f \in X,$$

with $Cf = e^{-|z|^2}f,\ \ f \in X$. We have

$$\|S(t)\| = \sup_{z \in \mathbb{C}} \left\{ e^{zt - |z|^2} \right\} = e^{\frac{t^2}{4}},$$

that is, S is not exponentially bounded. Note that in this example the operator $(\lambda I - A)$ is not invertible for any $\lambda \in \mathbb{C}$ and $\sigma(A) = \mathbb{C}$.

Example 1.4.4 (Local semigroups corresponding to the regularization methods)

Let

$$Au = -\frac{d^2 u}{ds^2} \text{ on } L^2[0, l] \text{ with}$$

$$D(A) = \left\{ u \in L^2[0, l] \ \Big| \ \frac{d^2 u}{ds^2} \in L^2[0, l], \ \ u(0) = u(l) = 0 \right\}.$$

then $\lambda_n = (\frac{\pi n}{l})^2$ are the eigenvalues for this operator and its eigenvectors $\alpha_n(s) = \sin \frac{\pi n}{l} s$ form an orthogonal basis in $L^2[0, l]$.

The operators

$$S_1(t)x := \sum_{k=1}^{\infty} e^{\lambda_k t - \varepsilon \lambda_k^2 T} x_k \alpha_k$$

and

$$S_2(t)x := \sum_{k=1}^{\infty} \frac{e^{\lambda_k t}}{1 + \varepsilon e^{\lambda_k T}} x_k \alpha_k,$$

where $x_k = (x, \alpha_k)$, $t < T$, $\varepsilon > 0$, are local C-regularized semigroups on $(0, T)$. The asymptotic C-resolvent for S_1 is

$$L_\tau(\lambda)x = \sum_{k=1}^{\infty} \frac{(e^{(\lambda_k - \lambda)\tau} - 1)e^{-\varepsilon \lambda_k^2 T}}{\lambda - \lambda_k} x_k \alpha_k.$$

1.5   Degenerate semigroups

In this section we consider degenerate integrated and C_0-semigroups connected with the degenerate Cauchy problem

$$Bu'(t) = Au(t), \quad t \geq 0, \quad u(0) = x, \quad \ker B \neq \{0\}, \qquad \text{(DP)}$$

where $B, A : X \to Y$ are linear operators in Banach spaces X, Y.

We assume that the set

$$\rho_B(A) :=$$
$$\left\{ \lambda \in \mathbb{C} \ \Big| \ R(\lambda) := (\lambda B - A)^{-1} B \text{ is a bounded operator on } X \right\}$$

is not empty. This set is called the B-resolvent set of operator A and the operator $R(\lambda)$ is called the B-resolvent of operator A.

We prove the equivalence of the uniform well-posedness of (DP) on the maximal correctness class

$$\Upsilon := \left\{ x \in D(A) \ \Big| \ Ax \in \operatorname{ran} B \right\} = R(\lambda)X,$$

the existence of a degenerate C_0-semigroup and MPFHY-type estimates together with the decomposition of X:

$$X = X_1 \oplus \ker B, \qquad (1.5.1)$$

where $X_1 := \overline{D_1}$ and $D_1 := R(\lambda)X$ with some $\lambda \in \rho_B(A)$. The decomposition (1.5.1) plays a role similar to the condition of density of a generator of a C_0-semigroup. We also consider the well-posedness of the degenerate Cauchy problem on subsets of $R^n(\lambda)X$.

1.5.1   Generators of degenerate semigroups

Definition 1.5.1 *A one-parameter family of bounded linear operators* $\{U(t),\ t \geq 0\}$ *on* X *is called a degenerate* C_0-*semigroup if the semigroup relation* (U1) *and the strong continuity condition* (U3) *of Definition 1.1.1 hold, and the operator* $U(0)$ *(and hence* $U(t)$ *for any* $t \geq 0$) *has a nonempty kernel.*

Definition 1.5.2 *Let* $n \in \mathbb{N}$. *An* n-*times integrated semigroup* $\{V(t),\ t \in [0, T)\}$ *is called a degenerate* n-*times integrated semigroup if conditions* (V1) – (V3) *of Definition 1.2.1 hold and* (V4) *does not.*

Remark 1.5.1 The exponential boundedness of a degenerate semigroup $\{U(t),\ t \geq 0\}$:

$$\exists K > 0,\ \omega \in \mathbb{R} : \quad \|U(t)\| \ \leq K e^{\omega t},\ t \geq 0,$$

follows from property (U1). Note that a degenerate n-times integrated semigroup, similarly to a nondegenerate semigroup, may be not exponentially bounded.

Remark 1.5.2 Property (U1) implies that the operator $U(0)$ is a projector on X, generating the following decomposition of X

$$X = \operatorname{ran} U(0) \oplus \ker U(0).$$

In this case, the restriction of $U(\cdot)$ on $\operatorname{ran} U(0)$ is a nondegenerate C_0-semigroup.

Let $V(\cdot)$ be a strongly continuous operator-function such that

$$\exists K > 0, \ \omega \in \mathbb{R}: \quad \|V(t)\| \leq K e^{\omega t},$$

for all $t \geq 0$, and

$$r(\lambda) = \int_0^\infty \lambda^n e^{-\lambda t} V(t) dt$$

for some $n \in \mathbb{N} \cup \{0\}$. It was proved in Propositions 1.1.2, 1.2.1 that $r(\lambda)$ satisfies the (pseudo) resolvent identity

$$(\mu - \lambda) r(\lambda) r(\mu) = r(\lambda) - r(\mu), \qquad \operatorname{Re} \lambda, \ \operatorname{Re} \mu > \omega,$$

if and only if V satisfies (V1) for $n \geq 1$ or (U1) for $n = 0$. Such function $r(\lambda)$ is called the pseudoresolvent. Moreover, if V is nondegenerate, then $r(\lambda)$ is invertible and is the resolvent of the generator A defined by

$$A := \lambda I - r^{-1}(\lambda).$$

Obviously, when dealing with single-valued operators, this definition of the generator cannot be used in the case of a degenerate semigroup V. We introduce the notion of a pair of generators A, B, which allows us to link degenerate semigroups with the degenerate Cauchy problem (DP).

Definition 1.5.3 *Let* $n \in \mathbb{N} \cup \{0\}$. *Linear operators* $A, B : X \to Y$ *are called the generators of an exponentially bounded degenerate n-times integrated semigroup* $\{V(t), \ t \geq 0\}$ *on* X *if* A *is closed,* B *is bounded and*

$$(\lambda B - A)^{-1} B = \int_0^\infty \lambda^n e^{-\lambda t} V(t) dt$$

for all $\lambda \in \mathbb{C}$ *with* $\operatorname{Re} \lambda > \omega$.

From Arendt-Widder's Theorem 1.2.1 we obtain the following result characterizing the generators of degenerate $(n+1)$-times integrated semigroups.

Theorem 1.5.1 *Let* $A, B : X \to Y$ *be linear operators,* A *be closed and* B *be bounded. Then the following statements are equivalent:*

(I) *the operators A and B are the generators of a degenerate $(n+1)$-times integrated semigroup $\{V(t),\ t \geq 0\}$ satisfying the condition*

$$\lim_{h \to 0} \sup \frac{1}{n}\|V(t+h) - V(t)\| \leq Ke^{\omega t}, \quad t \geq 0; \qquad (1.5.2)$$

(II) *the estimates*

$$\exists K > 0, \ \omega \in \mathbb{R}:$$

$$\left\| \frac{d^k}{d\lambda^k} \left[\frac{(\lambda B - A)^{-1}B}{\lambda^n} \right] \right\| \leq \frac{K\, k!}{(\lambda - \omega)^{k+1}} \qquad (1.5.3)$$

hold for all $\lambda > \omega$ and $k = 0, 1, \ldots$. $\square$

In the next subsection we use degenerate 1-time integrated semigroups and degenerate C_0-semigroups in the investigation of the uniform well-posedness of the Cauchy problem (DP).

1.5.2   Degenerate 1-time integrated semigroups

Let A and B be the generators of a degenerate 1-time integrated semigroup $\{V(t),\ t \geq 0\}$ satisfying (1.5.2). Then the B-resolvent set $\rho_B(A)$ is not empty. We consider the set $D_1 := R(\lambda)X$, where $\lambda > \omega$. As in the case of domains of generators of nondegenerate semigroups, the introduced set does not depend on $\lambda > \omega$. To show this we consider $\lambda, \mu > \omega$ and $x = R(\lambda)z$ with $z \in X$. Then $(\lambda B - A \pm \mu B)x = Bz$ and

$$x = R(\lambda)z = (\mu B - A)^{-1}B(z - (\mu - \lambda)x) \equiv (\mu B - A)^{-1}Bz_1 = R(\mu)z_1.$$

Proposition 1.5.1 *Let A and B be the generators of a degenerate 1-time integrated semigroup $\{V(t),\ t \geq 0\}$ satisfying (1.5.2). Then $\ker B \cap X_1 = \{0\}$ and $\ker B \oplus X_1$ is a subspace in X.*

Proof Recall that $X_1 := \overline{D}_1$. Consider the family of linear operators $\lambda R(\lambda)$, $\lambda > \omega$. By (1.5.3) they are bounded on X. We show that for any $x \in X_1$

$$\lambda R(\lambda)x \underset{\lambda \to \infty}{\longrightarrow} x. \qquad (1.5.4)$$

Let $x \in D_1$, then there exists $y \in X$ such that $x = R(\mu)y$ for some $\mu \in R_B(A)$. Using the resolvent identity for $R(\lambda)$ and estimates (1.5.3) with $n = 0$, we obtain

$$
\begin{aligned}
\|\lambda R(\lambda)x - x\| &= \|\lambda R(\lambda)R(\mu)y - R(\mu)y\| \\
&= \left\| \frac{\lambda}{(\mu - \lambda)}R(\lambda)y - \frac{\lambda}{(\mu - \lambda)}R(\mu)y - R(\mu)y \right\| \\
&\leq \frac{\lambda}{\mu - \lambda}\|R(\lambda)\| + \frac{\mu}{\mu - \lambda}\|R(\mu)y\| \underset{\lambda \to \infty}{\longrightarrow} 0.
\end{aligned}
$$

Furthermore, (1.5.3) implies that operators $\lambda R(\lambda)$ are uniformly bounded, and by the Banach-Steinhaus Theorem, (1.5.4) holds for any $x \in X_1$. Hence,

$$\ker B \cap X_1 = \{0\}.$$

The uniformly bounded family of operators $\lambda R(\lambda)$ converges on the set $\ker B \oplus X_1$ as $\lambda \to \infty$. Hence, by the Banach-Steinhaus theorem, it converges on $\overline{\ker B \oplus X_1}$ to a bounded linear operator P such that $Px = x$ for $x \in X_1$. Furthermore, $Px = 0$ for $x \in \ker B$, $\operatorname{ran} P = X_1$ and $P^2 = P$. Hence, P is a projector in $\overline{\ker B \oplus X_1}$ and

$$\overline{\ker B \oplus X_1} = \ker P \oplus \operatorname{ran} P = \ker B \oplus X_1. \ \square$$

The connection between the degenerate Cauchy problem (DP) and the generators of degenerate semigroups is established with the help of the following theorem.

Theorem 1.5.2 *Let A and B be the generators of a degenerate 1-time integrated semigroup $\{V(t),\ t \geq 0\}$ satisfying (1.5.2), then*

$$R(\lambda)V(t) = V(t)R(\lambda) \tag{1.5.5}$$

for all $\lambda \in \mathbb{C}$ with $\operatorname{Re}\lambda > \omega$ and all $t \geq 0$. Further

$$tBx = BV(t)x - A\int_0^t V(s)x\,ds \tag{1.5.6}$$

for all $x \in \ker B \oplus X_1$ and $t \geq 0$.

The operator-function $V'(\cdot)$ is a degenerate C_0-semigroup on the subspace $\ker B \oplus X_1$ that coincides with

$$F := \left\{x \in X \ \middle|\ V(t)x \in C^1\{(0,\infty),\ X\}\right\},$$

and

$$BV''(t)x = AV'(t)x \tag{1.5.7}$$

for all $x \in R(\lambda)X_1$ and $t \geq 0$.

Proof Let $\operatorname{Re}\lambda,\ \operatorname{Re}\mu > \omega$. Then for any $x \in X$,

$$\int_0^\infty \mu e^{-\mu t} V(t)R(\lambda)x\,dt \ =\ R(\mu)R(\lambda)x = R(\lambda)R(\mu)x$$

$$=\ \int_0^\infty \mu e^{-\mu t} R(\lambda)V(t)x\,dt.$$

Using the uniqueness of the Laplace transform, we obtain (1.5.5).

Now let $x \in X$ and $\lambda > \omega$, then

$$\int_0^\infty \lambda^2 e^{-\lambda t} t\, Bx\, dt \;=\; Bx = (\lambda B - A)(\lambda B - A)^{-1} Bx \qquad (1.5.8)$$

$$= \int_0^\infty \lambda^2 e^{-\lambda t} BV(t)x\, dt - A \int_0^\infty \lambda e^{-\lambda t} V(t)x\, dt.$$

Take $x \in D_1$, then $x = (\lambda_0 B - A)^{-1} By$ for some $y \in X$. For such x we have $V(t)x \in D(A)$ and

$$
\begin{aligned}
AV(t)x &= AV(t)(\lambda_0 B - A)^{-1} By = A(\lambda_0 B - A)^{-1} BV(t)y \\
&= \lambda_0(\lambda_0 B - A)^{-1} BV(t)y - BV(t)y \\
&= \lambda_0 BV(t)x - BV(t)y.
\end{aligned}
$$

Hence,

$$\|AV(t)x\| \;\leq\; K\|B\|\, e^{\omega t}\Big(|\lambda_0|\,\|x\| + \|y\|\Big).$$

As $AV(\cdot)x$ is exponentially bounded for $x \in D_1$, the Laplace transform of $AV(\cdot)x$ exists. Since the operator A is closed, by (1.5.8) we have

$$
\begin{aligned}
\int_0^\infty \lambda^2 e^{-\lambda t} tBx\, dt &= \int_0^\infty \lambda^2 e^{-\lambda t} BV(t)x\, dt - \int_0^\infty \lambda e^{-\lambda t} AV(t)\, dt \\
&= \int_0^\infty \lambda^2 e^{-\lambda t}\left[BV(t)x - \int_0^t AV(s)x\, ds \right] dt.
\end{aligned}
$$

Hence,

$$tBx = BV(t)x - \int_0^t AV(s)x\, ds$$

for all $x \in D_1$ and $t \geq 0$. Due to the closedness of A, this equality holds for all $x \in X_1 = \overline{D}_1$. Moreover, the equality holds for $x \in \ker B$, since $V(t)x = 0$ for all $x \in \ker B$ and $t \geq 0$. (1.5.6) is proved.

Since property (1.5.2) guarantees that $V'(t)$ is bounded:

$$\exists M > 0,\ \forall x \in F, \qquad \|V'(t)x\| \;\leq\; Me^{\omega t}\|x\|, \quad t \geq 0,$$

we conclude that the linear manifold F is closed. Let $x \in F$. Differentiating the equality (V1) with $n = 1$, we obtain

$$V(t+s)x - V(s)x = V(t)V'(s)x, \qquad t, s \geq 0. \qquad (1.5.9)$$

The left-hand side of this equality is differentiable with respect to t, hence for any $x \in F$ we have $V'(s)x \in F$, i.e. for any $s \geq 0$, $V'(s)$ is a bounded operator on F. Differentiating (1.5.9) we have

$$V'(t+s)x = V'(t)V'(s)x, \qquad t, s \geq 0,\ \ x \in F, \qquad (1.5.10)$$

which implies the semigroup property (U1) for V'. We now show that V' is a degenerate semigroup on F. From the definition of a generator of V, we have

$$(\lambda B - A)^{-1} B x = \int_0^\infty e^{-\lambda t} V'(t) x \, dt \qquad (1.5.11)$$

for all $x \in F$ and all $\lambda \in \mathbb{C}$ with $\operatorname{Re} \lambda > \omega$. From (1.5.10) and (1.5.11) we have $[V'(0)]^2 = V'(0)$. Now consider $x \in \ker B$, then the uniqueness of the Laplace transform guarantees that $V'(t)x = 0$ for $t \geq 0$. If $x \in \ker V'(0)$, then (1.5.10) implies $V'(t)x = 0$ for $t \geq 0$. Hence, by (1.5.11), $x \in \ker B$. Therefore, $\ker V'(0) = \ker B$.

Thus, $V'(0)$ is a projector, $\{V'(t),\ t \geq 0\}$ is a degenerate C_0-semigroup on F, and F is decomposed into the direct sum:

$$F = \ker V'(0) \oplus \operatorname{ran} V'(0) = \ker B \oplus \operatorname{ran} V'(0).$$

We now show that $\operatorname{ran} V'(0) = X_1$. By Remark 1.5.2 we have that the set

$$F_1 := \Big\{ x \in X \ \Big| \ V''(t)x \text{ exists for } t \geq 0 \Big\}$$

contains the domain of the generator of the nondegenerate semigroup $V'(t)\big|_{\operatorname{ran} V'(0)}$, which is dense in $\operatorname{ran} V'(0)$. Hence $\overline{F}_1 \supset \operatorname{ran} V'(0)$. Taking into account (1.5.8) and that $V(t)x = 0$ for $x \in \ker B$, we have the inclusion

$$\ker B \subset F_1 \qquad \text{and hence} \qquad \overline{F}_1 \supset F.$$

As $\overline{F}_1 \subset \overline{F} = F$, we have $\overline{F}_1 = F$. Let $x \in F_1$, then

$$\begin{aligned}
\lambda R(\lambda) x &= \int_0^\infty \lambda e^{-\lambda t} V'(t) x \, dt \\
&= V'(0)x + \int_0^\infty e^{-\lambda t} V''(t) x \, dt.
\end{aligned}$$

As $\lambda \to \infty$, the limit of the second term in the right-hand side of this equality is equal to zero, hence

$$V'(0)x = \lim_{\lambda \to \infty} \lambda R(\lambda) x.$$

Since the operators $\lambda R(\lambda)$ are uniformly bounded, the equality holds for any $x \in F$. Hence, $\operatorname{ran} V'(0) \subset \overline{D}_1 = X_1$.

Now we consider the set $R(\lambda) X_1$ for some $\lambda \in \mathbb{C}$ with $\operatorname{Re} \lambda > \omega$. It is not difficult to verify that $R(\lambda) X_1$ does not depend on λ. Let $x = R(\lambda)y$, where $y \in X_1$. We apply the operator $(\lambda B - A)^{-1}$ to the equality (1.5.6) with y. Using the property (1.5.5) and the equality

$$(\lambda B - A)^{-1} A = \lambda (\lambda B - A)^{-1} B - I,$$

which holds on $D(A)$, we obtain

$$tx = V(t)x - \int_0^t V(s) \left[\lambda(\lambda B - A)^{-1}B - I\right] y\, ds. \qquad (1.5.12)$$

Hence, $x \in F$ and $R(\lambda)X_1 \subset F$. From (1.5.4) we have $\overline{R(\lambda)X_1} = X_1$, therefore,

$$X_1 \subset \overline{F} = F, \text{ and } \ker V'(0) \oplus X_1 = \ker B \oplus X_1 \subset F.$$

This inclusion, together with already proved inclusions:

$$\operatorname{ran} V'(0) \subset X_1$$

and

$$F = \ker B \oplus \operatorname{ran} V'(0) \subset \ker B \oplus X_1,$$

gives the equality $F = \ker B \oplus X_1$.

To prove (1.5.7) we differentiate (1.5.12), where $y \in X_1$ and $x = R(\lambda)y$:

$$x = V'(t)x - V(t)\left[\lambda R(\lambda) - I\right]y.$$

Since $[\lambda R(\lambda) - I]y \in X_1$, the second term in the right-hand side is differentiable in t. Hence $V'(\cdot)x$ is also differentiable and

$$V''(t)x = V'(t)[\lambda R(\lambda) - I]y.$$

Applying operator B to both sides of this equality, we obtain

$$\begin{aligned} B\frac{d}{dt}V'(t)x &= B[\lambda R(\lambda) - I]V'(t)y = AR(\lambda)V'(t)y \\ &= AV'(t)R(\lambda)y = AV'(t)x. \quad \square \end{aligned}$$

From Theorems 1.5.1, 1.5.2 we obtain the result about the connection of the well-posedness of the degenerate Cauchy problem and estimates for the B-resolvent of A.

Theorem 1.5.3 *Let $A, B : X \to Y$ be linear operators. Suppose that A is closed, B is bounded, and estimates (1.5.3) are fulfilled for $n = 0$. Then (DP) is uniformly well-posed on $R(\lambda)X_1$, $\operatorname{Re}\lambda > \omega$.* $\square$

1.5.3   Maximal correctness class

Definition 1.5.4 *If, for any $x \in E \subseteq X$, a Cauchy problem (DP) has a unique stable solution, then the set E is called the correctness class for this problem.*

The set

$$\Upsilon := \left\{ x \in D(A) \ \middle| \ Ax \in \operatorname{ran} B \right\}$$

is the *maximal correctness class* for the Cauchy problem (DP) because if $u(\cdot)$ is a solution of (DP), then $u(t) \in \Upsilon$ for $t \geq 0$, and the condition $x \in \Upsilon$ is necessary for the existence of a solution.

Proposition 1.5.2 *Let $A, B : X \to Y$ be linear operators such that $\rho_B(A) \neq \emptyset$, then $\Upsilon = D_1 := R(\lambda)X$.*

Proof Let $x \in D_1$, then there exists $y \in X$ such that $x = R(\lambda)y$. Hence, $Ax = -By + \lambda Bx = B(\lambda x - y)$, and $x \in M$. Conversely, if $x \in M$, then there exists $z \in X$ such that $Ax = Bz$. Then $(\lambda B - A)x = B(\lambda x + z)$ and $x = R(\lambda)(\lambda x - z) \in D_1$. $\square$

Theorem 1.5.4 *Let $A, B : X \to Y$ be linear operators, A be closed, B be bounded, and the B-resolvent set of A be nonempty. Then the following statments are equivalent:*

(I) *the Cauchy problem (DP) is uniformly well-posed on D_1;*

(II) *the operators A and B are the generators of a degenerate C_0-semigroup;*

(III) *the estimates (1.5.3) hold with $n = 0$ and X admits the decomposition $X = \ker B \oplus X_1$.*

Proof (I) $\Longrightarrow$ (II). Let $u(\cdot)$ be a solution of (DP). Define on Υ the solution operators $\widetilde{U}(t)x := u(t)$ for $x \in D_1$ and $t \geq 0$. The uniform well-posedness of (DP) implies that operators $\widetilde{U}(t)$ are bounded on D_1 and therefore they can be extended to $X_1 = \overline{D}_1$. As in the nondegenerate case (Theorem 1.1.1) the operators $\widetilde{U}(t)$, $t \geq 0$ form a C_0-semigroup. This semigroup $\left\{ \widetilde{U}(t), \ t \geq 0 \right\}$ is nondegenerate and is exponentially bounded on X_1:

$$\exists K > 0, \ \omega \in \mathbb{R} : \qquad \left\| \widetilde{U}(t) \right\| \leq K\, e^{\omega t}, \quad t \geq 0.$$

Let $G : X_1 \to X_1$ be the generator of this semigroup, then G is a closed, densely defined linear operator on X_1. The function $\widetilde{U}(t)x$, $t \geq 0$, is differentiable with respect to $t \geq 0$ if and only if $x \in D(G)$. For any $x \in D(G)$, $\widetilde{U}(\cdot)x$ is the solution of the Cauchy problem

$$u'(t) = Gu(t), \quad t \geq 0, \ u(0) = x.$$

Since the semigroup $\widetilde{U}$ is exponentially bounded on X_1, the Laplace transform of $\widetilde{U}(\cdot)x$ is defined for $\operatorname{Re} \lambda > \omega$ and

$$\int_0^\infty e^{-\lambda t} \widetilde{U}(t)x\, dt = (\lambda I - G)^{-1}x, \quad x \in X_1.$$

By construction of $\widetilde{U}$ we have $D_1 \subset D(G)$. Let $x \in D_1$, then integrating by parts we have

$$\lambda \int_0^\infty e^{-\lambda t} \widetilde{U}(t) x dt = x + \int_0^\infty e^{-\lambda t} \widetilde{U}'(t) x dt.$$

for $\lambda \in \mathbb{C}$ with $\mathrm{Re}\,\lambda > \omega$. Applying B to this equality and taking into account that $\widetilde{U}(\cdot)$ is a solution of (DP) and that A is closed, we obtain

$$\lambda B \int_0^\infty e^{-\lambda t} \widetilde{U}(t) x dt = Bx + \int_0^\infty e^{-\lambda t} A\widetilde{U}(t) x dt, \quad x \in D_1,$$

and

$$(\lambda B - A) \int_0^\infty e^{-\lambda t} \widetilde{U}(t) x dt = Bx, \quad x \in X_1. \qquad (1.5.13)$$

Now we prove that $(\lambda B - A)$ is invertible for $\lambda \in \mathbb{C}$ with $\mathrm{Re}\,\lambda > \omega$. Let $x \in \ker(\lambda B - A)$ and consider $w(t) := e^{\lambda t} x$, then

$$Bw'(t) = \lambda B w(t) = A w(t), \qquad w(0) = x.$$

Since $x \in D_1$ and (DP) is well-posed on D_1, we have $w(\cdot) = U(\cdot)x$ and $\|w(t)\| \leq K e^{\omega t}\|x\|$ for $t \geq 0$. Therefore, if $0 \neq x \in \ker(\lambda B - A)$, then $\mathrm{Re}\,\lambda \leq \omega$. If $\mathrm{Re}\,\lambda > \omega$, then the operator $(\lambda B - A)$ is invertible. Hence, from (1.5.13) we obtain

$$R(\lambda)x = (\lambda B - A)^{-1} Bx = \int_0^\infty e^{-\lambda t} \widetilde{U}(t) x dt, \quad x \in X_1, \ \mathrm{Re}\,\lambda > \omega,$$

and

$$R(\lambda)x = (\lambda I - G)^{-1} x, \quad x \in X_1, \ \mathrm{Re}\,\lambda > \omega. \qquad (1.5.14)$$

Let $\lambda_0 \in \rho_B(A)$ and define the operator

$$Px := (\lambda_0 I - G) R(\lambda_0) x.$$

Since $D_1 \subset D(G)$, the domain of P coincides with X. This operator is closed since $R(\lambda_0)$ is bounded and $(\lambda_0 I - G)$ is closed. Hence, P is a bounded operator on X. From (1.5.14) we have $P^2 = P$, i.e. P is a projector on X.

Let $U(t) := \widetilde{U}(t)P$ for $t \geq 0$. Clearly, the family of bounded linear operators $\{U(t),\ t \geq 0\}$ is a degenerate C_0-semigroup and for any $x \in X$ we have the equality

$$R(\lambda)x = \int_0^\infty e^{-\lambda t} U(t) x dt, \quad \mathrm{Re}\,\lambda > \omega,$$

which means that operators A and B are the generators of $\{U(t),\ t \geq 0\}$.

(II) $\Longrightarrow$ (III). If A and B are the generators of a degenerate C_0- semi-group $\{U(t),\ t \geq 0\}$, then it is not difficult to verify that A and B are the generators of the degenerate 1-time integrated semigroup $V(t) := \int_0^t U(t)dt$, $t \geq 0$, which satisfies (1.5.2). Hence, by Theorem 1.5.1 the estimates (1.5.3) hold for $n = 0$, and by Theorem 1.5.2

$$X = F = \left\{ x \in X \ \middle| \ V(\cdot)x \subset C^1\{[0,\infty),\ X\} \right\} = \ker B \oplus X_1.$$

(III) $\Longrightarrow$ (I). If $X = \ker B \oplus X_1$, then the equality $R(\lambda)X_1 = R(\lambda)X$ holds. Hence, by Theorem 1.5.3 the Cauchy problem (DP) is uniformly well-posed on $D_1 = R(\lambda)X$. $\square$

1.5.4 (n,ω)-well-posedness of a degenerate Cauchy problem

Let $A, B : X \to Y$ be linear operators, A be closed, and B be bounded. Let the B-resolvent set of A be nonempty and consider $\lambda \in \rho_B(A)$. We introduce the sets $D_i = R^i(\lambda)X$, $i = 1, 2, \ldots$, which are obviously linear manifolds. At the begining of this section we proved that D_1 does not depend on $\lambda \in \rho_B(A)$. Similarly, one can prove that D_i do not depend on λ for all $i \geq 1$. We denote by $[D_n]$ sets D_n, $n \in \mathbb{N}$, equipped with the norm

$$\|x\|_{R^n} := \inf_{y : y = R^n(\lambda)x} \|y\|$$

for all $x \in D_n$.

Definition 1.5.5 *The Cauchy problem* (DP) *is called* (n,ω)*-well-posed on* $D \subset [D_n]$ *if for any* $x \in D$ *there exists a unique solution* $u(\cdot)$ *satisfying*

$$\|u(t)\| \leq Ke^{\omega t}\|x\|_{R^n}$$

for some $n \in \mathbb{N}$, $K > 0$ *and* $\omega \in \mathbb{R}$.

To obtain a sufficient condition for the (n,ω)-well-posedness of (DP) we need the following properties of degenerate integrated semigroups that can proved in a way similar to Proposition 1.2.2 and Theorem 1.5.2.

Theorem 1.5.5 *Let* $n \in \{0\} \cup \mathbb{N}$. *Let* A *and* B *be the generators of a degenerate* $(n+1)$*-times integrated semigroup* $\{V(t),\ t \geq 0\}$ *satisfying the condition (1.5.2). Then*

$$R(\lambda)V(t) = V(t)R(\lambda),\quad t \geq 0,\quad \mathrm{Re}\,\lambda > \omega$$

and

$$\frac{t^{n+1}}{(n+1)!}Bx = BV(t)x - A\int_0^t V(s)xds,\quad t \geq 0,\quad x \in X_1 = \overline{D_1}.$$

The family of operators $\{V'(t),\ t \geq 0\}$ is a degenerate exponentially bounded n-times integrated semigroup on the subspace $F = \overline{R(\lambda)X_1}$. For all $x \in R^k(\lambda)F,\ k = 0, 1, \ldots, n$, we have

$$BV^{n+1}(t)x = \frac{t^{n-k}}{(n-k)!}Bx - AV^k(t)x, \quad t \geq 0,$$

For all $x \in R^{n+1}(\lambda)F$ we have

$$B\frac{d}{dt}V^{n+1}(t)x = AV^{n+1}(t)x, \quad t \geq 0. \ \square$$

Proofs of this theorem and the next one can be obtained from Theorem 1.6.4.

Theorem 1.5.6 *Let $A, B : X \to Y$ be linear operators, A be closed, B be bounded and estimates (1.5.3) be fulfilled. Then (DP) is (n, ω)-well-posed on*

$$D = ((\lambda B - A)^{-1}B)^{n+1}\overline{(\lambda B - A)^{-1}B(X_1)} \subset D_{n+1}. \ \square$$

In the Section 1.6, assuming a decomposition of X, we prove the necessary and sufficient conditions for the (n, ω)-well-posedness and for the n-well-posedness of the degenerate Cauchy problem (DP).

1.5.5 Examples

Non-densely defined operators that generate integrated semigroups provide examples of degenerate semigroups.

Let A be the generator of an integrated semigroup $\{V(t),\ t \geq 0\}$ in X. Suppose that $\overline{D(A)}$ is complemented in X, i.e. $X = \overline{D(A)} \oplus Y$ for some subspace Y. Let $B = P_{\overline{D(A)}}$ be the projector from X to $\overline{D(A)}$. We have $\ker B = Y$, $\operatorname{ran} B = \overline{D(A)}$ and $B^2 = B$. Then

$$(\lambda I - A)x = (\lambda B - A)x, \quad x \in D(A),$$

and

$$(\lambda B - A)^{-1}x = (\lambda I - A)^{-1}x = R_A(\lambda)x, \quad x \in \overline{D(A)}.$$

Hence, if MFPHY condition is fulfilled for $R_A(\lambda)$, then it is also fulfilled for $(\lambda B - A)^{-1}B$.

Example 1.5.1 Let $X = C[0, \infty)$ and

$$A = -\frac{d}{dx} \quad \text{with} \quad D(A) = \{f \in X \mid f' \in X,\ f(0) = 0\}.$$

Then $Y = \{f \in X \mid f \equiv const\}$, $Bf = f(x) - f(0)$ and

$$D_1 = \{f \in X \mid Af \in \operatorname{ran} B\} = \{f \in X \mid f' \in X,\ f(0) = 0,\ f'(0) = 0\}.$$

The Cauchy problem

$$\frac{\partial u(x,t)}{\partial t} - \frac{\partial u(0,t)}{\partial t} + \frac{\partial u(x,t)}{\partial x} = 0, \quad x \in [0,\infty), \ t \geq 0,$$
$$u(x,0) = f(x),$$

is uniformly well-posed on D_1. For $f \in D_1$ we have

$$u(x,t) = (U(t)f)(x) = \begin{cases} f(x-t), & x \geq t \\ 0, & x \leq t \end{cases}.$$

We refer the reader to G. Da Prato and E. Sinestrari [50], where several examples of non-densely defined operators can be found. Many examples of degenerate Cauchy problems are discussed in A. Favini and A. Yagi [102].

1.6 The Cauchy problem for inclusions

The degenerate Cauchy problem

$$Bu'(t) = Au(t), \quad t \geq 0, \quad u(0) = x, \quad \ker B \neq \{0\}, \qquad \text{(DP)}$$

which we studied in the previous section, and the degenerate Cauchy problem

$$\frac{d}{dt} Bv(t) = Av(t), \quad t \geq 0, \quad Bv(0) = x, \qquad \text{(DP1)}$$

can be considered as particular cases of the Cauchy problem for an inclusion

$$u'(t) \in \mathcal{A}u(t), \quad t \geq 0, \quad u(0) = x, \qquad \text{(IP)}$$

with a multivalued linear operator $\mathcal{A}$. If we set $\mathcal{A} = B^{-1}A$ for problem (DP) or $\mathcal{A} = AB^{-1}$ and $u := Bv$ for (DP1), then $u(\cdot)$ is a solution of the Cauchy problem (IP). Conversely, if $u(\cdot)$ is a solution of (IP) with $\mathcal{A} = B^{-1}A$ or $\mathcal{A} = AB^{-1}$, then $u(\cdot)$ is the solution of (DP) or any $v(\cdot)$ from the set $Bv = u$ is the solution of (DP1), respectively.

In this section, we use the technique of degenerate semigroups with multivalued generators for studying the well-posedness of the Cauchy problem (IP). As a consequence, we obtain a criterion for the well-posedness of (DP) and a criterion for the existence of a solution for (DP1).

1.6.1 Multivalued linear operators

Let X be a complex Banach space. For subsets from X we define addition and scalar multiplication

$$F + G := \left\{ f + g \,\middle|\, f \in F, g \in G \right\},$$
$$\forall \lambda \in \mathbb{C}, \quad \lambda F := \left\{ \lambda f \,\middle|\, f \in F \right\}.$$

Definition 1.6.1 *A map* $\mathcal{A} : X \mapsto 2^X$ *is called a multivalued linear operator on* X *if*

$$D(\mathcal{A}) = \left\{ u \in X \ \middle| \ \mathcal{A}u \neq \emptyset \right\}$$

is a linear manifold in X *and*

$$\lambda \mathcal{A}u + \mu \mathcal{A}v \subset \mathcal{A}(\lambda u + \mu v) \tag{1.6.1}$$

for any $\lambda, \mu \in \mathbb{C}$ *and* $u, v \in D(\mathcal{A})$.

In the particular case of $\lambda = \mu = 0$ we have $0 \in \mathcal{A}0$.

By $\mathcal{M}(X)$ we denote the space of all multivalued linear operators on X.

Proposition 1.6.1 *Let* $\mathcal{A} \in \mathcal{M}(X)$, *then*

(a) $\forall u, v \in D(\mathcal{A}), \quad \mathcal{A}u + \mathcal{A}v = \mathcal{A}(u + v)$,

(b) $\forall u \in D(\mathcal{A}), \lambda \neq 0, \quad \lambda \mathcal{A}u = \mathcal{A}(\lambda u)$.

Proof Since $\mathcal{A}$ is a linear operator, it is sufficient to prove the inclusions $\mathcal{A}(u + v) \subset \mathcal{A}u + \mathcal{A}v$ and $\mathcal{A}(\lambda u) \subset \lambda \mathcal{A}u$. From (1.6.1) we have

$$\mathcal{A}(u + v) - \mathcal{A}v \subset (\mathcal{A}(u + v) - v) = \mathcal{A}u,$$

therefore,

$$\mathcal{A}(u + v) \subset \mathcal{A}u + \mathcal{A}v.$$

Also

$$\mathcal{A}(\lambda u) = \lambda(\lambda^{-1}\mathcal{A}(\lambda u)) \subset \lambda \mathcal{A}u. \ \square$$

Corollary 1.6.1 *For any* $u, v \in D(\mathcal{A})$ *and any* λ, μ *with* $|\lambda| + |\mu| \neq 0$, *we have*

$$\lambda \mathcal{A}u + \mu \mathcal{A}v = \mathcal{A}(\lambda u + \mu v). \ \square$$

Proposition 1.6.2 *Let* $\mathcal{A} \in \mathcal{M}(X)$, *then* $\mathcal{A}0$ *is a linear manifold in* X *and* $\mathcal{A}u = f + \mathcal{A}0$ *for any* $u \in D(\mathcal{A})$ *and* $f \in \mathcal{A}u$.

Proof By Proposition 1.6.1 we have $\mathcal{A}0 + \mathcal{A}0 = \mathcal{A}0$ and $\lambda \mathcal{A}0 = \mathcal{A}0$, hence $\mathcal{A}0$ is a linear manifold. Let $u \in D(\mathcal{A})$ and $f \in \mathcal{A}u$, then

$$f + \mathcal{A}0 \subset \mathcal{A}u + \mathcal{A}0 = \mathcal{A}u.$$

On the other hand, for any $g \in \mathcal{A}u$ we have $g - f \in \mathcal{A}u - \mathcal{A}u = \mathcal{A}0$, hence

$$g = f + (g - f) \in f + \mathcal{A}0.$$

This proves the inverse inclusion $\mathcal{A}u \supset f + \mathcal{A}0$, and hence the equality $\mathcal{A}u = f + \mathcal{A}0$. In particular, $\mathcal{A}$ is a single-valued operator if and only if $\mathcal{A}0 = \{0\}$. $\square$

Definition 1.6.2 *An operator $\mathcal{A}^{-1}$ is called the inverse of an operator $\mathcal{A}$ if*

$$D(\mathcal{A}^{-1}) := \operatorname{ran} \mathcal{A} = \bigcup_{u \in D(\mathcal{A})} \mathcal{A}u$$

and

$$\mathcal{A}^{-1}f := \left\{ u \in D(\mathcal{A}) \mid f \in \mathcal{A}u \right\}.$$

It is easy to see that $\mathcal{A}^{-1}$ is a linear and (in general) multivalued operator on X.

Definition 1.6.3 *The set*

$$\rho(\mathcal{A}) := \left\{ \lambda \in \mathbb{C} \mid (\lambda I - \mathcal{A})^{-1} \right.$$

$$\left. \textit{is a single-valued bounded linear operator on } X \right\}$$

is called the resolvent set of the operator $\mathcal{A}$.

Proposition 1.6.3 *Let $\mathcal{A} \in \mathcal{M}(X)$ and $\rho(\mathcal{A}) \neq \emptyset$, then for any $\lambda \in \rho(\mathcal{A})$, $\mathcal{A}0 = \ker (\lambda I - \mathcal{A})^{-1}$.*

Proof By the definition of kernel we have that $f \in \ker (\lambda I - \mathcal{A})^{-1}$ if and only if $(\lambda I - \mathcal{A})^{-1}f = 0$, that is $f \in (\lambda I - \mathcal{A})0 = \mathcal{A}0$. $\square$

Proposition 1.6.4 *Let $\mathcal{A} \in \mathcal{M}(X)$ and suppose that the decomposition $X = \mathcal{A}0 \oplus Y$ holds for some subspace $Y \subset X$. Then $\tilde{\mathcal{A}}u := \mathcal{A}u \cap Y$ is a single-valued linear operator with $D(\tilde{\mathcal{A}}) = D(\mathcal{A})$.*

Proof First we show that $D(\tilde{\mathcal{A}}) = D(\mathcal{A})$. From the definition of $\tilde{\mathcal{A}}$ we have $D(\tilde{\mathcal{A}}) \subset D(\mathcal{A})$. Let $u \in D(\mathcal{A})$, then there exists $f \in X$ such that $u = f + \mathcal{A}0$ and $f = f_1 + f_2$, where $f_1 \in \mathcal{A}0$, $f_2 \in Y$. We have

$$\tilde{\mathcal{A}}u = \mathcal{A}u \cap Y = (f_2 + \mathcal{A}0) \cap Y \supset \{f_2\} \neq \emptyset,$$

that is $u \in D(\tilde{\mathcal{A}})$. Hence $D(\tilde{\mathcal{A}}) \supset D(\mathcal{A})$ and $D(\tilde{\mathcal{A}}) = D(\mathcal{A})$.

Now let $u, v \in D(\tilde{\mathcal{A}}) = D(\mathcal{A})$ and $\lambda, \mu \in \mathbb{C}$, then

$$\begin{aligned}
\lambda \tilde{\mathcal{A}}u &+ \mu \tilde{\mathcal{A}}v \\
&= \left[(\lambda \mathcal{A}u) \cap Y \right] + \left[(\mu \mathcal{A}v) \cap Y \right] \subset \mathcal{A}(\lambda u + \mu v) \cap Y \\
&= \tilde{\mathcal{A}}(\lambda u + \mu v).
\end{aligned}$$

Since $\tilde{\mathcal{A}}0 = \mathcal{A}0 \cap Y = \{0\}$, we have that $\tilde{\mathcal{A}}$ is a single-valued linear operator. $\square$

The operator $\tilde{\mathcal{A}}$ is called the single-valued branch of $\mathcal{A}$.

Definition 1.6.4 *We say that an operator $A \in \mathcal{M}(X)$ is closed if for any sequences $\{u_n\} \subset D(A)$ and $f_n \in Au_n$ such that*

$$\lim_{n \to \infty} u_n = u \quad and \quad \lim_{n \to \infty} f_n = f,$$

we have $u \in D(A)$ and $f \in Au$.

It is clear that A is closed if and only if $\tilde{A}$ is closed. Furthermore, for a closed operator A we have

$$\forall \{u_n\} \subset D(A), \quad A(\lim_{n \to \infty} u_n) = \lim_{n \to \infty} Au_n := \lim_{n \to \infty} \tilde{A}u_n + A0$$

if all limits exist, and

$$\forall u \in C\{[0,T], D(A)\},$$

$$A \int_0^t u(\tau)d\tau = \int_0^t Au(\tau)d\tau =: \int_0^t f(\tau)d\tau + A0, \qquad t \in [0,T),$$

if integrals exist.

1.6.2   Uniform well-posedness

Definition 1.6.5 *The Cauchy problem*

$$u'(t) \in Au(t), \quad t \geq 0, \; u(0) = x, \tag{IP}$$

is called uniformly well-posed on $E \subset X$, if

(a) *for any $x \in E$, there exists a unique solution*

$$u(\cdot) \in C^1\{[0,\infty), X\} \cap C\{[0,\infty), D(A)\}$$

(b) *for any $T > 0$, $\sup_{0 \leq t \leq T} \|u(t)\| \leq \|x\|$.*

The following theorem connects the well-posedness of (IP) with the well-posedness of (CP) with operator $\tilde{A}$, the single-valued branch of A:

$$u'(t) = \tilde{A}u(t), \quad t \geq 0, \; u(0) = x, \tag{1.6.2}$$

and gives the MFPHY-type condition for the uniform well-posedness of (IP). Note that in contrast to Theorem 1.5.1, here we do not assume that $\rho(A) \neq \emptyset$.

Theorem 1.6.1 *Let A be a multivalued linear operator on X. Let*

$$X_1 := \overline{D(A)}, \; \tilde{A}u := Au \cap X_1, \; D(\tilde{A}) := \left\{ u \in X \; \middle| \; \tilde{A}u \neq \emptyset \right\}.$$

Then the following statements are equivalent:

(I) *the Cauchy problem* (IP) *is uniformly well-posed on* $D(A)$;

(II) *the operator $\widetilde{A}$ is a single-valued branch of A and it is the generator of a C_0-semigroup $\{U(t),\ t \geq 0\}$ on X_1;*

(III) *the operator $\widetilde{A}$ is a single-valued branch of A and the MFPHY condition holds for $R_{\widetilde{A}} = (\lambda I - \widetilde{A})^{-1}$:*

$$\exists K > 0,\ \omega \in \mathbb{R}:$$

$$\left\|\frac{d^k}{d\lambda^k}(\lambda I - \widetilde{A})^{-1}\right\| \leq \frac{K\,k!}{(\lambda - \omega)^{k+1}}$$

for all $\lambda > \omega$ and $k = 0, 1, \dots$.

In this case for any $x \in D(A)$, $u(\cdot) = U(\cdot)x$ is the unique solution of (IP).

Proof (I) $\Longrightarrow$ (II). Define on $D(A)$ the solution operators $U(t)$ by $U(t)x := u(t)$, $t \geq 0$, where $u(\cdot)$ is the solution of (IP). From the well-posedness of (IP) we have that for any $t \geq 0$, $U(t)$ is a bounded operator on $D(A)$. Hence it can be extended to $X_1 = \overline{D(A)}$.

In a way similar to the nondegenerate case (Theorem 1.1.1), one can prove that the operators $U(t)$ form a C_0-semigroup on X_1. Let $G : X_1 \mapsto X_1$ be the generator of this semigroup. Then G is a closed, densely defined linear operator with the property

$$U(\cdot)x \in C^1\{[0, \infty], X\} \iff x \in D(G).$$

Since for any $x \in D(A)$ we have $u(\cdot) = U(\cdot)x \in C^1\{[0, \infty], X\}$, then $D(A) \subset D(G)$. Since the operator A is closed, we have

$$\forall x \in D(A),$$

$$U(t)x - x = \int_0^t U'(\tau)x\,d\tau \in \int_0^t AU(\tau)x\,d\tau = A\int_0^t U(\tau)x\,d\tau.$$

Using the closedness of A we can extend the obtained inclusion to X_1

$$U(t)x - x \in A\int_0^t U(\tau)x\,d\tau, \quad x \in X_1,$$

and

$$\frac{U(t)x - x}{t} \in \frac{1}{t}A\int_0^t U(\tau)x\,d\tau, \quad t > 0,\ x \in X_1. \tag{1.6.3}$$

Consider the domain of the generator

$$D(G) := \left\{x \in X_1 \ \middle|\ \exists \lim_{t \to 0} \frac{U(t)x - x}{t}\right\},$$

then (1.6.3) implies $D(G) \subset D(\mathcal{A})$, i.e. $D(G) = D(\mathcal{A})$ and for any $x \in D(\mathcal{A})$, $Gx \in \mathcal{A}x$. Since $Gx \in X_1$, we have

$$\forall x \in D(\mathcal{A}) = D(G), \qquad Gx \in \tilde{\mathcal{A}}x := \mathcal{A}x \cap X_1, \qquad (1.6.4)$$

hence $D(\mathcal{A}) \subset D(\tilde{\mathcal{A}})$. By definition of $\tilde{\mathcal{A}}$, we have $D(\tilde{\mathcal{A}}) \subset D(\mathcal{A})$, and therefore $D(\mathcal{A}) = D(\tilde{\mathcal{A}})$.

Now we show that $\tilde{\mathcal{A}}$ is a single-valued operator. Let $y \in \tilde{\mathcal{A}}0 := \mathcal{A}0 \cap X_1$ and $z := (\lambda I - G)^{-1}y$ with $\lambda \in \rho(G)$, $\operatorname{Re}\lambda > \omega$. Note that since G is the generator of a C_0-semigroup, then any $\lambda \in \mathbb{C}$ with $\operatorname{Re}\lambda > \omega$ belongs to $\rho(G)$. For these $z \in D(G) = D(\mathcal{A})$ we have

$$(\lambda I - G)z = y$$

and

$$\lambda z = y + Gz \in \mathcal{A}0 + Gz \subset \mathcal{A}0 + \mathcal{A}z = \mathcal{A}z,$$

that is $\lambda z \in \mathcal{A}z$. Hence $(ze^{\lambda t})' = \lambda z e^{\lambda t} \in \mathcal{A}(ze^{\lambda t})$ and $u(t) = ze^{\lambda t}$, $t \geq 0$, is a solution of (IP) with the initial value z for any λ with $\operatorname{Re}\lambda > \omega$. From (1.6.4) we have

$$\forall x \in D(G) = D(\tilde{\mathcal{A}}), \qquad Gx \subset \tilde{\mathcal{A}}x,$$

and since $\tilde{\mathcal{A}}$ is a single-valued operator, we have $G = \tilde{\mathcal{A}}$.

(II) $\Longrightarrow$ (III) $\Longrightarrow$ (I). By assumption, $\tilde{\mathcal{A}}$ is the generator of a C_0-semigroup. By Theorem 1.1.1 it is equivalent to the MFPHY condition for $R_{\tilde{\mathcal{A}}}$ and to the uniform well-posedness on $D(\tilde{\mathcal{A}})$ of the Cauchy problem (1.6.2). Hence (IP) is uniformly well-posed on $D(\mathcal{A}) = D(\tilde{\mathcal{A}})$. $\square$

Corollary 1.6.2 If, in Theorem 1.6.1, we do not assume that the operator $\mathcal{A}$ is closed, then the result holds true for the single-valued branch of $\overline{\mathcal{A}}$, the closure of $\mathcal{A}$.

Let us show that under the condition $\rho(\mathcal{A}) \neq \emptyset$ the estimates for the resolvent of $\mathcal{A}$ are fulfilled. In Definition 1.5.1 we defined degenerate semigroups and now we need the notion of a generator of a degenerate semigroup.

Definition 1.6.6 *An operator $\mathcal{A}$ is called the generator of a degenerate (exponentially bounded) C_0-semigroup $\{U(t), \ t \geq 0\}$ if it satisfies the relation*

$$(\lambda I - \mathcal{A})^{-1}x = \int_0^\infty e^{-\lambda t}U(t)x\,dt \qquad (1.6.5)$$

for all $\lambda \in \mathbb{C}$ with $\operatorname{Re}\lambda > \omega$.

It follows from (1.6.5) that $\ker U = \ker(\lambda I - \mathcal{A})^{-1}$. On the other hand, by Proposition 1.6.3, we have $\ker(\lambda I - \mathcal{A})^{-1} = \mathcal{A}0$. That is, the generator of any degenerate semigroup is a multivalued operator, and conversely, if a multivalued operator is the generator of a semigroup, then this semigroup is degenerate.

Theorem 1.6.2 *Let $\mathcal{A}$ be a multivalued linear operator in a Banach space X with nonempty resolvent set. Then the following is equivalent:*

(I) *the Cauchy problem* (IP) *is uniformly well-posed on $D(\mathcal{A})$;*

(II) *the MFPHY condition holds for $R_{\mathcal{A}}(\lambda)$ and X admits the decomposition*

$$X = \mathcal{A}0 \oplus X_1. \tag{1.6.6}$$

Proof (I) $\Longrightarrow$ (II). In Theorem 1.6.1 it was proved that $\tilde{\mathcal{A}}$ is the single-valued branch of $\mathcal{A}$ and the generator of a C_0-semigroup $\{U(t),\ t \geq 0\}$. Consider the semigroup $\mathcal{U}$ equal to U on X_1 and zero on $\mathcal{A}0$. It is a degenerate exponentially bounded C_0-semigroup with the generator $\mathcal{A}$:

$$(\lambda I - \mathcal{A})^{-1}x = \int_0^\infty e^{-\lambda t}\mathcal{U}(t)x dt, \quad \operatorname{Re}\lambda > \omega,$$

hence the MFPHY condition holds for $R_{\mathcal{A}}(\lambda)$.

Let $x \in X$, $y := (\lambda I - \mathcal{A})^{-1}x$, $\lambda \in \rho(\mathcal{A})$ and $z := x - (\lambda I - \tilde{\mathcal{A}})y$. We have $y \in D(\mathcal{A}) = D(\tilde{\mathcal{A}})$ and $(\lambda I - \tilde{\mathcal{A}})y \in X_1$. Since for any $f \in X_1$, $(\lambda I - \mathcal{A})^{-1}f = (\lambda I - \tilde{\mathcal{A}})^{-1}f$, we obtain

$$(\lambda I - \mathcal{A})^{-1}z = (\lambda I - \mathcal{A})^{-1}x - (\hat{\lambda}I - \mathcal{A})^{-1}(\lambda I - \tilde{\mathcal{A}})(\lambda I - \mathcal{A})^{-1}x.$$

Hence $z \in \ker(\lambda I - \mathcal{A})^{-1} = \mathcal{A}0$ and for $x \in X$ we have the decomposition

$$x = z + (\hat{\lambda}I - \tilde{\mathcal{A}})y \in \mathcal{A}0 + X_1.$$

As $\tilde{\mathcal{A}}$ is a single-valued operator, we have $\tilde{\mathcal{A}}0 = \mathcal{A}0 \cap X_1 = \{0\}$. Hence, we obtain the decomposition of X into the direct sum $X = \mathcal{A}0 \oplus X_1$.

(II) $\Longrightarrow$ (I). Since we have the decomposition $X = \mathcal{A}0 \oplus X_1$, then $\tilde{\mathcal{A}}$ is a single-valued closed linear operator with $D(\tilde{\mathcal{A}}) = D(\mathcal{A})$ (see Proposition 1.6.4), and the MFPHY conditions for $R_{\mathcal{A}}(\lambda)$ and $R_{\tilde{\mathcal{A}}}(\lambda)$ are equivalent. By Theorem 1.1.1, the Cauchy problem (1.6.2) with operator $\tilde{\mathcal{A}}$ is uniformly well-posed on $D(\tilde{\mathcal{A}})$. Let us show that this problem is well-posed on $D(\tilde{\mathcal{A}})$ if and only if (IP) is well-posed on $D(\mathcal{A})$. If

$$u'(t) = \tilde{\mathcal{A}}u(t), \quad t \geq 0,\ u(0) = x \in D(\tilde{\mathcal{A}}),$$

then

$$u'(t) \in \mathcal{A}u(t), \quad t \geq 0, \ u(0) = x \in D(\mathcal{A}).$$

Conversely, if $u'(t) \in \mathcal{A}u(t)$, then $u'(t) \in \overline{D(\mathcal{A})} = X_1$. Hence, $u'(t) \in \tilde{\mathcal{A}}u(t) = \mathcal{A}u(t) \cap X_1$, and therefore $u'(t) = \tilde{\mathcal{A}}u(t)$. Thus, the Cauchy problem (IP) is uniformly well-posed on $D(\mathcal{A})$. $\square$

The following two theorems are corollaries of Theorem 1.6.2.

Theorem 1.6.3 *Let* $X_1 = \overline{D_1}$, *where*

$$D_1 = \Big\{ x \in D(A) \ \Big| \ Ax \in \operatorname{ran} B \Big\}.$$

Consider the Cauchy problem (DP) *with linear operators* A *and* B *such that* AB^{-1} *is closed (for example, one of them is closed and the other is bounded). Suppose*

$$\rho_B(A) := \Big\{ \lambda \in \mathbb{C} \ \Big| \ (\lambda B - A)^{-1}B \ is \ bounded \Big\} \neq \emptyset.$$

Then the following conditions are equivalent:

(I) *the Cauchy problem* (DP) *is uniformly well-posed on* D_1;

(II) *there exists a* C_0-*semigroup* $\{U(t), \ t \geq 0\}$ *on* X_1 *such that*

$$(\lambda B - A)^{-1}Bx = \int_0^\infty e^{-\lambda t} U(t)x \, dt$$

for all $x \in X_1$ *and* $\lambda \in \mathbb{C}$ *with* $\operatorname{Re}\lambda > \omega$;

(III) $X = X_1 \oplus \ker B$ *and*

$\exists K > 0, \ \omega \in \mathbb{R}$:

$$\left\| \frac{d^k}{d\lambda^k}(\lambda B - A)^{-1}B \right\| \leq \frac{K\,k!}{(\lambda - \omega)^{k+1}}$$

for all $\lambda > \omega$ *and* $k = 0, 1, \dots$.

And, in this case, the solution of (DP) *is* $u(t) = U(t)x, \ t \geq 0, \ x \in D_1$. $\square$

If B is bounded and A is closed, then this result gives the same well-posedness conditions as Theorem 1.5.4. In Section 1.5 we used a degenerate C_0-semigroup with generators A, B, and here we use a C_0-semigroup with a generator equal to a single-valued branch of operator AB^{-1}. In the case of $\ker B = \{0\}$ we obtain Theorem 1.1.1.

Theorem 1.6.4 *Let* $X_1 = \overline{\mathcal{D}_1}$, *where*

$$\mathcal{D}_1 := \left\{ x \in X \mid x \in \operatorname{ran} B(\lambda B - A)^{-1} \right\}.$$

Consider the Cauchy problem (DP1) *with linear operators* A *and* B *such that* $B^{-1}A$ *is closed (for example both operators are bounded). Suppose*

$$\left\{ \lambda \in \mathbb{C} \mid B(\lambda B - A)^{-1} \text{ is bounded} \right\} \neq \emptyset.$$

Then the following conditions are equivalent:

(I) *for any* $x \in \mathcal{D}_1$ *there exist solutions* v *of the Cauchy problem* (DP1) *such that* Bv *is uniquely defined and is stable with respect to* x;

(II) *there exists a* C_0-*semigroup* $\{U(t),\ t \geq 0\}$ *on* X_1 *such that*

$$B(\lambda B - A)^{-1} x = \int_0^\infty e^{-\lambda t} U(t) x\, dt$$

for all $x \in X_1$ *and* $\lambda \in \mathbb{C}$ *with* $\operatorname{Re}\lambda > \omega$;

(III) $X = X_1 \oplus A \ker B$, *and*

$$\exists K > 0,\ \omega \in \mathbb{R}:$$

$$\left\| \frac{d^k}{d\lambda^k} B(\lambda B - A)^{-1} \right\| \leq \frac{K\, k!}{(\lambda - \omega)^{k+1}}$$

for all $\lambda > \omega$ *and* $k = 0, 1, \dots$.

And in this case the solution of (DP1) *is* $Bv(t) = U(t)x,\ t \geq 0,\ x \in \mathcal{D}_1$. $\square$

1.6.3 (n, ω)-**well-posedness**

Definition 1.6.7 *The Cauchy problem*

$$u'(t) \in \mathcal{A}u(t), \quad t \geq 0,\ u(0) = x, \tag{IP}$$

is called (n, ω)-*well-posed on* $E \subset X$ *if for any* $x \in E$ *there exists a unique solution*

$$u(\cdot) \in C^1\{[0, \infty),\ X\} \cap C\{[0, \infty),\ D(\mathcal{A})\}$$

such that

$$\|u(t)\| \leq K e^{\omega t} \|x\|_{\mathcal{A}^n},$$

for some $K > 0$, *where*

$$\|x\|_{\mathcal{A}^n} := \sum_{k=0}^n \|\mathcal{A}^k x\|.$$

Here

$$\|A^k x\| = \inf_{f \in A^k x} \|f\|_X$$

is the factor-norm of $A^k x$ in $X/A^k 0$.

If the resolvent set $\rho(\mathcal{A})$ is not empty, then the norm $\|x\|_{A^n}$ is equivalent to

$$\|x\|_{R^n} := \inf_{y:R_{\mathcal{A}}^n(\lambda)y=x} \|y\|, \quad \lambda \in \rho(\mathcal{A}).$$

Now we investigate the connections between the (n,ω)-well-posedness of (IP) and the existence of an exponentially bounded degenerate $k(n)$-times integrated semigroup.

Definition 1.6.8 *An operator $\mathcal{A}$ is called the generator of an exponentially bounded degenerate n-times integrated semigroup $\{V(t), \ t \geq 0\}$ if it satisfies the relation*

$$(\lambda I - \mathcal{A})^{-1}x = \int_0^\infty \lambda^n e^{-\lambda t} V(t) x \, dt \tag{1.6.7}$$

for all $\lambda \in \mathbb{C}$ with $\operatorname{Re}\lambda > \omega$.

As in the case of degenerate C_0-semigroups, it follows from (1.6.7) that $\ker V = \ker(\lambda I - \mathcal{A})^{-1} = \mathcal{A}0$, and an exponentially bounded n-times integrated semigroup has a multivalued generator if and only if the semigroup is degenerate.

Theorem 1.6.5 *Let $n \in \mathbb{N}$, $\omega \in \mathbb{R}$, and let $\mathcal{A}$ be a closed linear multivalued operator on X. Suppose that $\rho(\mathcal{A}) \neq \emptyset$ is in the half-plane $\operatorname{Re}\lambda > \omega$. If the Cauchy problem (IP) is (n,ω)-well-posed on $D(\mathcal{A}^{n+1})$, then the MFPHY-type condition*
$$\exists K > 0:$$
$$\left\| \frac{d^k}{d\lambda^k} \frac{(\lambda I - \mathcal{A})^{-1}}{\lambda^{n+1}} \right\| \leq \frac{K\, k!}{(\lambda - \omega)^{k+1}} \tag{1.6.8}$$

holds for all $\lambda > \omega$ and $k = 0, 1, \dots$.

Proof Let $x \in D(\mathcal{A}^{n+1}) =: D_{n+1}$ and let $u(t)$, $t \geq 0$, be the unique solution of (IP) corresponding to x. By the definition of the (n,ω)-well-posedness for (IP), $u(t)$ satisfies the estimate

$$\|u(t)\| \leq Ke^{\omega t} \|x\|_{R^n}. \tag{1.6.9}$$

Now we introduce the solution operators $U(t)$ on D_{n+1} by

$$U(t)x := u(t), \quad t \geq 0.$$

Due to (1.6.9) we can extend $U(t)$ to the space $[D_{n+1}]_n$, the closure of D_{n+1} in the norm $\| \cdot \|_{R^n}$. As $U(\cdot)x = u(\cdot)$ is a solution of (IP), we have

$$U'(t)x \in \mathcal{A}U(t)x = \lambda U(t)x - (\lambda I - \mathcal{A})U(t)x. \tag{1.6.10}$$

Now we show that $R_{\mathcal{A}}(\lambda)U(\cdot)x = U(\cdot)R_{\mathcal{A}}(\lambda)x$, $x \in D_{n+1}$, is a solution of (IP) with the initial value $R_{\mathcal{A}}(\lambda)x$. Let $y = R_{\mathcal{A}}(\lambda)x \in D_{n+1}$. Then from (n, ω)-well-posedness of (IP) we have $U'(t)R_{\mathcal{A}}(\lambda)x \in \mathcal{A}U(t)R_{\mathcal{A}}(\lambda)x$. So $U(\cdot)R_{\mathcal{A}}(\lambda)x$ is a solution of (IP) satisfying the estimate

$$\|U(t)R_{\mathcal{A}}(\lambda)x\| \le Ke^{\omega t}\|R_{\mathcal{A}}(\lambda)x\|_{R^n} \le K_1 e^{\omega t}\|x\|_{R^{n-1}}.$$

Applying $R_{\mathcal{A}}(\lambda)$ to (1.6.10), we obtain

$$R_{\mathcal{A}}(\lambda)U'(t)x = R_{\mathcal{A}}(\lambda)\mathcal{A}U(t)x = \lambda R_{\mathcal{A}}(\lambda)U(t)x - U(t)x. \tag{1.6.11}$$

As $U(t)x \in D(\mathcal{A})$, we have

$$U(t)x = R_{\mathcal{A}}(\lambda)(\lambda I - \mathcal{A})U(t)x \in (\lambda I - \mathcal{A})R_{\mathcal{A}}(\lambda)U(t)x.$$

Hence,

$$
\begin{aligned}
R_{\mathcal{A}}(\lambda)\mathcal{A}U(t)x &= \lambda R_{\mathcal{A}}(\lambda)U(t)x - U(t)x \in \mathcal{A}R_{\mathcal{A}}(\lambda)U(t)x, \\
R_{\mathcal{A}}(\lambda)U'(t)x &= (R_{\mathcal{A}}(\lambda)U(t)x)' \in \mathcal{A}R_{\mathcal{A}}(\lambda)U(t)x,
\end{aligned}
$$

and $R_{\mathcal{A}}(\lambda)U(0)x = R_{\mathcal{A}}(\lambda)x$. That is, $R_{\mathcal{A}}(\lambda)U(\cdot)x$, $x \in D_{n+1}$ is a solution of (IP). The equality $R_{\mathcal{A}}(\lambda)U(t)x = U(t)R_{\mathcal{A}}(\lambda)x$ follows from the uniqueness of the solution. Integrating (1.6.11) from 0 to t, we obtain

$$R_{\mathcal{A}}(\lambda)U(t)x - R_{\mathcal{A}}(\lambda)x = \int_0^t \lambda R_{\mathcal{A}}(\lambda)U(s)x\,ds - \int_0^t U(s)x\,ds,$$

or

$$
\begin{aligned}
\int_0^t U(s)x\,ds &= -R_{\mathcal{A}}(\lambda)U(t)x + R_{\mathcal{A}}(\lambda)x + \lambda \int_0^t R_{\mathcal{A}}(\lambda)U(s)x\,ds \\
&= -U(t)R_{\mathcal{A}}(\lambda)x + R_{\mathcal{A}}(\lambda)x + \lambda \int_0^t U(s)R_{\mathcal{A}}(\lambda)x\,ds, \quad x \in D_{n+1}.
\end{aligned}
$$

On D_n we define

$$U_1(t)x := -U(t)R_{\mathcal{A}}(\lambda)x + R_{\mathcal{A}}(\lambda)x + \lambda \int_0^t U(s)R_{\mathcal{A}}(\lambda)x\,ds.$$

Applying $R_A(\lambda)$ to both sides of this equality, we obtain

$R_A(\lambda)U_1(t)x$

$$= -R_A(\lambda)U(t)R_A(\lambda)x + R_A(\lambda)R_A(\lambda)x + \lambda R_A(\lambda) \int_0^t U(s)R_A(\lambda)x ds.$$

Commutativity of $R_A(\lambda)$ and $U(t)$ on D_{n+1} implies

$$R_A(\lambda)U_1(t)x = U_1(t)R_A(\lambda)x, \quad x \in D_n.$$

Hence, $U_1(t)$ commutes with $R_A(\lambda)$ on D_n, satisfies the estimate

$$\|U_1(t)x\| \le Ke^{\omega t}\|x\|_{R^{n-1}},$$

and can be extended to $[D_n]_{n-1}$. In general,

$$U_k(t)x := \tag{1.6.12}$$

$$-U_{k-1}(t)R_A(\lambda)x + \frac{t^{k-1}}{(k-1)!}R_A(\lambda)x + \lambda \int_0^t U_{k-1}(s)R_A(\lambda)x ds.$$

The operators $U_k(t)$, $t \ge 0$, $k = 1, \ldots, n$, are defined on D_{n+1-k}, where they commute with $R_A(\lambda)$ and satisfy the estimate

$$\|U_k(t)x\| \le Kt^{k-1}e^{\omega t}\|x\|_{R^{n-k}} \le Ke^{\omega_1 t}\|x\|_{R^{n-k}}$$

for any $\omega_1 > \omega$. Therefore, we can extend $U_k(t)$ to $[D_{n+1-k}]_{n-k}$. In particular, the operators $U_n(t)$ commute with $R_A(\lambda)$ on $D(\mathcal{A})$, satisfy the estimate

$$\|U_n(t)x\| \le Ke^{\omega_1 t}\|x\|,$$

and can be extended to $\overline{D(A)}$. The operators $U_{n+1}(t)$ commute with $R_A(\lambda)$ on the whole space X and satisfy

$$\|U_{n+1}(t)x\| \le Ke^{\omega_1 t}\|x\|. \tag{1.6.13}$$

Now denote $V(t)x := U_{n+1}(t)x$ and we show that $\{V(t), \ t \ge 0\}$ is an $(n + 1)$-times integrated semigroup with the generator $\mathcal{A}$. From (1.6.12) we have that for any $x \in X$, $V(t)x$ is continuous in $t \ge 0$, so (V2) holds. We have (V3) from (1.6.13). To show that V satisfies (V1), by Proposition 1.2.1 it is sufficient to show that

$$\int_0^\infty \lambda^{n+1}e^{-\lambda t}V(t)dt$$

satisfies the pseudoresolvent identity. Hence, it is sufficient to prove the equality

$$R_A(\lambda)x = \int_0^\infty \lambda^{n+1}e^{-\lambda t}V(t)x dt \tag{1.6.14}$$

for $\lambda \in \mathbb{C}$ with $\mathrm{Re}\,\lambda > \omega$. Multiplying (1.6.12) by $\lambda^k e^{-\lambda t}$ with $k = n + 1$ and integrating from 0 to ∞ we get

$$\int_0^\infty \lambda^{n+1} e^{-\lambda t} U_{n+1}(t) x\, dt = R_A(\lambda) x = \int_0^\infty \lambda^{n+1} e^{-\lambda t} V(t) x\, dt.$$

This equation holds for all λ from some open set where $R_A(\lambda)$ exists by assumption. Using the resolvent identity for the function

$$\int_0^\infty \lambda^{n+1} e^{-\lambda t} V(t) x\, dt,$$

the analytical extension of $R_A(\lambda)$ to the half-plane $\mathrm{Re}\,\lambda > \omega$, we obtain (1.6.14) for $\mathrm{Re}\,\lambda > \omega$. Thus, $\{V(t),\ t \geq 0\}$ is an $(n+1)$-times degenerate integrated semigroup generated by A, and hence the estimate (1.6.8) holds. □

Theorem 1.6.6 *Let A be a multivalued linear operator on X. Suppose that the condition*
$$\exists K > 0,\ \omega \in \mathbb{R}:$$

$$\left\| \frac{d^k}{d\lambda^k} \frac{(\lambda I - A)^{-1}}{\lambda^n} \right\| \leq \frac{K\,k!}{(\lambda - \omega)^{k+1}} \tag{1.6.15}$$

holds for all $\lambda > \omega$ and $k = 0, 1, \dots$. Then the Cauchy problem (IP) *is (n, ω)-well-posed on $R_A^{n+1}(\lambda)\overline{D(A)}$.*

Proof If the estimate (1.6.15) holds for some $K > 0$ and $\omega \in \mathbb{R}$, then by Theorem 1.2.1 A is the generator of an $(n+1)$-times integrated semigroup $\{V(t),\ t \geq 0\}$ with the property

$$\|V(t+h) - V(t)\| \leq Kh, \quad t \geq 0,\ h \geq 0. \tag{1.6.16}$$

Note that in this case the semigroup can be degenerate. By the definition of the generator, for all $y \in X$ we have

$$\begin{aligned}
\int_0^\infty \lambda^{n+1} e^{-\lambda t} V(t) R_A(\mu) y\, dt &= R_A(\lambda) R_A(\mu) y = R_A(\mu) R_A(\lambda) y \\
&= \int_0^\infty \lambda^{n+1} e^{-\lambda t} R_A(\mu) V(t) y\, dt.
\end{aligned}$$

The uniqueness of the Laplace transform implies

$$V(t) R_A(\mu) y = R_A(\mu) V(t) y, \quad y \in X.$$

Let $y = (\mu I - A)x, x \in D(A)$, then we have

$$V(t)x = R_A(\mu) V(t)(\mu I - A)x,$$

and applying $(\mu I - A)$:

$$(\mu I - A)V(t)x = (\mu I - A)R_A(\mu)V(t)(\mu I - A)x.$$

From $V(t)(\mu I - A)x \in (\mu I - A)R_A(\mu)V(t)(\mu I - A)x$ we obtain

$$V(t)(\mu I - A)x \in (\mu I - A)V(t)x.$$

Hence for all $x \in D(A)$, $V(t)Ax \in AV(t)x$, and

$$\int_0^\infty \lambda^{n+2} e^{-\lambda t} \frac{t^{n+1}}{(n+1)!} x \, dt$$

$$= \quad x = \lambda R_A(\lambda)x - R_A(\lambda)Ax$$

$$= \quad \int_0^\infty \lambda^{n+2} e^{-\lambda t} V(t)x \, dt - \int_0^\infty \lambda^{n+1} e^{-\lambda t} V(t)Ax \, dt$$

$$= \quad \int_0^\infty \lambda^{n+2} e^{-\lambda t} V(t)x \, dt - \int_0^\infty \lambda^{n+2} e^{-\lambda t} \int_0^t V(s)Ax \, ds \, dt.$$

From the uniqueness of the Laplace transform we have

$$\frac{t^{n+1}}{(n+1)!} x = V(t)x - \int_0^t V(s)Ax \, ds, \quad x \in D(A),$$

hence

$$V'(t)x \quad = \quad \frac{t^n}{n!} x + V(t)Ax,$$

$$\tag{1.6.17}$$

$$V'(t)x \quad \in \quad \frac{t^n}{n!} x + AV(t)x, \quad x \in D(A).$$

Using property (1.6.16) and the closedness of A we can extend the inclusion in (1.6.17) to $\overline{D(A)}$. Now take $x \in D(A^2)$ (as $Ax \cap D(A) \neq \emptyset$), then from (1.6.17) we have

$$AV'(t)x \quad = \quad \frac{t^n}{n!} Ax + AV(t)Ax, \quad x \in D(A^2),$$

$$V'(t)Ax \quad \in \quad \frac{t^n}{n!} Ax + V(t)A^2 x$$

$$\subset \quad \frac{t^n}{n!} Ax + AV(t)Ax, \quad x \in D(A^2),$$

therefore $V'(t)Ax \in AV'(t)x$ for $x \in D(A^2)$, and

$$V''(t)x \quad = \quad \frac{t^{n-1}}{(n-1)!} x + V'(t)Ax,$$

$$\tag{1.6.18}$$

$$V''(t)x \quad \in \quad \frac{t^{n-1}}{(n-1)!} x + AV'(t)x, \quad x \in D(A^2).$$

Now we show that the equation from (1.6.18) holds for all $x \in R_A(\lambda)\overline{D(A)}$. Let $y \in \overline{D(A)}$. By the closedness of A, there exists a sequence $y_n \in D(A)$ such that $y_n \to y$. Take $x_n = R_A(\lambda)y_n$, then $x_n \in R_A(\lambda)D(A)$ and $x_n \to R_A(\lambda)y = x \in R_A(\lambda)\overline{D(A)}$. Moreover

$$Ax_n = AR_A(\lambda)y_n = -(\lambda I - A)R_A(\lambda)y_n + \lambda R_A(\lambda)y_n,$$

and the set $\{-(\lambda I - A)R_A(\lambda)y_n\}$ contains the sequence $-y_n$ convergent to $-y$. Therefore $Ax_n \cap D(A) \neq \emptyset$, and $-y_n + \lambda R_A(\lambda)y_n \in D(A)$ converges to $-y + \lambda R_A(\lambda)y \in \overline{D(A)}$. Hence

$$Ax_n \to Ax \quad \text{and} \quad V'(t)Ax_n \to V'(t)Ax,$$

as $V'(t)$ is bounded (property (1.6.16)). Since $\frac{t^{n-1}}{(n-1)!}x_n \to \frac{t^{n-1}}{(n-1)!}x$ and $V''(t)$ is closed, we have

$$V''(t)x = \frac{t^{n-1}}{(n-1)!}x + V'(t)Ax, \quad x \in R_A(\lambda)\overline{D(A)}. \tag{1.6.19}$$

From the inclusion in (1.6.18), the closedness of A and the boundedness of $V'(t)$, we obtain

$$V''(t)x \in \frac{t^{n-1}}{(n-1)}x + AV'(t)x, \quad x \in R_A(\lambda)\overline{D(A)}.$$

For $x \in R_A^2(\lambda)\overline{D(A)}$ there exists $V''(t)Ax$, therefore, after differentiating (1.6.18), we have

$$\begin{aligned}
V^{(3)}(t)x &= \frac{t^{n-1}}{(n-2)!}x + V''(t)Ax \\
&\in \frac{t^{n-2}}{(n-2)!}x + AV''(t)x, \quad x \in R_A^2(\lambda)\overline{D(A)}.
\end{aligned}$$

We continue the process and obtain

$$\begin{aligned}
V^{(n+2)}(t)x &\in AV^{(n+1)}(t)x, \quad x \in R_A^{n+1}(\lambda)\overline{D(A)}, \\
V^{(n+1)}(0)x &= x.
\end{aligned}$$

Therefore for any $x \in R_A^{n+1}(\lambda)\overline{D(A)}$, $V^{(n+1)}(\cdot)x$ is a solution of the Cauchy problem (IP).

Now we show that the solution is unique. Let $y(\cdot)$ be another exponentially bounded solution of (IP). Then we have

$$\begin{aligned}
\int_0^\infty e^{-\lambda t}y(t)dt &= \frac{1}{\lambda}\left\{x + \int_0^\infty e^{-\lambda t}y'(t)dt\right\} \\
&\in \frac{1}{\lambda}\left\{x + A\int_0^\infty e^{-\lambda t}y(t)dt\right\},
\end{aligned}$$

hence,

$$x \in (\lambda I - A) \int_0^\infty e^{-\lambda t} y(t) dt.$$

Applying $R_A(\lambda)$ to this inclusion, we obtain

$$R_A(\lambda) x = \int_0^\infty e^{-\lambda t} y(t) dt,$$

or

$$\int_0^\infty \lambda^{n+2} e^{-\lambda t} V(t) x \, dt = \int_0^\infty e^{-\lambda t} V^{(n+2)}(t) dt = \int_0^\infty e^{-\lambda t} y(t) dt.$$

Therefore $V^{(n+2)}(t) x = y(t)$. We have shown that if there exists a k-times integrated semigroup then any solution of (IP) coincides with the k-th derivative of the integrated semigroup.

As in Theorem 1.2.4, the unique solution $u(\cdot)$ of (IP) satisfies the estimate in Definition 1.6.7, thus the Cauchy problem (IP) is (n, ω)-well-posed on $R_A^{n+1}(\lambda)\overline{D(A)}$. □

Now we compare the obtained results with the results of Theorem 1.6.2 on the uniform well-posedness of the Cauchy problem (IP). Decomposition (1.6.6) generalizes in the degenerate case the property of the generator of a nondegenerate C_0-semigroup to be densely defined. Only in this case we have the projector

$$P = U(0) : X \to X_1 = \overline{D(A)},$$

such that $X_1 = PX$, $A0 = \ker P$. Due to the decomposition (1.6.6) the MFPHY condition for $(\lambda I - \tilde{A})^{-1}$ on X_1 can be written as the condition for $(\lambda I - A)^{-1}$ on X. In this subsection, without any decomposition of X in Theorem 1.6.5, we obtained the MFPHY-type estimates for $\frac{(\lambda I - A)^{-1}}{\lambda^n}$ only on X_1, and on X we obtain the estimates for $\frac{(\lambda I - A)^{-1}}{\lambda^{n+1}}$. And note that the estimates (1.6.15) guarantee (n, ω)-well-posedness of (IP) only on some subset of $D(A)$, not on the whole of it.

Recall the property discussed in Proposition 1.5.1: if an operator A satisfies the MFPHY estimates, then

$$\forall x \in D(A), \quad \lim_{\lambda \to \infty} \lambda R_A(\lambda) x \to x.$$

The following proposition justifies the generalization of the decomposition (1.6.6), which we will use later in this section.

Proposition 1.6.5 *Let A satisfy (1.6.8), then*

$$\forall x \in D_{n+1}, \quad \lim_{\lambda \to \infty} \lambda R_A(\lambda) x \to x.$$

Proof Let $x \in D_{n+1}$, then there exists $y \in X$ such that $x = (\lambda_0 I - A)^{n+1} y$ for some $\lambda_0 \in \rho(A)$ and

$$
\begin{aligned}
\|\lambda(\lambda I &- A)^{-1} x - x\| \\
&= \|\lambda(\lambda I - A)^{-1}[(\lambda_0 I - A)^{-1}]^{n+1} y - [(\lambda_0 I - A)^{-1}]^{n+1} y\| \\
&= \|\lambda(\lambda_0 - \lambda)^{-1}(\lambda I - A)^{-1}[(\lambda_0 I - A)^{-1}]^{n} y \\
&\qquad\qquad - \lambda_0(\lambda_0 - \lambda)^{-1}[(\lambda_0 I - A)^{-1}]^{n+1} y\| \\
&= \quad \cdots \\
&= \|\lambda(\lambda_0 - \lambda)^{-(n+1)}(\lambda I - A)^{-1} y - \cdots \\
&\qquad\qquad - \lambda_0(\lambda_0 - \lambda)^{-1}[(\lambda_0 I - A)^{-1}]^{n+1} y\| \\
&\longrightarrow \quad 0 \qquad \text{as} \qquad \lambda \to \infty.
\end{aligned}
$$

It follows that $\ker R_A(\lambda) \cap D_{n+1} = \{0\}$. $\square$

Thus we have, that in general, the estimates (1.6.8) do not imply

$$
A0 \cap X_1 = \{0\} \qquad \text{or} \qquad A^{k+1} 0 \cap X_{k+1} = \{0\}
$$

for some $k = 1, \ldots, n$, and consequently, they do not imply any decomposition of X.

Now we assume the decomposition

$$
X = X_{n+1} \oplus \ker R_A^{n+1} = X_{n+1} \oplus A^{n+1} 0, \tag{1.6.20}
$$

the generalization of decomposition (1.6.6) for the case of a degenerate n-times integrated semigroup. Having the decomposition (1.6.20), we can prove the necessary and sufficient condition for the (n, ω)-well-posedness of the Cauchy problem (IP).

Theorem 1.6.7 *Let A be a multivalued linear operator on X with $\rho(A) \neq \emptyset$. Suppose that the decomposition (1.6.20) holds for some $n \in \mathbb{N}$ and $\lambda \in \mathbb{C}$ with $\operatorname{Re} \lambda > \omega$, $\omega \in \mathbb{R}$. Then the Cauchy problem (IP) is (n, ω)-well-posed on D_{n+1} if and only if the condition (1.6.15) is satisfied.*

Proof Suppose the Cauchy problem (IP) is (n, ω)-well-posed on D_{n+1}. To show the estimates (1.6.15) we define $U_0(t)$, $t \geq 0$, by

$$
U_0(t)x := \begin{cases} u(t), & x \in [D_{n+1}]_n \\ 0, & x \in A0 \end{cases} .
$$

Since $U_0(t)$ are defined on $[D_{n+1}]_n \oplus A0$, we can extend $R_A(\lambda)U_0(t)$ to $[D_n]_{n-1}$ and $A^2 0$, respectively, and therefore to the closure of D_n in the norm $\| \cdot \|_{R^{n-1}}$. Hence, $U_k(t)x$, $k = 1, \ldots n$ are defined for $x \in D_{n+1-k}$, they are exponentially bounded in the norm $\| \cdot \|_{R^{n-k}}$, and we can extend them to $[D_{n+1-k}]_{n-k}$ and $A^{k+1} 0$, respectively. In particular, the operators

$U_n(t)$ are defined and bounded on $\overline{D(\mathcal{A})}$ and on $\mathcal{A}^{n+1}0$, and hence on $X = X_{n+1} \oplus \mathcal{A}^{n+1}0$. As in the proof in Theorem 1.6.5, we have that $\{V(t) := U_n(t),\ t \geq 0\}$ is an exponentially bounded n-times integrated semigroup degenerate on $\mathcal{A}0$. Its generator $\mathcal{A}$ is defined by

$$R_{\mathcal{A}}(\lambda)x = \int_0^\infty \lambda^n e^{-\lambda t} V(t) x \, dt.$$

Hence, the condition (1.6.15) holds.

Taking into account the decomposition (1.6.20), we now clarify the structure of the constructed degenerate semigroup V on X by considering it on X_{n+1} and on $\mathcal{A}^{n+1}0$. By construction we have that all operator-functions $U_k(\cdot)$, $k \geq 1$, are equal to the primitives of $U_0(\cdot)$ of the corresponding order, which are extended from D_{n+1} to $X_{n+1} = \overline{D_{n+1}}$. They are equal to zero on $\mathcal{A}0 = \ker R_{\mathcal{A}}(\lambda)$, hence they are independent of λ on $\mathcal{A}0$. Now we show that $U_k(t)x$ are independent of λ on the sets $\mathcal{A}^{k+1}0$.

$$
\begin{aligned}
U_1(t)x &= R_{\mathcal{A}}(\lambda)x, \quad x \in \mathcal{A}^2 0 = \ker R_{\mathcal{A}}^2(\lambda), \\
U_2(t)x &= [\lambda t - 1] R_{\mathcal{A}}^2(\lambda)x + t R_{\mathcal{A}}(\lambda)x, \quad x \in \mathcal{A}^3 0, \\
U_3(t)x &= \left[\lambda^2 \frac{t^2}{2} - 2\lambda t + 1 \right] R_{\mathcal{A}}^3(\lambda)x + t \left[\lambda \frac{t}{2} - 1 \right] R_{\mathcal{A}}^2(\lambda)x \\
&\quad + \frac{t^2}{2} R_{\mathcal{A}}(\lambda)x, \quad x \in \mathcal{A}^4 0,
\end{aligned}
$$

$$\cdots$$

$$
U_n(t)x = \sum_{i=1}^n a_i(t,\lambda) R_{\mathcal{A}}^i(\lambda)x, \tag{1.6.21}
$$

$$
a_i = \sum_{k=o}^{i-1} (-1)^{k+i-1} C_{i-1}^k \lambda^k \frac{t^k}{k!}, \quad x \in \mathcal{A}^{n+1}0.
$$

The equality

$$\frac{d}{d\lambda} U_1(t)x = R_{\mathcal{A}}'(\lambda)x = -R_{\mathcal{A}}^2(\lambda)x = 0, \quad x \in \mathcal{A}^2 0,$$

implies that $U_1(t)$ is independent of λ. By induction one can prove

$$\frac{d}{d\lambda} U_k(t)x = 0, \quad x \in \mathcal{A}^{k+1}0, \ k = 2, 3, \ldots .$$

Suppose that the estimates (1.6.15) hold. Then by Theorem 1.6.6 the Cauchy problem (IP) is (n, ω)-well-posed on $R_{\mathcal{A}}^{n+1}(\lambda)\overline{D(\mathcal{A})} \subset D_{n+1}$. By the decomposition (1.6.20)

$$D_{n+1} = R_{\mathcal{A}}^{n+1}(\lambda)X = R_{\mathcal{A}}^{n+1}(\lambda)X_{n+1}.$$

On the other hand, we have

$$R_A^{n+1}(\lambda)X_{n+1} = R_A^{n+1}(\lambda)\overline{R_A^{n+1}(\lambda)X} \subset R_A^{n+1}(\lambda)\overline{D(A)},$$

and hence, $D_{n+1} \subset R_A^{n+1}(\lambda)\overline{D(A)}$. So $R_A^{n+1}(\lambda)\overline{D(A)} = D_{n+1}$, therefore, the Cauchy problem (IP) is (n, ω)-well-posed on D_{n+1}. $\square$

From this theorem we obtain conditions for the (n, ω)-well-posedness of the degenerate Cauchy problems (DP) and (DP1).

Corollary 1.6.2 Consider the Cauchy problem (DP) with linear operators A and B such that AB^{-1} is closed. Let

$$\rho_B(A) := \left\{ \lambda \in \mathbb{C} \mid (\lambda B - A)^{-1}B \text{ is bounded} \right\} \neq \emptyset,$$

and suppose that the decomposition (1.6.20) holds. Then the Cauchy problem (DP) is (n, ω)-well-posed on D_{n+1} if and only if the condition
$\exists K > 0, \ \omega \in \mathbb{R}:$

$$\left\| \frac{d^k}{d\lambda^k} \frac{R_A(\lambda)}{\lambda^n} \right\| \leq \frac{K\, k!}{(\lambda - \omega)^{k+1}} \tag{1.6.22}$$

holds for all $\lambda > \omega$ and $k = 0, 1, \ldots$.

Corollary 1.6.3 Consider the Cauchy problem (DP1) with linear operators A and B such that $B^{-1}A$ is closed. Let the set

$$\left\{ \lambda \in \mathbb{C} \mid B(\lambda B - A)^{-1} \text{ is bounded} \right\} \neq \emptyset,$$

and suppose that the decomposition (1.6.20) holds. Then for any $x \in D_{n+1}$ the Cauchy problem (DP1) has solutions v such that Bv is uniquely defined and $\|Bv(t)\| \leq Ke^{\omega t}\|x\|_{R^n}$ if and only if the condition (1.6.22) holds.

1.7 Second order equations

Let X, Y be Banach spaces and A, B, Q be linear operators from X to Y. We suppose that Q is bounded and $\ker Q \neq \{0\}$. Consider the abstract Cauchy problems

$$u''(t) = Au'(t) + Bu(t), \quad t \geq 0, \ u(0) = x, \ u'(0) = y, \tag{1.7.1}$$

and

$$Qu''(t) = Au'(t) + Bu(t), \quad t \geq 0, \ u(0) = x, \ u'(0) = y, \tag{1.7.2}$$

In this section we use semigroup methods for studying the well-posedness of these problems. First, we construct $\mathcal{M}, \mathcal{N}$-functions, the solution operators for problem (1.7.1), which generalize semigroups and cosine and sine operator-functions. Second, we study (1.7.1) and (1.7.2) by reducing them to a first order Cauchy problem in a product space. Using the results of Sections 1.2 and 1.6 on integrated semigroups we obtain necessary and sufficient conditions for well-posedness of (1.7.1) and (1.7.2) in terms of resolvent operators

$$R_r(\lambda^2) := (\lambda^2 I - \lambda A - B)^{-1} \quad \text{and} \quad R_d(\lambda^2) := (\lambda Q^2 - \lambda A - B)^{-1},$$

respectively.

The relation between operators A and B plays a very important role in studying the well-posedness of second order problems. We show that for the equivalence of the well-posedness of (1.7.1) and the existence of an integrated semigroup for the Cauchy problem

$$v'(t) = \Phi v(t), \quad t \geq 0, \ v(0) = v_0, \tag{1.7.3}$$

where

$$\Phi = \begin{pmatrix} 0 & I \\ B & A \end{pmatrix}, \ v(t) = \begin{pmatrix} u(t) \\ u'(t) \end{pmatrix}, \ v_0 = \begin{pmatrix} x \\ y \end{pmatrix},$$

operators A, B have to be biclosed. With the help of the theory of integrated semigroups we prove that the MFPHY-type condition:

$$\exists K > 0, \ \omega \geq 0 :$$

$$\tag{1.7.4}$$

$$\left\| \frac{d^k}{d\lambda^k} R_r(\lambda^2) \right\|, \ \left\| \frac{d^k}{d\lambda^k} \overline{R_r(\lambda^2)\left(I - \frac{A}{\lambda}\right)} \right\| \leq \frac{K \, k!}{(\operatorname{Re}\lambda - \omega)^{k+1}},$$

for all $\lambda \in \mathbb{C}$ with $\operatorname{Re}\lambda > \omega$ and $k = 0, 1, \ldots,$

is necessary and sufficient for well-posedness of (1.7.1) with biclosed operators A, B. Using the $\mathcal{M}, \mathcal{N}$-functions theory, we obtain the following MFPHY-type well-posedness condition:

$$\exists K > 0, \ \omega \geq 0 :$$

$$\tag{1.7.5}$$

$$\left\| \frac{d^k}{d\lambda^k} R_r(\lambda^2) \right\|, \ \left\| \frac{d^k}{d\lambda^k} \left(I - \frac{A}{\lambda}\right) R_r(\lambda^2) \right\| \leq \frac{K \, k!}{(\operatorname{Re}\lambda - \omega)^{k+1}},$$

for all $\lambda \in \mathbb{C}$ with $\operatorname{Re}\lambda > \omega$ and $k = 0, 1, \ldots,$

for problem (1.7.1) with commuting operators A, B. In Subsection 1.7.3, using the theory of degenerate integrated semigroups, we prove MFPHY-type necessary and sufficient conditions for well-posedness of the degenerate Cauchy problem (1.7.2).

1.7.1 $\mathcal{M},\mathcal{N}$-functions method

Consider the Cauchy problem

$$u''(t) = Au'(t) + Bu(t), \quad t \geq 0, \ u(0) = x, \ u'(0) = y, \qquad (1.7.1)$$

where A and B are closed linear operators in a Banach space X.

Definition 1.7.1 *The Cauchy problem (1.7.1) is called ω-well-posed on E_1, E_2 if for any $x \in E_1, y \in E_2$ there exists a unique exponentially bounded solution:*

$$\exists K > 0, \ \omega \geq 0: \quad \|u(t)\| \leq K e^{\omega t}\Big(\|x\| + \|y\|\Big), \quad t \geq 0.$$

Definition 1.7.2 *Let $A, B : X \to X$ be closed linear operators. A one-parameter family of bounded commuting operators $\{\mathcal{M}(t), \mathcal{N}(t), \ t \geq 0\}$ is called a family of $\mathcal{M}, \mathcal{N}$-functions generated by operators A, B if*

(M1) $\mathcal{M}(t + h) = \mathcal{M}(t)\mathcal{M}(h) + B\mathcal{N}(t)\mathcal{N}(h),$
$\mathcal{N}(t + h) = \mathcal{M}(t)\mathcal{N}(h) + \mathcal{M}(h)\mathcal{N}(t) + A\mathcal{N}(t)\mathcal{N}(h), \quad t, h \geq 0;$

(M2) $\mathcal{N}(0) = 0, \ \mathcal{M}(0) = I, \ \mathcal{N}'(0) = I, \ \mathcal{M}'(0) = 0;$

(M3) $\mathcal{M}(t)$ *and* $\mathcal{N}(t)$ *are strongly continuous in* $t \geq 0;$

(M4) $\exists K > 0, \ \omega \geq 0: \quad \|\mathcal{M}(t)\| \leq K e^{\omega t} \ \text{and} \ \|\mathcal{N}(t)\| \leq K e^{\omega t}, \ t \geq 0.$

The operators A, B are called the generators of the family of $\mathcal{M}, \mathcal{N}$-functions.

If $\mathcal{M}(t)$ and $\mathcal{N}(t)$ commute with A and B on $D(A)$ and $D(B)$ respectively, then it follows from (M1) that operators

$$\widetilde{\mathcal{M}}''(0)x \ := \ \lim_{h \to 0} \frac{\mathcal{M}(2h) - 2\mathcal{M}(h) + I}{h^2} \, x,$$

$$\widetilde{\mathcal{N}}''(0)x \ := \ \lim_{h \to 0} \frac{\mathcal{N}(2h) - 2\mathcal{N}(h)}{h^2} \, x,$$

defined for those $x \in X$ where limits exist, are the generators of the family of $\mathcal{M}, \mathcal{N}$-functions:

$$\widetilde{\mathcal{M}}''(0) = B, \quad \widetilde{\mathcal{N}}''(0) = A. \qquad (1.7.6)$$

Definition 1.7.3 *Operators A, B are called ω-closed if the operator $\lambda A + B$ is closed for any $\lambda \in \mathbb{C}$ with $\operatorname{Re} \lambda > \omega$. Operators A, B are called biclosed, if for any sequences $x_n \in D(A)$, $y_n \in D(B)$ such that*

$$x_n \to x \quad and \quad y_n \to y,$$

we have

$$A x_n + B y_n \to z \implies x \in D(A), \ y \in D(B)$$

and $Ax + By = z$.

It is not difficult to see that if A, B are biclosed, then A, B are ω-closed, and both A and B are closed. The following two examples illustrate that if A and B are closed operators, they may be not ω-closed, and if A, B are ω-closed, they may be not biclosed.

1. Consider closed operators $A = d^2/ds^2 + d/ds$, $B = -cd^2/ds^2$ on $C[a, b]$. Let x_n be such that

$$D(A) = D(B) \ni x_n(s) \to x(s), \ x_n'(s) \to y(s),$$

and $x_n''(s)$ is not convergent in $C[a, b]$. Then $(cA + B)x_n \to cy$, but $x \notin D(A) \cap D(B) = D(cA + B)$. Therefore, operators A, B are not ω-closed for $\omega < c$.

2. Consider operators $A = d/ds$, $B = d^2/ds^2$ on $C[a, b]$. They are ω-closed for any ω. Let sequences $x_n(s) \in D(A)$ and $y_n(s) = -\int_0^s x_n(t)dt$ be such that $x_n \to x$ but x_n' is not convergent in $C[a, b]$. Then A, B are not biclosed.

All pairs of operators where one is closed and the other is bounded, and pairs of closed operators with ranges in orthogonal subspaces, are examples of biclosed operators.

Denote $d_1 = D(AB)$, $d_2 = D(A^2) \cap D(B)$. Using the $\mathcal{M}, \mathcal{N}$-functions we obtain the following necessary and sufficient conditions for well-posedness of (1.7.1).

Theorem 1.7.1 *Let A, B be commuting closed linear operators with domains such that $\bar{d}_1 = \bar{d}_2 = X$. Then the following statements are equivalent:*

(I) *the Cauchy problem (1.7.1) is ω-well-posed on d_1, d_2;*

(II) *the operators A, B are the generators of a family of $\mathcal{M}, \mathcal{N}$-functions, and $\tilde{N}''(0) = A$, $\widetilde{M}''(0) = B$. In this case, the solution of (1.7.1) has the form*

$$u(\cdot) = \mathcal{M}(\cdot)x + \mathcal{N}(\cdot)y, \quad x \in d_1, \ y \in d_2;$$

(III) *for* $\operatorname{Re}\lambda > \omega$ *there exists* $R_r(\lambda^2)$ *and*

$$R_r(\lambda^2)x = \int_0^\infty e^{-\lambda t} \mathcal{N}(t)x dt, \qquad (1.7.7)$$

$$(\lambda I - A)R_r(\lambda^2)x = \int_0^\infty e^{-\lambda t} \mathcal{M}(t)x dt \qquad (1.7.8)$$

for all $x \in X$;

(IV) *the condition (1.7.5) if fulfilled for* $R_r(\lambda^2)$:

$$\exists K > 0, \ \omega \geq 0:$$

$$\left\| \frac{d^k}{d\lambda^k} R_r(\lambda^2) \right\|, \ \left\| \frac{d^k}{d\lambda^k} \left(I - \frac{A}{\lambda} \right) R_r(\lambda^2) \right\| \leq \frac{K\,k!}{(\operatorname{Re}\lambda - \omega)^{k+1}}$$

for all $\lambda \in \mathbb{C}$ *with* $\operatorname{Re}\lambda > \omega$ *and* $k = 0, 1, \dots$.

Proof (I) $\Longrightarrow$ (II). Bounded operators $\mathcal{M}(t)$ and $\mathcal{N}(t)$ are defined as the solution operators for initial values $(x, 0)$ and $(0, y)$, respectively:

$$\mathcal{M}''(t) = A\mathcal{M}'(t) + B\mathcal{M}(t), \ \ t \geq 0, \ \mathcal{M}(0) = x, \ \mathcal{M}'(0) = 0,$$

$$\mathcal{N}''(t) = A\mathcal{N}'(t) + B\mathcal{N}(t), \ \ t \geq 0, \ \mathcal{N}(0) = 0, \ \mathcal{N}'(0) = y.$$

Commutativity of A and B and the uniqueness of a solution imply commutativity of $\mathcal{M}(t)$ and $\mathcal{N}(h)$ for any $t, h \geq 0$, commutativity of operator-functions $\mathcal{M}, \mathcal{N}$ with operators A, B, property (M1), and equality (1.7.6). Property (M2) follows from the initial conditions. Continuity and exponential boundedness of a solution imply (M3) and (M4).

From (1.7.6) and the commutativity of $\mathcal{M}, \mathcal{N}$-functions we obtain the following relations for derivatives

$$
\begin{aligned}
\forall u \in D(B), \ \mathcal{M}'(t)u &= \mathcal{N}(t)Bu = B\mathcal{N}(t)u, \\
\forall u \in D(A), \ \mathcal{N}'(t)u &= \mathcal{M}(t)u + \mathcal{N}(t)Au \\
&= \mathcal{M}(t)u + A\mathcal{N}(t)u, \\
\forall u \in D(AB), \ \mathcal{M}''(t)u &= B\mathcal{N}(t)u + AB\mathcal{N}(t)u \\
&= A\mathcal{M}'(t)u + B\mathcal{M}(t)u, \\
\forall u \in D(A^2) \cap D(B), \ \mathcal{N}''(t)u &= A\mathcal{N}'(t)u + B\mathcal{N}(t)u
\end{aligned}
$$

for all $t \geq 0$. These relations imply that

$$u(\cdot) = \mathcal{M}(\cdot)x + \mathcal{N}(\cdot)y, \ x \in d_1, y \in d_2$$

is a solution of the Cauchy problem (1.7.1). Since $\mathcal{M}, \mathcal{N}$-functions are exponentially bounded, the solution is stable.

(II) $\Longrightarrow$ (III). As in the proof of the closedness of a generator of a C_0-semigroup in Proposition 1.1.1, one can prove the following property of generators of a family of $\mathcal{M}, \mathcal{N}$-functions:

$$\forall x_n \in D(A) \cap D(B), \tag{1.7.9}$$
$$x_n \to x \in D(B) \,\&\, P x_n \to y \in D(P) \Longrightarrow x \in D(P) \,\&\, P x = y,$$

where $P := \lambda A + B = \lambda \widetilde{\mathcal{N}}''(0) + \widetilde{\mathcal{M}}''(0)$, $\operatorname{Re} \lambda > \omega$.

Note that property (1.7.9) is weaker than the ω-closedness and hence, the biclosedness of A, B, but it allows us to conclude that

$$(\lambda^2 I - \lambda A - B) \int_0^\infty e^{-\lambda t} \mathcal{N}(t) x \, dt = \int_0^\infty e^{-\lambda t} (\lambda^2 I - \lambda A - B) \mathcal{N}(t) x \, dt$$

for all $x \in D(A^2) \cap D(B)$. Integrating by parts and using the formulae for the derivatives of $\mathcal{M}, \mathcal{N}$-functions, we obtain

$$(\lambda^2 I - \lambda A - B) \int_0^\infty e^{-\lambda t} \mathcal{N}(t) x \, dt = x$$

for all $x \in X = \overline{D(A^2) \cap D(B)}$. Applying $\int_0^\infty e^{-\lambda t} \mathcal{N}(t) dt$ to $(\lambda^2 I - \lambda A - B) x$ for $x \in D(A^2) \cap D(B)$, we obtain

$$\int_0^\infty e^{-\lambda t} \mathcal{N}(t) (\lambda^2 I - \lambda A - B) x \, dt = x$$

for all $x \in D(A) \cap D(B)$. Thus (1.7.7) is proved. The equality (1.7.8) can be proved by similar arguments.

(I) $\Longrightarrow$ (IV) follows from (1.7.7), (1.7.8) and exponential estimates (M4). For proof of (IV) $\Longrightarrow$ (I) see our book [128], where the detailed proof of this theorem is given. $\square$

1.7.2 Integrated semigroups method

We consider the Cauchy problem (1.7.3):

$$v'(t) = \Phi v(t), \quad t \geq 0, \ v(0) = v_0,$$

on the Banach space $\mathcal{X} := X \times X$, equipped with the norm

$$\|(x, y)\|_{\mathcal{X}} = \max \left\{ \|x\|_X, \|y\|_X \right\}.$$

Here

$$\Phi = \begin{pmatrix} 0 & I \\ B & A \end{pmatrix}, \quad v(t) = \begin{pmatrix} u(t) \\ u'(t) \end{pmatrix}, \quad v_0 = \begin{pmatrix} x \\ y \end{pmatrix},$$

Suppose that Φ is the generator of an exponentially bounded n-times integrated semigroup V on $\mathcal{X}$. We recall (see Proposition 1.2.1) that the operator Ψ defined by

$$\Psi x = \lambda x - \left(\int_0^\infty \lambda^n e^{-\lambda t} V(t) dt \right)^{-1} x$$

is called the generator of V.

Let $\rho(\Phi)$ be the resolvent set of Φ and $\rho(A, B)$ be the set of all $\lambda \in \mathbb{C}$ such that the operator $R_r(\lambda^2) : X \to X$ is bounded.

Theorem 1.7.2 *Let $A, B : X \to X$ be biclosed linear operators such that $\overline{d_1} = \overline{d_2} = X$ and $\rho(A, B) \neq \emptyset$. Then the following statements are equivalent:*

(I) *the operator $\Phi = \begin{pmatrix} 0 & I \\ B & A \end{pmatrix}$ is the generator of a 1-time integrated semigroup on $\mathcal{X}$;*

(II) *the condition (1.7.4) is fulfilled:*

$$\exists K > 0, \ \omega \geq 0 :$$

$$\left\| \frac{d^k}{d\lambda^k} R_r(\lambda^2) \right\|, \ \left\| \frac{d^k}{d\lambda^k} \overline{R_r(\lambda^2) \left(I - \frac{A}{\lambda} \right)} \right\| \leq \frac{K \, k!}{(\text{Re} \, \lambda - \omega)^{k+1}}$$

for all $\lambda \in \mathbb{C}$ with $\text{Re} \, \lambda > \omega$ and $k = 0, 1, \ldots$;

(III) *the Cauchy problem (1.7.1) is ω-well-posed on d_1, d_2.*

Proof (I) $\Longrightarrow$ (II). The operator $\Phi = \begin{pmatrix} 0 & I \\ B & A \end{pmatrix}$ is closed if and only if A, B are biclosed. So Φ is closed and densely defined: $\overline{D(\Phi)} = \mathcal{X}$. Since Φ is the generator of a 1-time integrated semigroup, by Theorem 1.2.3, the MFPHY-type condition holds for the resolvent of Φ:

$$\exists K > 0, \ \omega \geq 0 :$$

$$\left\| \left[\frac{R_\Phi(\lambda)}{\lambda} \right]^{(k)} \right\| \leq \frac{K \, k!}{(\lambda - \omega)^{k+1}} \tag{1.7.10}$$

for all $\lambda > \omega$ and $k = 0, 1, \ldots$, where

$$R_\Phi(\lambda) = (\lambda I - \Phi)^{-1} = \begin{pmatrix} \overline{R_r(\lambda^2)(\lambda I - A)} & R_r(\lambda^2) \\ R_r(\lambda^2) B & \lambda R_r(\lambda^2) \end{pmatrix}. \tag{1.7.11}$$

Note that (1.7.11) implies that estimates (1.7.10) are equivalent to estimates (1.7.4) for $R_r(\lambda^2)$.

(II) $\Longrightarrow$ (III). We have that the condition (1.7.4) is equivalent to (1.7.10) and the condition $\overline{d_1} = \overline{d_2} = X$ is equivalent to the condition $\overline{D(\Phi)} = \mathcal{X}$. Therefore the operator Φ is densely defined and (1.7.10) holds. Thus Φ is the generator of a 1-time integrated semigroup V, and the Cauchy problem (1.7.3) is $(1, \omega)$-well-posed. That is, for all

$$v_0 \in D(\Phi^2) = D(AB) \times \left[D(A^2) \cap D(B) \right]$$

there exists the unique solution $v(\cdot) = V'(\cdot)v_0$ with

$$\|v(t)\| = \|(u(t), u'(t))\| \leq K e^{\omega t} \|v_0\|_\Phi.$$

Then $w(\cdot) = R_\Phi(\lambda)v(\cdot)$ is a solution of (1.7.3) with the initial value $w(0) = R_\Phi(\lambda)v_0$ and with the stability property

$$\|w(t)\| \leq K e^{\omega t} \|R_\Phi(\lambda)v_0\|_\Phi \leq K e^{\omega t} \|v_0\|.$$

Applying $(\lambda I - \Phi)^{-1}$ to the equation (1.7.3) and integrating it, we obtain the equality

$$\int_0^t v(s)ds = -w(t) + w(0) + \lambda \int_0^t w(s)ds. \qquad (1.7.12)$$

From (1.7.12) we have

$$\left\| \int_0^t v(s)ds \right\| \leq K \left(2 + |\lambda|t \right) e^{\omega t} \|v_0\|,$$

hence

$$\|u(t)\| \leq K e^{\omega_1 t} \left(\|x\| + \|y\| \right)$$

for some $\omega_1 > \omega$. Thus, for all $x \in d_1, y \in d_2$ the Cauchy problem (1.7.1) has the unique solution $u(\cdot)$ stable in X. It is equal to the first coordinate of $v(\cdot)$, the unique solution of (1.7.3).

(III) $\Longrightarrow$ (I). If $\rho(A, B) \neq \emptyset$ then $\rho(\Phi) \neq \emptyset$ and the ω-well-posedness of the Cauchy problem (1.7.1) on d_1, d_2 is equivalent to the $(1, \omega)$-well-posedness of (1.7.3). In particular, $\int_0^t v(s)ds$ is stable in $\mathcal{X}$ if and only if $u(t)$ is stable in X. Since (1.7.3) is $(1, \omega)$-well-posed then Φ is the generator of a 1-time integrated semigroup (see Theorem 1.2.4). $\square$

Comparing the $\mathcal{M}, \mathcal{N}$-functions method and the integrated semigroups method, we note that by Theorem 1.7.1, if A, B are commuting operators satisfying the density property and (1.7.5), then (1.7.1) is ω-well-posed on d_1, d_2 and A, B are biclosed. By Theorem 1.7.2, if A, B are biclosed operators satisfying the density property and (1.7.4), then (1.7.1) is well-posed on d_1, d_2. The condition (1.7.5) is the same as (1.7.4) if and only if A and B commute.

1.7.3 The degenerate Cauchy problem

Let X, Y be Banach spaces. We consider the Cauchy problem

$$Qu''(t) = Au'(t) + Bu(t), \quad t \geq 0, \quad u(0) = x, \quad u'(0) = y, \qquad (1.7.2)$$

where $A, B : X \to Y$ are biclosed linear operators and operator $Q : X \to Y$ is linear and bounded.

Let $\mathcal{X} = X \times X$ and $\mathcal{Y} = Y \times Y$. Consider the closed linear operator

$$\Phi = \begin{pmatrix} 0 & I \\ B & A \end{pmatrix} : \mathcal{X} \to \mathcal{Y}$$

and the bounded linear operator

$$\Psi = \begin{pmatrix} I & 0 \\ 0 & Q \end{pmatrix} : \mathcal{X} \to \mathcal{Y}.$$

As in the regular case it is not difficult to prove that the existence of a unique solution

$$u(\cdot) \in C\{[0, \infty), \ D(B)\} \cap C^1\{[0, \infty), \ D(A)\} \cap C^2\{[0, \infty), \ X\}$$

of (1.7.2) is equivalent to the existence of a unique solution

$$v(\cdot) = \begin{pmatrix} u(\cdot) \\ u'(\cdot) \end{pmatrix} \in C\{[0, \infty), \ D(\Phi)\} \cap C^1\{[0, \infty), \ \mathcal{X}\}$$

of the degenerate Cauchy problem

$$\Psi v'(t) = \Phi v(t), \quad t \geq 0, \quad v(0) = \begin{pmatrix} x \\ y \end{pmatrix}. \qquad (1.7.13)$$

Using Theorem 1.6.5 and Corollary 1.6.4 about the connection between the well-posedness of (1.7.13) and existence of a degenerate n-times integrated exponentially bounded semigroup V with the pair of generators Ψ, Φ:

$$\mathcal{R}_d(\lambda)\xi := (\lambda\Psi - \Phi)^{-1}\Psi\xi = \int_0^\infty \lambda^n e^{-\lambda t} V(t)\xi dt, \quad \xi \in \mathcal{X}, \ \operatorname{Re}\lambda > \omega,$$

and the MFPHY-type condition for $\mathcal{R}_d(\lambda)$, we obtain the following theorem on the well-posedness of (1.7.2).

Theorem 1.7.3 *Suppose operators A, B are biclosed and the operator $(\lambda\Psi - \Phi)^{-1}$ is bounded for some λ. Assume that $\mathcal{X}$ admits the decomposition:*

$$\mathcal{X} = \overline{\mathcal{D}}_{n+1} \oplus \ker \mathcal{R}_d^{n+1} \qquad (1.7.14)$$

where $\mathcal{D}_{n+1} := \mathcal{R}_d^{n+1}(\lambda)\mathcal{X}$. Then the following statements are equivalent:

(I) *the Cauchy problem (1.7.13) is* (n, ω)*-well-posed on* $\mathcal{D}_{n+1}$;

(II) *the MFPHY-type condition holds for* $\mathcal{R}_d(\lambda) = (\lambda\Psi - \Phi)^{-1}\Psi$:

$\exists K > 0, \ \omega \in \mathbb{R}$:

$$\left\| \left[\frac{\mathcal{R}_d(\lambda)}{\lambda^n} \right]^{(k)} \right\| \leq \frac{K\,k!}{(\lambda - \omega)^{k+1}} \qquad (1.7.15)$$

for all $\lambda > \omega$ *and* $k = 0, 1, \ldots$;

(III) *the MFPHY-type condition holds for* $\mathcal{R}_d(\lambda^2) := (\lambda^2 Q - \lambda A - B)^{-1}$:

$\exists K > 0, \ \omega \in \mathbb{R}$:

$$\left\| \left[\frac{\mathcal{R}_d(\lambda^2)}{\lambda^{n-1}} \right]^{(k)} \right\| \leq \frac{K\,k!}{(\operatorname{Re}\lambda - \omega)^{k+1}},$$

$$\left\| \left[\frac{\mathcal{R}_d(\lambda^2)(\lambda Q - A)}{\lambda^n} \right]^{(k)} \right\| \leq \frac{K\,k!}{(\operatorname{Re}\lambda - \omega)^{k+1}}$$

or all $\lambda \in \mathbb{C}$ *with* $\operatorname{Re}\lambda > \omega$ *and* $k = 0, 1, \ldots$;

(IV) *the Cauchy problem (1.7.2) is* $(n - 1, \omega)$*-well-posed on the sets of initial values* E_1 *and* E_2 *such that* $\begin{pmatrix} x \\ y \end{pmatrix} \in \mathcal{D}_{n+1}$.

Proof (I) $\Longleftrightarrow$ (II) follows from Corollary 1.6.4.

(II) $\Longleftrightarrow$ (III) follows from the equality

$$\mathcal{R}_d(\lambda) = (\lambda\Psi - \Phi)^{-1}\Psi = \begin{pmatrix} \overline{\mathcal{R}_d(\lambda^2)(\lambda Q - A)} & \mathcal{R}_d(\lambda^2) \\ \overline{\mathcal{R}_d(\lambda^2)B} & \lambda\mathcal{R}_d(\lambda^2) \end{pmatrix} \begin{pmatrix} I & 0 \\ 0 & Q \end{pmatrix}.$$

(I) $\Longleftrightarrow$ (IV). Unique solutions of (1.7.13) and (1.7.2), stable with respect to corresponding initial data, are connected by the relation $v(t) = \begin{pmatrix} u(t) \\ u'(t) \end{pmatrix}$. $\square$

Corollary 1.7.1 Suppose operators A, B are biclosed and the operator $(\lambda\Psi - \Phi)^{-1}$ is bounded for some λ. Assume that $\mathcal{X}$ admits the decomposition:

$$\mathcal{X} = \overline{\mathcal{D}_2} \oplus \ker \mathcal{R}_d^2.$$

Then the Cauchy problem (1.7.2) is well-posed on the sets of initial values E_1 and E_2 such that $\begin{pmatrix} x \\ y \end{pmatrix} \in \mathcal{D}_2$ if and only if the MFPHY-type condition holds for $\mathcal{R}_d(\lambda^2) := (\lambda^2 Q - \lambda A - B)^{-1}$:

$\exists K > 0,\ \omega \in \mathbb{R}:$

$$\left\| \left[R_d(\lambda^2) \right]^{(k)} \right\| \leq \frac{K\,k!}{(\operatorname{Re}\lambda - \omega)^{k+1}},$$

$$\left\| \left[R_d(\lambda^2)\left(Q - \frac{A}{\lambda} \right) \right]^{(k)} \right\| \leq \frac{K\,k!}{(\operatorname{Re}\lambda - \omega)^{k+1}}$$

or all $\lambda \in \mathbb{C}$ with $\operatorname{Re}\lambda > \omega$ and $k = 0, 1, \dots$.

Chapter 2

Abstract Distribution Methods

2.1 The Cauchy problem

In the previous chapter we showed that under different conditions on the resolvent of an operator $A : D(A) \subset X \to X$, the Cauchy problem

$$u'(t) = Au(t), \quad t \in [0, T),\ T \leq \infty,\ u(0) = x, \qquad \text{(CP)}$$

has a unique solution for x from different correctness classes stable with respect to x in different norms. In this section, we consider the Cauchy problem (CP) for any $x \in X$ in the space of distributions. Our main aim is to obtain necessary and sufficient conditions for well-posedness in the space of distributions in terms of the resolvent of A. We show that they are the same as in the case when A is the generator of an integrated semi-group. That is, only the Cauchy problem with a generator of an integrated semigroup can be well-posed in the space of distributions.

In the first subsection, we consider abstract distributions and distribution semigroups. In the second, we prove necessary and sufficient conditions for the well-posedness of (CP) in the space of distributions. As in the well-posedness theorems in Chapter 1, it is done via intermediate results about the existence of some semigroups. Here, they coincide with solution operator distributions and distribution semigroups. In the third subsection, we consider the particular case of exponential distributions.

2.1.1 Abstract (vector-valued) distributions

Let X be a Banach space and $\mathcal{D}$ be the Schwartz space of infinitely differentiable functions $\varphi : \mathbb{R} \to \mathbb{R}$ with compact supports and with the following convergence: $\varphi_n \to \varphi$ in $\mathcal{D}$ if there exists a compact set $K \subset \mathbb{R}$ such that $\operatorname{supp} \varphi_n \subset K$ and

$$\sup_{t \in K} |\varphi_n^{(i)}(t) - \varphi^{(i)}(t)| \underset{n \to \infty}{\longrightarrow} 0, \quad i = 0, 1, \ldots .$$

Let $\mathcal{D}_+, \mathcal{D}_-, \mathcal{D}_0, \mathcal{E}$ be subspaces of functions from $\mathcal{D}$ with supports bounded from the left, from the right, with support in $[0, \infty)$, and with any support, respectively. Let $\mathcal{S}$ be the countably normed space of rapidly decreasing functions with the system of norms

$$\|\varphi\|_{j,k} := \sup_{0 \le i \le k} \sup_t (1 + |t|)^j |\varphi^{(i)}(t)|. \tag{2.1.1}$$

Definition 2.1.1 *The space of distributions on X is the space of linear continuous operators from $\mathcal{D}$ to X:*

$$\mathcal{D}'(X) := \mathcal{L}(\mathcal{D}, X), \qquad \mathcal{D}'_\pm(X) := \mathcal{L}(\mathcal{D}_\mp, X);,$$

$\mathcal{D}'_0(X)$ *is the subspace of distributions vanishing on* $(-\infty, 0)$, $\mathcal{S}'(X) := \mathcal{L}(\mathcal{S}, X)$ *is the space of tempered distributions,* $\mathcal{E}'(X) := \mathcal{L}(\mathcal{E}, X)$ *is the space of distributions with compact supports,* $\mathcal{E}'_0(X)$ *is the subspace of distributions from* $\mathcal{E}'(X)$ *vanishing on* $(-\infty, 0)$.

We denote by $U(\varphi)$ or $\langle U, \varphi \rangle$ or $\langle U(\hat{t}), \varphi(\hat{t}) \rangle$ the value of a distribution U for a function φ from a space of test functions. For a locally integrable function $f : \mathbb{R} \to X$ we write

$$f(\varphi) = \langle f, \varphi \rangle = \langle f(\hat{t}), \varphi(\hat{t}) \rangle := \int_{\mathbb{R}} f(t)\varphi(t)dt, \quad f \in \mathcal{D}'(\mathbb{R}).$$

Definition 2.1.2 *A sequence U_n is called convergent to U in $\mathcal{D}'(X)$ if $\|(U_n - U)(\varphi)\| \longrightarrow_{n \to \infty} 0$ uniformly with respect to φ from a bounded set $K \subset \mathcal{D}$. A set $K \subset \mathcal{D}$ is said to be bounded if for any $\{\varphi_n\} \subset K$ and $\varepsilon_n \to 0$ $(\varepsilon_n \in \mathbb{R})$, we have $\varepsilon_n \varphi_n \to 0$ in $\mathcal{D}$.*

The following proposition for abstract distributions is an analogue of the well-known fact that every linear continuous operator on a normed space is bounded.

Proposition 2.1.1 *Let $U \in \mathcal{D}'(X)$ and K be a compact set from $\mathbb{R}$. Then there exist $p \ge 0$, $C > 0$ such that for any $\varphi \in \mathcal{D}$ with $\operatorname{supp} \varphi \subset K$*

$$\|U(\varphi)\| \le C\|\varphi\|_p, \quad \text{where} \quad \|\varphi\|_p := \sup_{0 \le j \le p} \sup_{t \in K} |\varphi^{(j)}(t)|. \ \square \tag{2.1.2}$$

The number p in (2.1.2) is called the local order of a distribution U.

Let $\mathcal{D}^p(K)$ be the Banach space of p-times continuously differentiable functions with supports in a compact set $K \subset \mathbb{R}$ and with the norm (2.1.2), then any function from $\mathcal{D}^p(K)$ can be approximated by an infinitely differentiable function.

Proposition 2.1.2 *For any $\varphi \in \mathcal{D}^p(K)$ and $C > 0$ there exists $\psi \in \mathcal{D}$ such that*

$$\operatorname{supp}\psi \subseteq K_\varepsilon = \left\{t \in \mathbb{R} \,\Big|\, dist(t, K) \leq \varepsilon\right\}$$

and $\|\varphi - \psi\|_p \leq \varepsilon$. $\square$

Thus, every distribution from $\mathcal{D}'(X)$ can be locally extended to a smooth function. The following structure theorems are based on this fact.

Theorem 2.1.1 *Let $U \in \mathcal{D}'(X)$ and Ω be an open bounded set from $\mathbb{R}$. Then there exists a continuous function $f : \mathbb{R} \to X$ and $m > 0$ such that*

$$U(\varphi) = f^{(m)}(\varphi)$$

for all $\varphi \in \mathcal{D}$ with $\operatorname{supp}\varphi \subset \Omega$.
If $U = 0$ on $(-\infty, a)$, then $f(t) = 0$ for $t < a$. $\square$

The number m in the structure theorems is equal to the local order of U plus two.

For the space $\mathcal{S}'(X)$ the structure theorem is global. The primitive of a distribution U is defined by $\langle U, \varphi_t \rangle$ for some $\varphi_t \in \mathcal{D}^p(K)$. Later on, we use this construction of the m-th continuous primitive for constructing an m-times integrated semigroup related to U.

Theorem 2.1.2 *Let $U \in \mathcal{S}'(X)$. Then there exist $m, r > 0$ and a continuous function $f : R \to X$ such that*

$$U(\varphi) = f^{(m)}(\varphi)$$

for all $\varphi \in \mathcal{S}$, and $|f(t)| =_{|t| \to \infty} \mathcal{O}(|t|^r)$.
If $U = 0$ on $(-\infty, a)$, then $f(t) = 0$ for $t < a$. $\square$

For distributions with a point support t_0 the structure theorems give the following representations.

Proposition 2.1.3 *Let U be a distribution with a point support t_0, then*

$$\exists p > 0, \qquad U = \sum_{k \leq p} \alpha_k \delta^{(k)}(t - t_0). \quad \square$$

Here $\delta(t - t_0) = \delta_{t_0}$ is the Dirac distribution (delta function) concentrated at t_0.

Now we define a convolution of distributions. Let Z, X, Y be Banach spaces such that the bilinear mapping $(u, v) \to uv : Z \times X \to Y$ is defined. If $Z = X = Y = \mathbb{R}$, then this mapping is the usual multiplication. We consider the particular case of $Z = \mathcal{L}(X, Y)$ which is relevant to the study of the Cauchy problem. Let $U \in \mathcal{D}'_0(Z)$, $V \in \mathcal{D}'_0(X)$ and $c > 0$. By

Theorem 2.1.1 there exist m, $p \geq 0$ and continuous functions $f : \mathbb{R} \to Z$ and $g : \mathbb{R} \to X$ vanishing for $t < 0$ such that

$$U(\varphi) = f^{(m)}(\varphi), \quad V(\varphi) = g^{(p)}(\varphi).$$

Then we define

$$\langle (U * V), \varphi \rangle := \langle (f * g)^{(m+p)}, \varphi \rangle \qquad (2.1.3)$$

for any φ with $\operatorname{supp} \varphi \subset (-\infty, c)$. This definition does not depend on m, p, c, f, g. If $U \in \mathcal{S}'_{\pm}(\mathcal{L}(X,Y))$ and $V \in \mathcal{S}'_{\pm}(X)$, then $U * V \in \mathcal{S}'_{\pm}(Y)$.

Example 2.1.1 Let $S \subset \mathcal{D}'_0(\mathcal{L}(X,Y))$ and $x \in X$, then

$$\langle S * (\delta \otimes x), \varphi \rangle = \langle S, \varphi \rangle x = \langle \delta \otimes I * S, \varphi \rangle x, \qquad (2.1.4)$$

$$\langle S * (\delta^{(k)} \otimes x), \varphi \rangle = \langle S^{(k)}, \varphi \rangle x = \langle \delta^{(k)} \otimes I * S, \varphi \rangle x, \qquad (2.1.5)$$

here $\langle \delta^{(k)} \otimes x, \varphi \rangle := \langle \delta^{(k)}, \varphi \rangle x$, $\langle \delta^{(k)} \otimes I, \varphi \rangle := \langle \delta^{(k)}, \varphi \rangle I$, and $I = I_X$ is the identity operator on X.

It follows from Theorem 2.1.1 that for any $\varphi \in \mathcal{D}$ there exists a continuous function f such that $f(t) = 0$ for $t < 0$, and $m \geq 0$ such that

$$\langle S, \varphi \rangle = (-1)^m \int_0^\infty \varphi^{(m)}(t) f(t) dt.$$

Let

$$\eta(t) = \begin{cases} t, & t \geq 0 \\ 0, & t < 0 \end{cases},$$

then $\eta'' = \delta$. Suppose $\operatorname{supp} \varphi \in (-\infty, c)$. By definition of the convolution we have

$$\begin{aligned} \langle S * (\delta \otimes x), \varphi \rangle &= (-1)^{m+2} \int_0^c \varphi^{(m+2)}(t) dt \int_0^t (t-s) f(s) ds \\ &= (-1)^m \int_0^c \varphi^{(m)}(t) f(t) x \, dt = \langle S, \varphi \rangle x. \end{aligned}$$

The second equality in (2.1.4) and the equalities (2.1.5) can be proved in the similar way.

To define the Laplace transform for distributions, one considers the space $\mathcal{S}'_\omega(X)$ of exponential distributions of type ω. A distribution U belongs to $\mathcal{S}'_\omega(X)$ if $e^{-\omega \hat{t}} U(\hat{t}) \in \mathcal{S}'(X)$. The Laplace transform of $U \in \mathcal{S}'_\omega(X)$ is the function

$$\hat{U}(\lambda) = \mathcal{L}U(t) := \langle U, e^{-\lambda \hat{t}} \rangle := \langle e^{-\omega \hat{t}} U(\hat{t}), \chi(\hat{t}) e^{-(\lambda-\omega)\hat{t}} \rangle, \qquad (2.1.6)$$

where

$$\chi(t) = \begin{cases} 1, & t \geq 0 \\ 0, & t \leq c < 0 \end{cases}$$

and $\chi(\cdot) \in C^\infty(\mathbb{R})$. The following properties of Laplace transforms for distributions are similar to the properties of classical Laplace transforms.

Proposition 2.1.4 *Let* $U, V \in \mathcal{S}'_\omega(X)$, *then* $\mathcal{L}U(\lambda)$ *is an analytic function for* $\operatorname{Re}\lambda > \omega$ *and*

$$(\mathcal{L}U)'(\lambda) = -\mathcal{L}(\hat{t}U), \quad \mathcal{L}U'(\lambda) = \lambda\mathcal{L}U(\lambda),$$

$$\mathcal{L}(U * V)(\lambda) = \mathcal{L}U(\lambda)\mathcal{L}V(\lambda), \quad \operatorname{Re}\lambda > \omega. \ \square$$

Now we consider the Cauchy problem in spaces of distributions. It is not difficult to verify that if u is a classical solution of (CP), then the corresponding distribution

$$U(\varphi) := \langle uH, \varphi \rangle = \int_0^\infty \varphi(t)u(t)dt, \quad \varphi \in \mathcal{D},$$

(here $H(\cdot)$ is the Heaviside function) is a solution for the equation

$$P * U = \delta \otimes x, \qquad (\delta P)$$

where

$$P := \delta' \otimes I - \delta \otimes A \ \in \mathcal{D}'_0(\mathcal{L}(X_A, X)),$$

$$\langle \delta \otimes A, \varphi \rangle := \langle \delta, \varphi \rangle A$$

and

$$X_A := \left\{ D(A), \ \|x\|_A := \|x\| + \|Ax\| \right\}.$$

Indeed, we have

$$\begin{aligned}
\int_0^\infty \varphi(t)u'(t)dt &= -\varphi(0)x - \int_0^\infty \varphi'(t)u(t)dt \\
&= -(\delta \otimes x)(\varphi) + U'(\varphi)
\end{aligned}$$

and

$$\int_0^\infty \varphi(t)u'(t)dt = A \int_0^\infty \varphi(t)u(t)dt = AU(\varphi).$$

Taking into account Example 2.1.1, we obtain

$$\begin{aligned}
\langle P * U, \varphi \rangle &= \langle (\delta' \otimes I) * U, \varphi \rangle - \langle (\delta \otimes A) * U, \varphi \rangle \\
&= \langle U', \varphi \rangle - \langle AU, \varphi \rangle = \langle \delta \otimes x, \varphi \rangle.
\end{aligned}$$

Definition 2.1.3 *A distribution* $U \in \mathcal{D}'_0(X_A)$ *is called a solution of the Cauchy problem* (CP) *in the sense of distributions if it is a solution of* (δP).

Definition 2.1.4 *The Cauchy problem* (CP) *is called well-posed in the sense of distributions (or* (δP) *is well-posed) if*

(a) *for any $x \in X$ there exists a unique solution $U \in \mathcal{D}_0'(X_A)$ of* (δP);

(b) *for any sequence of initial values $\{x_n\} \subset X$ such that $x_n \to 0$, the corresponding sequence of solutions $\{U_n\}$ converges to 0 in $\mathcal{D}_0'(X_A)$.*

    The Cauchy problem (CP) *is called ω-well-posed in the sense of distributions if*

(a1) *for any $x \in X$ there exists a unique solution $U \in \mathcal{S}_\omega'(X_A)$ of* (δP);

(b1) *for any $x_n \to 0$ we have $U_n \to 0$ in $\mathcal{S}_\omega'(X_A)$.*

Similarly to the case of a uniformly well-posed Cauchy problem, the solution of a well-posed problem (δP) can be constructed with the help of a semigroup, namely a distribution semigroup .

Definition 2.1.5 *We say that a distribution $Q \in \mathcal{D}_0'(\mathcal{L}(X))$ is a distribution semigroup if*

(i) $\forall \varphi, \psi \in \mathcal{D}_0, \qquad \langle Q, \varphi * \psi \rangle = \langle Q, \varphi \rangle \langle Q, \psi \rangle.$

A distribution semigroup is said to be nondegenerate if

(ii) $\forall \varphi \in \mathcal{D}_0, \qquad Q(\varphi)x = 0 \Longrightarrow x = 0.$

A distribution semigroup is said to be regular on

$$\operatorname{ran} Q := \left\{ Q(\varphi)x \ \middle|\ \varphi \in \mathcal{D}_0, \ x \in X \right\}$$

or simply regular if

(iii) *for any $y = Q(\varphi)x$ with $\varphi \in \mathcal{D}_0$ and $x \in X$, the distribution Qy is a continuous function $u(\cdot)$ such that $u(t) = 0$ for $t < 0$ and $u(0) = y$.*

A distribution semigroup is said to be with dense range if

(iv) $\overline{\operatorname{ran} Q} = X.$

A distribution semigroup Q is said to be an exponential distribution semigroup of type ω if $Q \in \mathcal{S}_\omega'(\mathcal{L}(X))$.

The generator of distribution semigroup is defined by $A := \overline{Q(-\delta')}$.

To justify the definition of the generator, for any distribution E with a compact support in $[0, \infty)$, we give the definition of the operator $Q(E)$. Let $E \in \mathcal{E}'$ and let $\{\delta_n\} \subset \mathcal{D}_0$ be any sequence convergent to δ (Dirac δ-function). Then

$$D(Q(E)) := \tag{2.1.7}$$
$$\left\{ x \in X \mid Q(\delta_n)x \to x \text{ and } \exists y \text{ such that } Q(E * \delta_n)x \to y \right\},$$

and $Q(E)x := y$. It is not difficult to prove that this definition does not depend on the choice of the sequence δ_n.

Now we state the properties of distribution semigroups. In Theorem 2.1.4 we use these properties to show that every distribution semigroup coincides with a solution operator distribution.

Proposition 2.1.5 *Let $E \in \mathcal{E}'_0$ and Q be a distribution semigroup with dense range. Then $Q(E)$ has the following properties*

(Q1) $\forall \varphi \in \mathcal{D}_0$, $x \in X$,

$$Q(\varphi)x \in D(Q(E)) \quad and \quad Q(E)Q(\varphi)x = Q(E * \varphi)x;$$

(Q2) $\overline{D(Q(E))} = X$;

(Q3) $\forall \varphi \in \mathcal{D}_0$, $x \in D(Q(E))$,

$$Q(E)Q(\varphi)x = Q(\varphi)Q(E)x;$$

(Q4) *There exists* $\overline{Q(E)}$, *the closure of $Q(E)$, and*
$\forall \varphi \in \mathcal{D}_0$, $x \in X$,

$$\overline{Q(E)}Q(\varphi)x = Q(E)Q(\varphi)x.$$

If Q is regular, then

(Q5) $\forall \psi \in \mathcal{D}$,
$$\overline{Q(\psi_+)} = Q(\psi),$$

where $\psi_+(\cdot) \in \mathcal{E}'_0$ is defined by

$$\psi_+(t) = \begin{cases} \psi(t), & t \geq 0 \\ 0, & t < 0 \end{cases}.$$

Proof (Q1). Taking into account the fact that $\delta_n * \varphi \in \mathcal{D}_0$ and property (i), we have

$$Q(\delta_n)Q(\varphi)x = Q(\delta_n * \varphi)x \to Q(\varphi)x,$$
$$Q(E * \delta_n)Q(\varphi)x = Q(E * \delta_n * \varphi)x \to Q(E * \varphi)x, \quad x \in X.$$

Hence $Q(\varphi)x \in D(Q(E))$ and $Q(E)Q(\varphi)x = Q(E * \varphi)x$. (Q1) and (iv) imply (Q2).

(Q3). For x, y from (2.1.7) we have

$$Q(\varphi)Q(E * \delta_n)x \to Q(\varphi)y = Q(\varphi)Q(E)x,$$
$$Q(\varphi)Q(E * \delta_n)x = Q(E * \delta_n * \varphi)x \to Q(E * \varphi)x = Q(E)Q(\varphi)x.$$

(Q4). It is well-known that a linear operator A has a closure if and only if $\forall x_n \in D(A): x_n \to 0$,

$$Ax_n \to y \implies y = 0.$$

Let $x_n \in D(Q(E))$ be such that $x_n \to 0$ and $Q(E)x_n \to y$, then
$\forall \varphi \in \mathcal{D}_0$,

$$Q(\varphi)Q(E)x_n = Q(E * \varphi)x_n \to Q(\varphi)y.$$

Since $Q(E * \varphi) \in \mathcal{L}(X, X)$, we obtain $Q(\varphi)y = 0$. It follows from (ii) that $y = 0$.

(Q5). Let $\psi \in \mathcal{D}$, then by (Q4), for $\psi_+ \in \mathcal{E}'_0$ we have

$$\overline{Q(\psi_+)}Q(\varphi)x = Q(\psi_+)Q(\varphi)x = Q(\psi_+ * \varphi)x$$

for all $\varphi \in \mathcal{D}_0$ and $x \in X$. If we show that for any $x \in X$

$$Q(\psi_+)Q(\varphi)x = Q(\psi)Q(\varphi)x, \qquad (2.1.8)$$

then by (Q2) we have (Q5). By the structure theorem, in a general case there exists $m \geq 0$ and a continuous function $f(\cdot) \in C\{\mathbb{R}, \mathcal{L}(X)\}$ such that $f(t) = 0$, $t < 0$ and

$$Q(\psi)x = (-1)^m \int_0^\infty f(t)\psi^{(m)}(t)x\,dt.$$

for all $x \in X$. Due to the regularity of Q, for any $y \in \operatorname{ran} Q$ there exists $f \in C\{\mathbb{R}, \mathcal{L}(X)\}$ such that

$$Q(\psi)y = \int_0^\infty f(t)\psi(t)y\,dt.$$

Note that in this case $m = 0$, and the proof is also valid if $m = 1$.

Hence for any $y = Q(\varphi)x$, where $x \in X$, and any sequence $\{\delta_j\} \subset \mathcal{D}_0$ such that $\delta_j \to \delta$, we have

$$
\begin{aligned}
Q(\psi_+)y &= \lim_{j \to \infty} Q(\psi_+ * \delta_j)y \\
&= \lim_{j \to \infty} \int_0^\infty f(t)\,dt \int_0^\infty \psi(\tau)\delta_j(t - \tau)y\,d\tau \\
&= \int_0^\infty f(t)\psi(t)y\,dt.
\end{aligned}
$$

Therefore (2.1.8) and (Q5) hold. $\square$

2.1.2 Well-posedness in the space of distributions

Theorem 2.1.3 *Let A be a closed linear operator on a Banach space X. Then*

(W) *the Cauchy problem* (CP) *is well-posed in the sense of distributions*

if and only if

(S) *there exists a solution operator distribution $S \in \mathcal{D}'_0(\mathcal{L}(X, X_A))$ such that*

$$P * S = \delta \otimes I, \tag{2.1.9}$$

$$S * P = \delta \otimes J, \tag{2.1.10}$$

where $I = I_X, J = I_{X_A}$. In this case

$$\langle U, \varphi \rangle = \langle S, \varphi \rangle x,$$

for all $\varphi \in \mathcal{D}$ and $x \in X$.

Proof (W) $\Longrightarrow$ (S). Consider the distribution S defined by

$$S(\varphi)x = Sx(\varphi) := U(\varphi)$$

for all $\varphi \in \mathcal{D}$, where U is the unique solution of (δP). Then $Sx \in \mathcal{D}'_0(X_A)$. The well-posedness of the Cauchy problem implies that for every $\varphi \in \mathcal{D}$

$$\|x_n\| \to 0 \implies \|S(\varphi)x_n\|_A = \|U_n(\varphi)\|_A \to 0,$$

therefore $S(\varphi) \in \mathcal{L}(X, X_A)$. We show that $S \in \mathcal{D}'_0(\mathcal{L}(X, X_A))$, i.e.,
$$\forall \{\varphi_n\} \subset \mathcal{D},$$
$$\varphi_n \to 0 \implies \|S(\varphi_n)\|_{\mathcal{L}(X, X_A)} \to 0.$$

This implication follows from the estimate
$$\exists \, m, \, C > 0:$$
$$\|\langle Sx, \varphi_n \rangle\| \le C \|\varphi_n\|_m$$

which holds on the set of all Sx, where x are from a bounded set M, and for all sequences $\{\varphi_n\}$ such that $\varphi_n \to 0$. This estimate is the analogue of Proposition 2.1.1 for bounded sets. Here m depends on M and on a compact set K containing $\operatorname{supp} \varphi_n$.

Now, using the structure theorem, we show (2.1.9). Let $\varphi \in \mathcal{D}_-$ and $\operatorname{supp} \varphi \subset (-\infty, a)$. There exist

$$g \in C\{\mathbb{R}, \mathcal{L}(X_A, X)\}, \quad f \in C\{\mathbb{R}, \mathcal{L}(X, X_A)\}, \quad v \in C\{\mathbb{R}, X_A\},$$

such that

$$P = g^{(p)}, \quad S = f^{(m)}, \quad Sx = v^{(q)}$$

on $(-\infty, a)$ and

$$
\langle P * S, \varphi \rangle x = \langle g * f^{(m+p)}, \varphi \rangle x
$$

$$
= \lim_{n \to \infty} (-1)^{m+p} \frac{1}{n} \sum_{k=-\infty}^{\infty} g\left(\frac{k}{n}\right) \int_R f\left(t - \frac{k}{n}\right) x \varphi^{(m+p)}(t)\, dt
$$

$$
= \lim_{n \to \infty} (-1)^{m+p} \frac{1}{n} \sum_{k=-\infty}^{\infty} g\left(\frac{k}{n}\right) \int_R f(t) \varphi^{(m+p)}\left(t + \frac{k}{n}\right) x\, dt
$$

$$
= \lim_{n \to \infty} (-1)^p \frac{1}{n} \sum_{k=-\infty}^{\infty} g\left(\frac{k}{n}\right) \left\langle Sx, \varphi^{(p)}\left(\hat{t} + \frac{k}{n}\right) \right\rangle
$$

$$
= (-1)^p \int_R g(s) \langle Sx, \varphi^{(p)}(\hat{t} + s) \rangle ds
$$

$$
= (-1)^{(p+q)} \int_R g(s) \int_R v(t) \varphi^{(p+q)}(t + s)\, dt\, ds
$$

$$
= (-1)^{(p+q)} \int_R \varphi^{(p+q)}(t) \int_R g(s) v(t - s)\, ds\, dt
$$

$$
= \langle P * Sx, \varphi \rangle = \langle P * U, \varphi \rangle = \langle \delta \otimes I, \varphi \rangle x.
$$

Let us prove (2.1.10). Equality (2.1.9) implies the following equalities

$$
\begin{aligned}
(P * S')x &= (P * S)'x = \delta' \otimes x, \quad x \in X, \\
(P * S)Ax &= \delta \otimes Ax, \quad x \in D(A).
\end{aligned}
$$

By definition of P we have

$$
(P * \delta \otimes x) = \delta' \otimes x - \delta \otimes Ax, \quad x \in D(A).
$$

Hence

$$
P * (S'x - SAx - \delta \otimes x) = 0
$$

and since the Cauchy problem is well-posed, we have

$$
S'x - SAx = \delta \otimes x, \quad x \in D(A).
$$

Therefore,

$$
(S * P) = S' - SA = \delta \otimes J.
$$

(S) $\Longrightarrow$ (W). Let $S \in \mathcal{D}'_0(\mathcal{L}(X, X_A))$. For any $\varphi \in \mathcal{D}_0$ and $x \in X$, we define

$$
U(\varphi) := S(\varphi)x = Sx(\varphi),
$$

then $U \in \mathcal{D}'_0(X_A)$ is a solution of (δP):

$$
\langle \delta' \otimes I * Sx, \varphi \rangle - \langle \delta \otimes A * Sx, \varphi \rangle = \langle \delta \otimes x, \varphi \rangle.
$$

The associativity of convolution implies the uniqueness of the solution

$$U = \delta \otimes J * U = S * P * U = S * (\delta \otimes x) = Sx.$$

Let $\|x_n\| \to 0$. Since for any $\varphi \in \mathcal{D}$, $S(\varphi) \in \mathcal{L}(X, X_A)$, we have

$$\|Sx_n(\varphi)\|_A \le \|S(\varphi)x_n\| + \|S'(\varphi)x_n\| + \|\varphi(0)x_n\| \to 0.$$

We show that this convergence is uniform with respect to φ from any bounded set $\mathcal{K} \subset \mathcal{D}$. Since $\mathcal{K}$ is a bounded set in $\mathcal{D}$, it follows that
a) there exists a compact set K such that $\operatorname{supp}\varphi \subset K$ for any $\varphi \in \mathcal{K}$;
b) for any $i \ge 0$, there exists C_i such that

$$\forall \varphi \in \mathcal{K}, \quad \sup_{t \in R} |\varphi^{(i)}(t)| \le C_i.$$

By Theorem 2.1.1 there exist $m \ge 0$ and $f \in C\{\mathbb{R}, \mathcal{L}(X, X_A)\}$ such that $S = f^{(m)}$ on K, therefore

$$\forall \varphi \in \mathcal{K}, \quad \|S(\varphi)\|_{\mathcal{L}(X,X_A)} = \left\| \int\limits_K f(t)\varphi^{(m)}(t)\,dt \right\|_{\mathcal{L}(X,X_A)}$$

$$\le C_m \left\| \int\limits_K f(t)\,dt \right\|_{\mathcal{L}(X,X_A)} = C,$$

and

$$\|S(\varphi)x_n\|_A \le C\|x_n\|. \quad \square$$

Thus we have shown that S is a solution operator distribution for the Cauchy problem (δP). Now we prove that every solution operator distribution coincides with a distribution semigroup generated by A.

Theorem 2.1.4 *Let A be a closed densely defined linear operator on a Banach space X. Then*

(S) *there exists a solution operator distribution $S \in \mathcal{D}'_0(\mathcal{L}(X, X_A))$ satisfying (2.1.9) and (2.1.10)*

if and only if

(D) *A is the generator of a regular distribution semigroup Q with dense range.*

In this case $S = Q$.

Proof (D) $\Longrightarrow$ (S). Let Q be a distribution semigroup with the generator A. For any $\psi \in \mathcal{D}$ and the corresponding function ψ_+ introduced in (Q5) of Proposition 2.1.5, we have $\psi_+ \in \mathcal{E}'_0$ and

$$\delta' * \psi_+ = \psi'_+ + \psi(0)\delta.$$

Hence, by (Q1) and by the definition of the generator of distribution semi-group, we write

$\forall x \in X, \ \varphi \in \mathcal{D}_0,$
$$\begin{aligned} Q(\delta' * \psi_+ * \varphi)x &= Q(\psi'_+ * \varphi)x + \psi(0)Q(\varphi)x = Q(\delta')Q(\psi_+ * \varphi)x \\ &= -AQ(\psi_+ * \varphi)x = -AQ(\psi_+)Q(\varphi)x \\ &= Q(\psi_+)Q(\delta' * \varphi)x = Q(\psi_+)(-A)Q(\varphi)x. \end{aligned}$$

From (Q5) we have

$$\begin{aligned} -AQ(\psi)Q(\varphi)x &= Q(\psi')Q(\varphi)x + \psi(0)Q(\varphi)x \\ &= -Q(\psi)AQ(\varphi)x, \end{aligned} \tag{2.1.11}$$

or for $y = Q(\varphi)x$

$$-AQ(\psi)y - Q(\psi')y = \psi(0)y, \tag{2.1.12}$$

$$-Q(\psi)Ay - Q(\psi')y = \psi(0)y. \tag{2.1.13}$$

Since the range of a distribution semigroup is dense in X and the operator A is closed, we have $Q(\psi)y \in D(A)$, and (2.1.12) holds for every $y \in X$. Let $\|y_n\| \to 0$ in (2.1.12), then

$$\|AQ(\psi)y_n\| \to 0 \implies Q(\psi) \in \mathcal{L}(X, X_A).$$

If $\psi_n \to 0$ $(\psi_n \in \mathcal{D})$ in (2.1.12), then $\|Q(\psi'_n)\| \to 0$ and $\|AQ(\psi_n)\| \to 0$. Hence, since $Q \in \mathcal{D}'_0(\mathcal{L}(X, X))$, we have $Q \in \mathcal{D}'_0(\mathcal{L}(X, X_A))$.

To complete the proof it remains to show (2.1.13) for $y \in D(A)$. Let $x \in D(Q(-\delta'))$ and $\varphi_n \to \delta$ in (2.1.11), then $Q(\varphi_n)x \to x$, $Q(\delta' * \varphi_n)x \to -Ax$, and we have

$$-Q(\psi)Ax = Q(\psi')x + \psi(0)x, \quad x \in D(Q(-\delta')) \tag{2.1.14}$$

Since
$\forall x \in D(A), \ \exists x_n \in D(Q(-\delta')):$

$$x_n \to x \quad \text{and} \quad Ax_n \to Ax,$$

(2.1.14) holds for $x \in D(A)$. Thus, (2.1.9), (2.1.10) are valid for $S = Q$, i.e. the implication (D) $\implies$ (S) is proved.

(S) $\implies$ (D). Let $Q \in \mathcal{D}'_0(\mathcal{L}(X, X_A))$ be a solution of (2.1.9), (2.1.10). For any $F \in \mathcal{D}'_+(X)$, $U = Q * F$ is a solution of the equation

$$P * U = F. \tag{2.1.15}$$

It follows from (2.1.10) that the solution is unique. Hence the well-posedness of (2.1.15) is equivalent to the well-posedness of (δP).

Now, using the fact that (2.1.15) has a unique solution, we prove properties (i) – (iv) for Q.

(i). Let $\varphi, \psi \in \mathcal{D}_0$. Define

$$\psi_1(t) := \psi(-t), \quad \varphi_1(t) := \varphi(-t), \quad t \geq 0,$$

and let $u(\hat{t})$ be a solution of (2.1.15) with $F = \varphi_1 \otimes x$:

$$u' - Au = \varphi_1 \otimes x. \tag{2.1.16}$$

Then $u(\cdot) = (Q * \varphi_1)x$ is an infinitely differentiable function with values in $D(A)$. Let $w(\hat{t}), v(\hat{t})$ be solutions of (2.1.15) with $F = (\psi_1 * \varphi_1) \otimes x$ and $F = \psi_1 \otimes u(0)$, respectively. Then $w(t), v(t)$ are also infinitely differentiable in t. These functions have the following initial values

$$u(0) = Q(\varphi)x, \quad w(0) = Q(\varphi * \psi)x,$$

$$v(0) = Q(\psi)u(0) = Q(\psi)Q(\varphi)x.$$

To prove (i) one needs to show $v(0) = w(0)$. From (2.1.16) we have

$$-A(u * \varphi_1) + (u * \varphi_1)' = (\varphi_1 * \psi_1) \otimes x,$$

hence $w(t) = u * \varphi_1$ and $w(0) = u(\psi)$.

Let $H(\cdot)$ be the Heaviside function, then $H(t)u(t) \in \mathcal{D}_0'(X_A)$. Since A does not depend on t, we have

$$-A(Hu) + (Hu)' = H(t)(-Au + u') + \delta \otimes u(0).$$

Since

$$\forall \varphi \in \mathcal{D}_0, \qquad H(t)(-Au + u') = (H\varphi_1) \otimes x = 0,$$

we obtain

$$-A(Hu) + (Hu)' = \delta \otimes u(0) \tag{2.1.17}$$

and

$$-A((Hu) * \psi_1) + ((Hu) * \psi_1)' = \psi_1 \otimes u(0).$$

Therefore,

$$u = (Hu) * \psi_1, \quad v(0) = \int_0^\infty u(t)\psi(t)\, dt,$$

and

$$\forall \psi \in \mathcal{D}_0, \qquad v(0) = u(\psi) = w(0).$$

Thus

$$\forall \varphi, \psi \in \mathcal{D}_0, \qquad Q(\varphi * \psi) = Q(\varphi)Q(\psi).$$

(iii). Let Z be a solution of (2.1.15) with $F = \delta \otimes y$, where $y = Q(\varphi)x$, $\varphi \in \mathcal{D}_0$ and $x \in X$. Since $u(0) = y$, by (2.1.17) we have

$$Z(t) = Qy = H(t)u(t),$$

and Qy is a continuous function on $(0, \infty)$.

(ii). Let x be such that

$$\forall \varphi \in \mathcal{D}_0, \qquad Q(\varphi)x = 0. \tag{2.1.18}$$

We show that $x = 0$. Since $Q \in \mathcal{D}_0'(\mathcal{L}(X, X_A))$, we have $Qx = 0$ on $(-\infty, 0)$. Furthermore, (2.1.18) implies that $Qx = 0$ on $(0, \infty)$. Hence, by Proposition 2.1.3

$$Qx = \sum_{i=1}^{k} \delta^{(i)} x_i, \quad x_i \in X_A.$$

Since $U = Qx$ is a solution of the equation $-AU + U' = \delta \otimes x$, we have

$$\delta^{(k+1)} \otimes x_k \quad + \quad \delta^{(k)} \otimes (x_{k-1} - Ax_k) + \dots$$
$$+ \quad \delta' \otimes (x_1 - Ax_2) + \delta \otimes (-x - Ax_1) = 0,$$

and $x_k = \dots = x_1 = x = 0$.

(iv). Let X^* and X_A^* be the dual spaces of X and X_A. Define $Q^*(\varphi) := (Q(\varphi))^*$, then

$$Q^*(\varphi) \in \mathcal{L}(X_A^*, X^*), \quad Q^* \in \mathcal{D}_0'(\mathcal{L}(X_A^*, X^*)), \quad A^* \in \mathcal{L}(X^*, X_A^*).$$

Since (2.1.12), (2.1.13) hold for Q, we have

$$Q^* \left(\frac{d}{dt} - A^* \right) = \delta \otimes I_{X^*}$$

and

$$\left(\frac{d}{dt} - A^* \right) Q^* = \delta \otimes I_{X_A^*}.$$

Therefore $Q^* \in \mathcal{D}_0'(\mathcal{L}(X^*, X^*))$ is a distribution semigroup which is nondegenerate, as Q is nondegenerate. Hence if $z^* \in X_A^*$ is such that

$$\forall \varphi \in \mathcal{D}_0, \ x \in X, \quad \langle Q(\varphi)x, z^* \rangle = \langle x, Q^*(\varphi)z^* \rangle = 0,$$

then

$$Q^*(\varphi)z^* = 0 \implies z^* = 0 \implies \overline{\{Q(\varphi)x\}} = X.$$

To complete the proof we show that if A_1 is a generator of Q then $A_1 = A$. From the well-posedness of (2.1.15) we have that the mapping $F \to Q * F$ is an isomorphism from $\mathcal{D}_0'(X)$ to $\mathcal{D}_0'(X_A)$. Since A_1 is a generator of Q, the mapping $F \to Q * F$ is an isomorphism from $\mathcal{D}_0'(X)$ to $\mathcal{D}_0'(X_{A_1})$. Hence $D(A) = D(A_1)$.

Since

$$-AQx + Q'x = \delta \otimes Ix = -A_1Qx + Q'x,$$

we have

$$AQ(\varphi)x = A_1Q(\varphi)x$$

for all $\varphi \in \mathcal{D}_0$ and $x \in X$. Taking into account that $\overline{\{Q(\varphi)x\}} = X$, we conclude $A = A_1$. $\square$

Theorem 2.1.5 *Let A be a closed linear operator on a Banach space X. Then*

(S) *there exists a solution operator distribution $S \in \mathcal{D}_0'(\mathcal{L}(X, X_A))$ satisfying (2.1.9) and (2.1.10)*

if and only if

(R) *there exist $M > 0$ and $n \in \{0\} \cup \mathbb{N}$ such that*

$$\|R_A(\lambda)\| \leq \frac{M|\lambda|^n}{\log(1+|\lambda|)}$$

for all λ from

$$\Lambda = \left\{ \lambda \in \mathbb{C} \;\middle|\; \operatorname{Re}\lambda > \frac{n}{\tau}\log(1+|\lambda|) + \frac{1}{\tau}\log\frac{C}{\gamma} \right\},$$

where

$$\tau \in (0,T), \; C > 0, \; 0 < \gamma < 1.$$

Proof (R) $\Longrightarrow$ (S). For any $m \geq 2n + 2$ we define the function

$$T_m(t) := \frac{1}{2\pi i} \int_{\partial\Lambda} \lambda^{-m} e^{\lambda t} R_A(\lambda)\, d\lambda,$$

where $\partial\Lambda$ is the boundary of Λ. On $\partial\Lambda$ the function under the integral sign grows not faster than $|\lambda|^{\frac{nt}{\tau}+n-m}/\log|\lambda|$. Hence, this integral defines a continuous function on $(-\infty, \tau_m)$ with values in $\mathcal{L}(X, X_A)$, where $\tau_m = \frac{\tau}{n}(m - n - 1)$. In this case, by the abstract Cauchy theorem $T_m(t) = 0$ for $t \leq 0$.

We now define $S := T_m^{(m)}$ on $(-\infty, \tau_m)$. Since for any $\varphi \in \mathcal{D}$, $\operatorname{supp}\varphi \subset (-\infty, \tau_m)$, and

$$\forall m' > m, \quad \langle T_{m'}^{(m'-m)}(t), \varphi \rangle = \langle T_m(t), \varphi \rangle$$

for all $t \in (-\infty, \tau_m)$, we have

$$\int_{\mathbb{R}} \varphi^{(m)} T_m(t)\, dt = \int_{\mathbb{R}} \varphi^{(m')} T_{m'}(t)\, dt,$$

and thus S is well-defined.

Let us show that S is a solution of (2.1.10). Let $x \in X$. For any $\varphi \in \mathcal{D}$ with $\operatorname{supp} \varphi \in (-\infty, a)$ we can choose $\tau_m = a$, then

$$
\begin{aligned}
\langle S * P, \varphi \rangle x &= S'(\varphi)x - S(\varphi)Ax \\
&= \frac{(-1)^{m+1}}{2\pi i} \int_0^a \varphi^{(m+1)}(t) \int_{\partial\Lambda} \lambda^{-m} e^{\lambda t} R_A(\lambda) x\, d\lambda\, dt \\
&\quad - \frac{(-1)^m}{2\pi i} \int_0^a \varphi^{(m)}(t) \int_{\partial\Lambda} \lambda^{-m} e^{\lambda t} R_A(\lambda) Ax\, d\lambda\, dt \\
&= \frac{(-1)^m}{2\pi i} \int_0^a \varphi^{(m)}(t) \int_{\partial\Lambda} \lambda^{-m} e^{\lambda t} x\, d\lambda\, dt.
\end{aligned}
$$

By the residue theorem, for any $t \in [0, \tau_m]$ we obtain

$$\frac{1}{2\pi i} \int_{\partial\Lambda} \lambda^{-m} e^{\lambda t}\, d\lambda = \frac{t^{m-1}}{(m-1)!}.$$

This yields that

$$
\begin{aligned}
\langle S * P, \varphi \rangle x &= (-1)^m \int_0^a \varphi^{(m)}(t) \frac{t^{m-1}}{(m-1)!}\, dt\, x \\
&= \varphi(0)x = \langle \delta \otimes x, \varphi \rangle.
\end{aligned}
$$

In the same way, one can show that S is a solution of (2.1.9).

(S) $\Longrightarrow$ (R). Let $S \in \mathcal{D}_0'(\mathcal{L}(X, X_A))$ be a solution of (2.1.9), (2.1.10). By Propositions 2.1.1, 2.1.2 we have

$$\forall T > 0,\ \varepsilon > 0,\ \exists p \in \mathbb{Z}_+,\ C > 0: \ \forall \varphi \in \mathcal{D},\ \operatorname{supp} \varphi \in [-1 - \varepsilon, T + \varepsilon],$$

$$\|S(\varphi)\| \le C\|\varphi\|_p,$$

and S can be extended to $\mathcal{D}^p[-1, T]$. We denote the extension by the same symbol. Using this extension we can construct a continuous primitive of order $p + 2$ for S. Let $t \in [0, T]$, and consider the function

$$\psi_{t,p}(s) = \chi(s)\eta_p(t - s) \in \mathcal{D}^p[-1, T],$$

where $\chi(\cdot) \in C^\infty(\mathbb{R})$ is defined by

$$\chi(s) = \begin{cases} 0, & s \leq -1 \\ 1, & s \geq 0 \end{cases} \tag{2.1.19}$$

and

$$\eta_p(t) = \begin{cases} \frac{t^{p+1}}{(p+1)!}, & t \geq 0 \\ 0, & t < 0 \end{cases}.$$

Define

$$V(t) := \langle S, \psi_{t,p} \rangle. \tag{2.1.20}$$

We now show that $\{V(t), \; 0 \leq t < T\}$ is a local $(p+2)$-times integrated semigroup.

By Proposition 2.1.2 there is $\{\varphi_n\} \in \mathcal{D}$ such that

$$\text{supp } \varphi_n \in [-1 - \varepsilon, T + \varepsilon]$$

for all $n \in \mathbb{N}$, and

$$\|\varphi_n - \psi_{t,p+1}\|_{p+1} \to 0$$

for all $t \in [0, T]$. Hence

$$\|\varphi_n - \psi_{t,p+1}\|_p \to 0 \quad \text{and} \quad \|\varphi'_n - \psi'_{t,p+1}\|_p \to 0.$$

Consider (2.1.9), (2.1.10) with $\varphi = \varphi_n$. Let $n \to \infty$, then

$$-\langle S, \psi'_{t,p+1} \rangle x - A\langle S, \psi_{t,p+1} \rangle x = \psi_{t,p+1}(0)x$$

for all $x \in X$ and $0 \leq t \leq T$. Furthermore

$$-\langle S, \psi'_{t,p+1} \rangle x - \langle S, \psi_{t,p+1} \rangle Ax = \psi_{t,p+1}(0)x, \tag{2.1.21}$$

for all $x \in D(A)$. Since

$$\psi'_{t,p+1}(s) = \chi'(s)\frac{(t-s)^{p+2}}{(p+2)!} + \chi(s)\eta_p(t-s)$$

$$+\chi'(s)\frac{(t-s)^{p+2}}{(p+2)!} + \psi'_{t,p}(s)$$

and

$$\text{supp } \chi'(s)\frac{(t-s)^{p+2}}{(p+2)!} \in [-1, 0],$$

for defined by (2.1.20) $V(t)$ we have

$$\forall t \in [0, T], \quad V(t) = -\langle S, \psi'_{t,p+1} \rangle = \langle S, \psi_{t,p} \rangle, \tag{2.1.22}$$

and $V(t) \in \mathcal{L}(X)$. Since the mapping $t \to \psi_{t,p} \; : \; [0,T] \to \mathcal{D}^p[-1,T]$ is continuous, the function $V(\cdot)$ is continuous in $t \in [0,T]$. Since $\operatorname{supp} \psi_{0,p} \subset [-1,0]$ and $\operatorname{supp} S \subset [0,\infty)$ it follows that $V(0) = 0$. Furthermore,

$\forall t \in (0,T), \; k = 1,2,\ldots,p+1,$

$$\frac{\psi_{t+h,k}(s) - \psi_{t,k}(s)}{h} \to \psi_{t,k-1}(s) \tag{2.1.23}$$

uniformly with respect to $s \in [-1,T]$. Hence we have the convergence in $\mathcal{D}^p[-1,T]$. Therefore

$$
\begin{aligned}
\frac{d}{dt} S(\psi_{t,p+1}) &= \lim_{h \to 0} \frac{S(\psi_{t+h,p+1}) - S(\psi_{t,p+1})}{h} \\
&= \lim_{h \to 0} S\left(\frac{\psi_{t+h,p+1} - \psi_{t,p+1}}{h} \right) = S(\psi_{t,p}).
\end{aligned}
$$

Since $S(\psi_{0,p+1}) = 0$, we have

$$S(\psi_{t,p+1}) = \int_0^t V(s)\, ds. \tag{2.1.24}$$

From (2.1.21) – (2.1.24) we obtain

$$V(t)x = \frac{t^{p+2}}{(p+2)!} x + \int_0^t V(s) A x\, ds, \quad t \in [0,T), \; x \in D(A), \tag{2.1.25}$$

$$V(t)x = \frac{t^{p+2}}{(p+2)!} x + A \int_0^t V(s) x\, ds, \quad t \in [0,T), \; x \in X. \tag{2.1.26}$$

As it was proved in Theorem 1.2.5 for A satisfying (2.1.25) – (2.1.26), there exists the resolvent of A and (R) holds for $R_A(\lambda)$ with $n = p+2$, where p is the local order of S.

Remark 2.1.1 It follows from the obtained Theorems 2.1.3–2.1.5 and Propositions 1.2.3, 1.2.4, that for a closed linear operator A condition (W) and polynomial estimates (R) with some n are equivalent to k-well-posedness of (CP), with some k.

2.1.3 Well-posedness in the space of exponential distributions

Now we consider (CP) in the space of exponential distributions $\mathcal{S}'_\omega(\mathcal{L}(X, X_A))$. We show that the MFPHY-type condition:

$\exists M > 0, \; \omega \in \mathbb{R} :$

$$\left\| \frac{d^k}{d\lambda^k} \left[\frac{(\lambda I - A)^{-1}}{\lambda^n} \right] \right\| = \frac{M\, k!}{(\operatorname{Re}\lambda - \omega)^{k+1}} \tag{2.1.27}$$

for all $\lambda \in \mathbb{C}$ with $\operatorname{Re}\lambda > \omega$ and all $k = 0, 1, \dots ,$

is necessary and sufficient for the ω-well-posedness of (δP) with a closed densely defined operator A. Hence, by Theorem 1.2.1, (δP) with such A is ω-well-posed if and only if A is the generator of an exponentially bounded n-times integrated semigroup with the same n as in (2.1.27). In the following theorem we give the constructive proof of these connections.

Theorem 2.1.6 *Let A be a linear closed densely defined operator on a Banach space X. Then the following statements are equivalent:*

($\mathbf{W}_{\exp}$) *the Cauchy problem* (CP) *is ω-well-posed in the sence of distributions;*

($\mathbf{S}_{\exp}$) *there exists a solution operator distribution $S \in \mathcal{S}'_\omega(\mathcal{L}(X, X_A))$ such that*

$$P * S = \delta \otimes I, \quad S * P = \delta \otimes J, \tag{2.1.28}$$

In this case

$$\langle U, \varphi \rangle = \langle S, \varphi \rangle x$$

for all $\varphi \in \mathcal{D}$ and $x \in X$;

($\mathbf{D}_{\exp}$) *A is the generator of an exponentially bounded distribution semigroup Q with dense range. In this case $S = Q$;*

($\mathbf{I}_{\exp}$) *A is the generator of an exponentially bounded n-times integrated semigroup $\{V(t), \ t \geq 0\}$ for some n;*

($\mathbf{R}_{\exp}$) *the MFPHY-type condition (2.1.27) holds.*

Proof Implications ($\mathbf{W}_{\exp}$) $\Longleftrightarrow$ ($\mathbf{S}_{\exp}$) $\Longleftrightarrow$ ($\mathbf{D}_{\exp}$) follow from the corresponding implications in Theorems 2.1.3 – 2.1.4. If a solution S of (2.1.28) belongs to $\mathcal{S}'_\omega(\mathcal{L}(X, X_A))$, then we have that $Q = S \in \mathcal{S}'_\omega(\mathcal{L}(X, X_A))$ and $U = Qx \in \mathcal{S}'_\omega(X_A)$. The converse is also valid.

($\mathbf{I}_{\exp}$) $\Longrightarrow$ ($\mathbf{S}_{\exp}$). Let $n \in \mathbb{N}$ and A be the generator of an exponentially bounded n-times integrated semigroup V. Since $V(t)$ is defined for all $t \geq 0$, $S \in \mathcal{S}'_\omega(\mathcal{L}(X, X_A))$ can be defined as $S = V^{(n)}$. Let

$$S(\varphi) := (-1)^n \int_0^\infty \varphi^{(n)}(t) V(t) \, dt, \quad \varphi \in \mathcal{D}.$$

Taking into account the following equality for an n-times integrated semigroup:

$$V'(t)x = \frac{t^{n-1}}{(n-1)!}x + AV(t)x, \quad t \geq 0, \ x \in D(A),$$

we have

$$(-1)^{n+1} \int_0^\infty \varphi^{(n+1)}(t)V(t)x\,dt$$

$$= (-1)^n \int_0^\infty \varphi^{(n)}(t)\frac{t^{n-1}}{(n-1)!}x\,dt + (-1)^n A \int_0^\infty \varphi^{(n)}(t)V(t)x\,dt$$

for any $\varphi \in \mathcal{D}$ and $x \in D(A)$. Since A is closed and all other operators in this equality are bounded, it is valid for all $x \in X$. Integrating the first term in the right-hand side by parts, we obtain

$$AS(\varphi)x = S'(\varphi)x - \varphi(0)x, \quad x \in X. \qquad (2.1.29)$$

Hence S is a solution of the first equation in (2.1.28). Using the commutativity of $V(t)$ and A on $D(A)$, we obtain the second one. From (2.1.29) we have

$$\|S(\varphi)x\|_A \leq \|S(\varphi)x\| + \|Q'(\varphi)x\| + |\varphi(0)|\,\|x\|.$$

Therefore $S(\varphi) \in \mathcal{L}(X, X_A)$ for any φ, and $S \in \mathcal{D}'_0(\mathcal{L}(X, X_A))$. Moreover, exponential boundedness of V implies

$$\exists \omega \in \mathbb{R} : \quad e^{-\omega t}S \in \mathcal{S}'(\mathcal{L}(X, X_A)),$$

hence $S \in \mathcal{S}'_\omega(\mathcal{L}(X, X_A))$.

($S_{\exp}$) $\Longrightarrow$ ($R_{\exp}$) $\Longrightarrow$ ($I_{\exp}$). Let $S \in \mathcal{S}'_\omega(\mathcal{L}(X, X_A))$ be a (unique) solution of (2.1.29). Applying the Laplace transform to (2.1.29), we have

$$(\lambda I - A)\mathcal{L}S(\lambda) = I$$

and

$$\mathcal{L}S(\lambda)(\lambda I - A) = J$$

for all $\lambda \in \mathbb{C}$ with $\operatorname{Re}\lambda > \omega$. From these equalities we have that the half-plane $\operatorname{Re}\lambda > \omega$ belongs to the resolvent set of operator A, and $(\lambda I - A)^{-1} = \mathcal{L}S(\lambda)$. Since $e^{-\omega t}S \in \mathcal{S}'(X)$, by the analogue of Proposition 2.1.1 for exponentially bounded distributions, we have

$$\exists\, p \geq 0, C > 0 : \ \forall \psi \in \mathcal{S}, \ \|(e^{-\omega t}S)(\psi)\| \leq C\|\psi\|_{p,p}, \qquad (2.1.30)$$

where

$$\|\psi\|_{j,k} := \sup_{0 \leq i \leq k} \sup_t (1 + |t|)^j |\psi^{(i)}(t)|.$$

Using the density of $\mathcal{S}$ in the space $\mathcal{S}^p(\mathbb{R})$ of p-times continuously differentiable functions with the norm $\|\cdot\|_{j,k}$ and the estimate (2.1.30), we can extend S to $\mathcal{S}^p(\mathbb{R})$. Let η_p and χ be defined by (2.1.19), then $\chi(s)\eta_p(t - s) \in \mathcal{S}^p(\mathbb{R})$. Introduce the function

$$V(t) = S\big(\chi(s)\eta_p(t - s)\big) := (e^{-\omega s}S)\big(e^{\omega s}\chi(s)\eta_p(t - s)\big). \qquad (2.1.31)$$

It is not difficult to verify that the mapping $t \mapsto e^{\omega s}\chi(s)\eta_p(t-s)$ from $\mathbb{R}$ to $\mathcal{S}^p(\mathbb{R})$ is continuous, hence $V(\cdot)$ is a continuous function with values in $\mathcal{L}(X, X_A)$. Since $\operatorname{supp} S \subseteq [0, +\infty)$ and $\operatorname{supp}(e^{\omega s}\chi(s)\eta_p(t-s)) \subseteq [-1, t]$, we obtain that $V(t) = 0$ for $t \leq 0$. Moreover, (2.1.30) implies the following estimate

$$
\begin{aligned}
\|V(t)\| &= \|(e^{-\omega s}S)(e^{\omega s}\chi(s)\eta_p(t-s))\| \\
&\leq C\|e^{\omega s}\chi(s)\eta_p(t-s)\|_{p,p} \leq M e^{\omega t}(1+|t|)^{p+1} \\
&\leq M e^{\omega' t}
\end{aligned}
$$

for $\omega' > \omega$ and $t \geq 0$.

Let us show that S is n-th derivative of V for $n = p+2$. For any $\varphi \in \mathcal{D}$ we have

$$
\begin{aligned}
\langle V^{p+2}, \varphi \rangle &= (-1)^{p+2} \int_0^\infty \varphi^{(p+2)}(t)V(t)\,dt \\
&= (-1)^{p+2} \int_0^\infty \varphi^{(p+2)}(t)\langle S(\hat{s}), \chi(\hat{s})\eta_p(t-\hat{s})\rangle\,dt \\
&= \langle S(\hat{s}), \chi(\hat{s})(-1)^{p+2} \int_0^\infty \varphi^{(p+2)}(t)\eta_p(t-\hat{s})\,dt \rangle \\
&= \langle S, \chi(\hat{s})\varphi(\hat{s})\rangle = \langle S, \varphi \rangle.
\end{aligned}
$$

Using properties of the Laplace transform we have

$$
(\lambda I - A)^{-1} = \lambda^{p+2} \int_0^\infty e^{-\lambda t}V(t)\,dt, \quad \operatorname{Re}\lambda > \omega'.
$$

Hence, for $n = p+2$ and all $k = 0, 1, \ldots$, we obtain

$$
\begin{aligned}
\left\| \frac{d^k}{d\lambda^k}\left[\frac{(\lambda I - A)^{-1}}{\lambda^n}\right] \right\| &= \left\| \frac{d^k}{d\lambda^k}\mathcal{L}V(\lambda) \right\| = \|\mathcal{L}(t^k V)(\lambda)\| \\
&\leq \int_0^\infty e^{-t\operatorname{Re}\lambda}\, t^k \|V(t)\|\,dt \\
&\leq M \int_0^\infty t^k e^{-(\operatorname{Re}\lambda-\omega')t}\,dt \\
&= \frac{M\,k!}{(\operatorname{Re}\lambda - \omega')^{k+1}}, \quad \operatorname{Re}\lambda > \omega'.
\end{aligned}
$$

By Theorem 1.2.1, these estimates are the necessary and sufficient condition for a closed densely defined operator A to be the generator of an exponentially bounded n-times integrated semigroup V. $\quad\square$

2.2   The degenerate Cauchy problem

Let X, Y be Banach spaces. In this section we consider the degenerate Cauchy problem

$$Bu'(t) = Au(t), \quad t \geq 0, \ u(0) = x, \qquad \text{(DP)}$$

with linear operators $B, A : X \to Y$ such that A is closed and B is bounded with $\ker B \neq \{0\}$. Suppose

$$\rho_B(A) := \qquad (2.2.1)$$
$$\left\{ \lambda \in \mathbb{C} \ \middle| \ (\lambda B - A)^{-1}B =: R(\lambda) \text{ is bounded in } X \right\} \neq \emptyset,$$

Recall that the problem (DP) is equivalent to the Cauchy problem for the inclusion

$$u'(t) \in \mathcal{A}u(t), \quad t \geq 0, \ u(0) = x, \qquad \text{(IP)}$$

with $\mathcal{A} = B^{-1}A$.

We prove the necessary and sufficient conditions for the well-posedness of (DP) and (IP) in spaces of distributions, generalizing the conditions obtained in Section 2.1. Naturally, the semigroups connected with these problems are degenerate on X. The important fact is that an n-times integrated semigroup with the generator $\mathcal{A}$ is degenerate on $\ker B = \ker R(\lambda) = \mathcal{A}0$, but the corresponding solution operator distribution and distribution semigroup are degenerate on $\ker R^{n+1} := \ker R^{n+1}(\lambda) = \mathcal{A}^{n+1}0$. We assume that X admits the decomposition

$$X = X_{n+1} \oplus \ker R^{n+1}, \qquad (2.2.2)$$

where

$$X_{n+1} := \overline{D}_{n+1}, \quad D_{n+1} := R^{n+1}(\lambda)X,$$

which generalizes the decomposition $X = X_1 \oplus \mathcal{A}0$ and the density condition for a generator of a C_0-semigroup.

In the first subsection we describe the structure of $\ker R^{n+1}$. It is used in the main theorem (second subsection) where we discuss the structure of solution operator distributions and distribution semigroups on $\ker R^{n+1}$. We prove that (DP) is well-posed in the space of distributions if and only if $\hat{A}$ (the part of operator $\mathcal{A}$ in X_{n+1}) is the generator of a distribution semigroup, or equivalently, $\hat{A}$ is the generator of a local k-times integrated semigroup for some k. In the third subsection, we consider (DP) in the space of exponential distributions. The distinctive feature of the exponential case is that the parameter k of the integrated semigroup coincides with the parameter in the estimate for the resolvent of the generator and with the parameter n in the decomposition (2.2.2).

2.2.1 *A*-associated vectors and degenerate distribution semigroups

Definition 2.2.1 *The sets*

$$K_i = \Big\{ x \in X \mid \exists y \in K_{i-1} : Ay = Bx, \ y \neq 0 \Big\}, \ i \geq 1,$$
$$K_0 = \ker B,$$

are called the sets of i-th A-associated with $\ker B$ *vectors.*

From this definition we have that for any $\varphi_i \in K_i$ there exists a sequence of A-associated vectors $\{\varphi_0, \varphi_1, \ \dots \ \varphi_{i-1}\}$ such that $\varphi_k \in K_k$ and $A\varphi_k = B\varphi_{k+1}$ for $k = 0, 1, \dots, i-1$. This sequence is unique up to $\ker A$.

Proposition 2.2.1 *Consider linear operators* $B, A : X \to Y$ *with* $\rho_B(A) \neq \emptyset$. *Then*

(I) $\ker B \cap \ker A = \{0\}$; $K_i \cap \ker A = \emptyset$, $i \geq 1$; $K_i \cap K_j = \emptyset$, $j \neq i$;

(II) $K_i \cup \{0\}$ *is a closed linear manifold in* X *for all* $i \geq 0$;

(III) *let* $\Lambda \subset \mathbb{C}$ *be a region containing some sequence convergent to infinity. If there exists* $M > 0$ *such that*

$$\|R(\lambda)\| \leq \frac{M|\lambda^n|}{|f(\lambda)|}$$

for all $\lambda \in \Lambda$ *and some* f *such that* $|f(\lambda)| \longrightarrow_{\lambda \to \infty} \infty$, *then*

$$\forall x \in D_{n+1}, \quad \lim_{\lambda \to \infty} \lambda(\lambda B - A)^{-1} Bx = x,$$

$$\overline{D}_{n+1} \cap \ker B = \{0\}, \ \text{and} \ K_{n+1} = \emptyset;$$

(IV) *any* $x \in \ker R^{n+1}$ *can be represented in the form*

$$x = x_1 + x_2 + \dots + x_n, \quad \text{where} \quad x_i \in K_i.$$

If $\ker B$ *is complemented in* $\ker R^{n+1}$, *then the decomposition*

$$\ker R^{n+1} = \ker B \oplus \mathcal{K}_1 \oplus \ \dots \ \oplus \mathcal{K}_n,$$

holds, where $\mathcal{K}_i \cong (K_i \cup \{0\})/\ker B$ *and* $\mathcal{K}_i \subset X$.

Proof (I). Condition (2.2.1) implies that $\ker A \cap \ker B = \{0\}$. For φ_i and for $\lambda \in \rho_B(A)$ the following relations hold

$$(\lambda B - A)^{-1} B \varphi_i = -\varphi_{i-1} - \lambda \varphi_{i-2} \ - \dots \ - \lambda^{i-1} \varphi_0,$$

$$\left[(\lambda B - A)^{-1} B\right]^{j} \varphi_i = \tag{2.2.3}$$
$$(-1)^{j} \left[C_{j-1}^{0} \varphi_{i-j} + C_{j}^{1} \lambda \varphi_{i-j-1} + \ldots + C_{i}^{i-j} \lambda^{i-j} \varphi_0\right],$$
$$j = 1, \ldots, i.$$

In particular,

$$\left[(\lambda B - A)^{-1} B\right]^{i} \varphi_i = (-1)^{i} \varphi_0,$$

for any $\varphi_i \in K_i$, $\varphi_0 \in \ker B$. To prove these relations we write

$$
\begin{aligned}
(\lambda B - A)^{-1} B \varphi_i &= (\lambda B - A)^{-1} (A \pm \lambda B) \varphi_{i-1} \\
&= -\varphi_{i-1} + \lambda (\lambda B - A)^{-1} B \varphi_{i-1} \\
&= -\varphi_{i-1} - \lambda \varphi_{i-2} - \ldots - \lambda^{i-1} \varphi_0.
\end{aligned}
\tag{2.2.4}
$$

Let $\varphi_i \in K_i \cap \ker A$, then from (2.2.3) we have

$$
\begin{aligned}
A\varphi_0 &= A(-1)^{i} \left[(\lambda B - A)^{-1} B\right]^{i} \varphi_i \\
&= B(-1)^{i} \left[(\lambda B - A)^{-1} B\right]^{i-1} (\lambda B - A)^{-1} A \varphi_i = 0.
\end{aligned}
$$

Since $\ker A \cap \ker B = \{0\}$, we have $\varphi_0 = 0$. Therefore

$$K_i \cap \ker A = \emptyset \quad \text{for any} \quad i \geq 1.$$

Let $j = 0$. Take $\varphi_i \in K_i$ and $\varphi_{i-1} \in K_{i-1}$ such that $A\varphi_{i-1} = B\varphi_i$. Since $K_i \cap \ker A = \emptyset$, then $A\varphi_{i-1} \neq 0$. Hence, $\varphi_i \notin \ker B$, and $K_0 \cap K_i = \emptyset$.

Now let $j > 0$ and $i > j$. Consider any $x \in K_j \cap K_i$. We have $\left[(\lambda B - A)^{-1} B\right]^{i} x \neq 0$ since $x \in K_i$, and $\left[(\lambda B - A)^{-1} B\right]^{i} x = 0$ since $x \in K_j$. Hence, $K_j \cap K_i = \emptyset$.

(II). For any $i \geq 1$, $K_i \cup \{0\}$ are obviously linear manifolds. We now show that they are closed. Let $\{\varphi_i^k\}$ be a sequence from K_i convergent to f_i, then

$$\left[(\lambda B - A)^{-1} B\right]^{j} \varphi_i^k \to \left[(\lambda B - A)^{-1} B\right]^{j} f_i$$

for all $j = 1, \ldots, i$ and $\lambda \in \rho_B(A)$. In particular,

$$\varphi_0^k = (-1)^{i} \left[(\lambda B - A)^{-1} B\right]^{i} \varphi_i$$

is convergent. Furthermore, (2.2.3) implies that all φ_j^k, $j = 1, \ldots, i$, are convergent. Denote their limits by f_j. Considering (2.2.4) for $\varphi_0^k, \varphi_1^k, \ldots \varphi_i^k$ and letting $k \to \infty$, we obtain

$$(\lambda B - A)^{-1} B f_i = (\lambda B - A)^{-1} (A \pm \lambda B) f_{i-1},$$

i.e. $Bf_i = Af_{i-1}$. Similarly, $Bf_j = Af_{j-1}$ for any $j = 1, 2, \ldots, i-1$. Hence $f_i \in K_i$.

(III). Let $x \in D_{n+1}$, then there exists $y \in X$ such that

$$x = \left[(\lambda_0 B - A)^{-1} B\right]^{n+1} y$$

for some $\lambda_0 \in \rho_B(A)$, and

$$\|\lambda(\lambda B - A)^{-1} Bx - x\|$$
$$= \left\|\lambda(\lambda B - A)^{-1} B\left[(\lambda_0 B - A)^{-1} B\right]^{n+1} y - \left[(\lambda_0 B - A)^{-1} B\right]^{n+1} y\right\|$$
$$= \left\|\lambda(\lambda_0 - \lambda)^{-1}(\lambda B - A)^{-1} B\left[(\lambda_0 B - A)^{-1} B\right]^{n} x \right.$$
$$\left. -\lambda_0(\lambda_0 - \lambda)^{-1}\left[(\lambda_0 B - A)^{-1} B\right]^{n+1} y\right\|$$
$$= \left\|\lambda(\lambda_0 - \lambda)^{-(n+1)}(\lambda B - A)^{-1} By - \ldots \right.$$
$$\left. -\lambda_0(\lambda_0 - \lambda)^{-1}\left[(\lambda_0 B - A)^{-1} B\right]^{n+1} y\right\| \xrightarrow[\lambda \to \infty]{} 0.$$

Hence, if $x \in X_{n+1} = \overline{D}_{n+1} \cap \ker B$, then $x = 0$.

Consider $x \in K_{n+1}$. Let $x_0 \in \ker B$ be the first vector in the sequence of A-associated vectors $\{x_0, x_1, x_{n-1}, x\}$, then $\|x_0\| = \|R(\lambda_0)^n R(\lambda)x\|$. Applying the resolvent identity, we obtain that $x_0 = 0$, therefore $K_{n+1} = \emptyset$.

(IV). Let $x \in \ker R^{n+1}$, then we have $\left[(\lambda B - A)^{-1} B\right]^{n+1} x = 0$ and either

$$\forall j = 1, 2, \ldots n, \quad \left[(\lambda B - A)^{-1} B\right]^{j} x = 0 \iff x \in \ker B$$

or

$$\exists\, 1 \le j \le n, \quad \left[(\lambda B - A)^{-1} B\right]^{j+1} x = 0 \quad \text{and} \quad \left[(\lambda B - A)^{-1} B\right]^{j} x \ne 0.$$

If $j = 1$, we have $y = (\lambda B - A)^{-1} Bx \in \ker B$. Hence $Bx = -Ay$ and $x \in K_1$. If $j = 2$, then

$$y = (\lambda B - A)^{-1} B(\lambda B - A)^{-1} Bx \in \ker B$$
$$\implies y_1 = (\lambda B - A)^{-1} Bx \in K_1,$$

and $-x + \lambda y_1 \in K_2$, and so on. Thus, for any $x \in \ker R^{n+1}$ we have $x = x_1 + x_2 + \ldots + x_n$, where $x_i \in K_i$, $1 \le i \le n$. If $\ker B$ complemented in $\ker R^{n+1}$ (i.e., if there exists Y such that $\ker R^{n+1} = Y \oplus \ker B$), then

$$x = \tilde{x}_0 + \tilde{x}_1 + \tilde{x}_2 + \ldots + \tilde{x}_n, \quad \tilde{x}_i \in \mathcal{K}_i, \ 0 \le i \le n.$$

We now prove by contradiction that this representation is unique. Suppose that it is not unique, then we have $0 = z_1 + z_2 + \ldots + z_n$, where $z_i \in \mathcal{K}_i$, and there exists $z_j \ne 0$ such that $z_i = 0$ for all $i > j$. Applying $\left[(\lambda_0 B - A)^{-1} B\right]^{j}$ to this equality, we have for x_0 that the vector from $\ker B$ for which z_j is the j-th A-associated vector, is equal to 0. Hence the decomposition in (iv) holds true. $\square$

Corollary 2.2.1 If the decomposition (2.2.2) holds, and $\ker B$ is complementable in $\ker R^{n+1}$, then we have the decomposition

$$X = X_{n+1} \oplus \ker B \oplus \mathcal{K}_1 \oplus \ \ldots\ \oplus \mathcal{K}_n =: \mathcal{X} \oplus \ker B, \qquad (2.2.5)$$

and $\mathcal{K}_{n+1} = \emptyset$.

Taking into account the discussions in the previous section we can give the following definition of a solution for (DP) in spaces of distributions.

Definition 2.2.2 *An abstract distribution $U \in \mathcal{D}'_0(X_{A,B})$ is a solution of* (DP) *in the sense of distributions if*

$$\forall \varphi \in \mathcal{D}, \qquad\qquad\qquad\qquad (2.2.6)$$
$$B\langle U, \varphi' \rangle + A\langle U, \varphi \rangle = -\langle \delta, \varphi \rangle Bx, \quad x \in X,$$

or equivalently

$$P * U = \delta \otimes Bx, \quad x \in X, \qquad\qquad (\delta\text{DP})$$

where

$$P := \delta' \otimes B - \delta \otimes A,$$
$$\langle \delta' \otimes B, \varphi \rangle := \langle \delta', \varphi \rangle B,$$
$$\langle \delta \otimes A, \varphi \rangle := \langle \delta, \varphi \rangle A$$

and

$$X_{A,B} = \Big\{ x \in D_1\,,\ \|x\|_R = \inf_{y:R(\lambda)y=x} \|y\| \Big\}.$$

A solution U is called degenerate on $Z \subset X$, if for U corresponding to $x \in Z$ we have $U(\varphi) = 0$ for any $\varphi \in \mathcal{D}_0$.

For the degenerate Cauchy problem, the set

$$\begin{aligned} D_1 \ &:= \ R(\lambda)X \\ &= \ \Big\{ x \in D(A) \ \Big|\ \text{there exists } y \text{ such that } By = Ax \Big\} \end{aligned}$$

is the generalization of $D(A)$ in the nondegenerate case.

It follows from Definition 2.2.2 that $U \in \mathcal{D}'_0(X_{A,B})$ is a solution of (δDP) if and only if $U \in \mathcal{D}'_0(X_A)$ is a solution of

$$\mathcal{P} * U \in \delta \otimes x, \quad x \in X, \qquad\qquad (\delta\text{IP})$$

where $\mathcal{P} := \delta' \otimes I - \delta \otimes \mathcal{A}$, $\mathcal{A} = B^{-1}A$, and

$$X_A := \Big\{ D(\mathcal{A}),\ \|x\|_A = \|x\| + \|\mathcal{A}x\| \Big\}.$$

Here $\|\mathcal{A}x\|$ is the factor-norm on $X/\mathcal{A}0$.

If $\ker B$ is complemented in X:

$$X = \mathcal{X} \oplus \ker B = \mathcal{X} \oplus \mathcal{A}0,$$

then there exists $\tilde{\mathcal{A}}$, the single-valued branch of $\mathcal{A}$ (i.e., $\tilde{\mathcal{A}}$ is the single-valued operator with $D(\tilde{\mathcal{A}}) = D(\mathcal{A})$) and $X_{A,B}$ is homeomorphic to the space

$$X_{\tilde{\mathcal{A}}} := [D(\tilde{\mathcal{A}})] = \left\{ D(\tilde{\mathcal{A}}), \ \|x\|_{\tilde{\mathcal{A}}} = \|x\| + \|\tilde{\mathcal{A}}x\| \right\}.$$

If, moreover, the decomposition (2.2.2) holds, then $\hat{\mathcal{A}}$, the part of $\mathcal{A}$ in X_{n+1}, is the part of $\tilde{\mathcal{A}}$ in X_{n+1}.

Definition 2.2.3 *The problem* (DP) *is called well-posed in the sense of distributions (or (δDP) well-posed) if for any $x \in X$ there exists a unique solution of (δDP), and $U_j \to 0$ for any sequence $x_j \to 0$. If $U \in S'_\omega(X_{\hat{\mathcal{A}}})$, then ($\delta$DP) is called w-well-posed.*

If $\mathcal{A} = B^{-1}A$, then (δIP) is well-posed if and only if (δDP) is well-posed. In this case $U \in \mathcal{D}'_0(X_A)$.

As in the nondegenerate case, the solution of well-posed (δDP) can be constructed with the help of a semigroup of solution operators. In this case it is a distribution semigroup (see Definition 2.1.5), which is degenerate on $\ker R^{n+1}$.

Here we say that a distribution semigroup Q is degenerate on

$$Z = \ker Q := \left\{ x \in X \ \Big| \ \forall \varphi \in \mathcal{D}_0, \ Q(\varphi)x = 0 \right\},$$

if Z contains at least one non-zero element, and we say that Q is nondegenerate if $\ker Q = \{0\}$.

For any $E \in \mathcal{E}'_0$ we define an operator $Q(E)$. Let $E \in \mathcal{E}'_0$ and let $\{\delta_n\} \subset \mathcal{D}_0$ be any sequence covergent to δ. Then for any x from

$$D(Q(E)) := \tag{2.2.7}$$
$$\left\{ x \ \Big| \ Q(\delta_n)x \to x \ \text{ and } \ \exists y \text{ such that } Q(E * \delta_n)x \to y \right\},$$

we have $Q(E)x := y$.

Now we state the properties of degenerate distribution semigroups. In the main theorem we use these properties to show that the distribution semigroup connected with the well-posed (δDP) is regular on X_{n+1} and degenerate on $\ker R^{n+1}$.

Proposition 2.2.2 *Let Q be a distribution semigroup degenerate on Z and let $E \in \mathcal{E}'_0$. Then $Q(E)$ has the following properties*

(Q1) $\forall \varphi \in \mathcal{D}_0,\ x \in X,$

$$Q(\varphi)x \in D(Q(E)) \quad and \quad Q(E)Q(\varphi)x = Q(E*\varphi)x;$$

(Q2) $D(Q(E)) \cap Z = \{0\};$

(Q3) $\forall\ \varphi \in \mathcal{D}_0,\ x \in D(Q(E)),$

$$Q(E)Q(\varphi)x = Q(\varphi)Q(E)x;$$

(Q4) *if* $X = \mathcal{X} \oplus Z$ *and* $\operatorname{ran} Q \subset \mathcal{X}$, *then there exists a closure of* $Q(E)$
and

$\forall \varphi \in \mathcal{D}_0,\ x \in X,$

$$\overline{Q(E)}Q(\varphi)x = Q(E)Q(\varphi)x;$$

(Q5) *if* Q *is regular and* $\operatorname{ran} Q \subset \mathcal{X}$, *then for any* $x \in \overline{\operatorname{ran} Q}$ *and* $\psi \in \mathcal{D}$,
we have

$$\overline{Q(\psi_+)}x = Q(\psi)x,$$

where $\psi_+(\cdot) \in \mathcal{E}'_0$ *is defined by*

$$\psi_+(t) = \begin{cases} \psi(t), & t \geq 0 \\ 0, & t < 0 \end{cases}.$$

Proof In addition to the proof of the Proposition 2.1.3 for the nondegenerate case, we need to show only (Q2), (Q4).

(Q2). Consider $x \in Z$ and a sequence $\{\delta_j\}$ from (2.2.7), then $Q(\delta_j)x = 0$. Hence, for any $E \in \mathcal{E}'_0$ we have $x \notin D(Q(E))$, i.e.

$$D(Q(E)) \cap Z = \{0\}.$$

(Q4). A linear operator A has a closure if and only if
$\forall \{x_j\} \subset D(A)$ such that $x_j \to 0$, we have

$$Ax_j \to y \implies y = 0.$$

Consider $\{x_j\} \subset D(Q(E))$ such that $x_j \to 0$ and $Q(E)x_j \to y$, then $y \in \overline{\operatorname{ran} Q} \subset \mathcal{X}$. By (Q1) we have
$\forall \varphi \in \mathcal{D}_0,$

$$Q(\varphi)Q(E)x_j = Q(E*\varphi)x_j \to Q(\varphi)y = 0 \implies y \in Z,$$

and $y \in \overline{\operatorname{ran} Q} \subset \mathcal{X}$. Since $Z \cap \mathcal{X} = \{0\}$, we have $y = 0$. $\square$

Due to (Q4), the operator $\overline{Q(-\delta')}$ is defined for any distribution semi-group Q with $\operatorname{ran} Q \subset \mathcal{X}$, where $X = \mathcal{X} \oplus Z$. This operator is called the generator of Q.

2.2.2 Well-posedness in the sense of distributions

In the following theorem, assuming the decomposition (2.2.5), we show the connection between the well-posedness of (DP), the existence of a degenerate distribution semigroup, the existence of a degenerate local k-times integrated semigroup with the generator $\hat{A}$, polynomial estimates for $R_{\hat{A}}(\lambda)$ in some region Λ containing a right semiaxis, and the existence of a unique classical solution for local (CP) on D_{k+1}.

We recall that the operator $\overline{A_0}$, where $A_0 x := \lim_{t \to 0} t^{-1}[V^{(k)}(t)x - x]$ is defined for those x where the limit exists, is called the generator of a k-times integrated semigroup V. A local k-times integrated semigroup $\{V(t),\ t \in [0, T)\}$ is called degenerate on Z if

$$\forall t \in [0, T),$$
$$V(t)x = 0 \implies x \in Z,$$

and V is called nondegenerate if $Z = \{0\}$.

Theorem 2.2.1 *Let $A, B : X \to Y$ be linear operators, A be closed and invertible, B be bounded. Suppose that the decomposition (2.2.5) holds. Then the following statements are equivalent:*

(W) *the problem (δDP) is well-posed. Its solution belongs to $\mathcal{D}_0'(X_{\hat{A}})$ for $x \in X_{n+1}$ and is degenerate for $x \in \ker R^{n+1}$;*

(S) *there exists a solution operator distribution*

$$S \in \mathcal{D}_0'(\mathcal{L}(X, X_{A,B})) \cap \mathcal{D}_0'(\mathcal{L}(X_{n+1}, X_{\hat{A}}))$$

such that

$$(P * S)x = \delta \otimes Bx, \quad x \in X, \tag{2.2.8}$$

and

$$(S * \widetilde{\mathcal{P}})x = \delta \otimes x, \quad x \in D(\mathcal{A}), \tag{2.2.9}$$

where $\widetilde{\mathcal{P}} := \delta' \otimes I - \delta \otimes \widetilde{A}$, $\hat{A}$ is the part of A in X_{n+1}, and $\widetilde{A}$ is the single-valued branch of $\mathcal{A}$.

In this case, for any $x \in X$,

$$\langle U, \varphi \rangle = \langle S, \varphi \rangle x$$

for all $\varphi \in \mathcal{D}$. Furthermore, S is degenerate on $\ker R^{n+1}$;

(D) *there exists a regular distribution semigroup Q degenerate on $\ker R^{n+1}$ with $\operatorname{ran} Q$ dense in X_{n+1}. Its generator is equal to $\hat{A}$. In this case $S = Q$;*

(I) *for any $T > 0$, the operator $\hat{A}$ is the generator of a local k-times integrated semigroup $\{V(t),\ t \in [0, T)\}$ on X_{n+1}, for some $k = k(T) \geq n$;*

(R) *for the resolvent of* $\hat{A}$, *the polynomial estimate*

$$\|R_{\hat{A}}(\lambda)\| \leq \frac{M\,|\lambda|^l}{\log(1+|\lambda|)} \leq M_1|\lambda|^l \qquad (2.2.10)$$

holds for some $M > 0$, $l \geq 0$, *and any*

$$\lambda \in \Lambda = \left\{ \lambda \in \mathbb{C} \;\middle|\; \operatorname{Re}\lambda > \frac{l}{\tau}\,\log(1+|\lambda|) + \frac{1}{\tau}\,\log\frac{C}{\gamma} \right\},$$

where

$$C > 0,\; 0 < \gamma < 1,\; \tau > 0;$$

(Wk) *for any* $T > 0$, *the local Cauchy problem*

$$u'(t) = \hat{A}u(t), \quad 0 \leq t < T,\; u(0) = x, \qquad \text{(CP)}$$

is k-well-posed on D_{k+1} *for some* $k = k(T) \geq n$.

Proof (W) $\Longrightarrow$ (S). Consider the distribution S defined by

$$S(\varphi)x = Sx(\varphi) := U(\varphi), \quad \varphi \in \mathcal{D},\; x \in X,$$

where U is a unique solution of (δDP). Then $Sx \in \mathcal{D}'_0(X_{A,B}) = \mathcal{D}'_0(X_{\tilde{A}})$ and $Sx \in \mathcal{D}'_0(X_{\hat{A}})$ for $x \in X_{n+1}$.

The well-posedness of (δDP) implies that for every $\varphi \in \mathcal{D}$,

$$\|x_n\| \to 0 \Longrightarrow \|S(\varphi)x_n\|_{\tilde{A}} = \|U_n(\varphi)\|_{\tilde{A}} \to 0.$$

Therefore $S(\varphi) \in \mathcal{L}(X, X_{\tilde{A}})$. We now show that $S \in \mathcal{D}'_0(\mathcal{L}(X, X_{\tilde{A}}))$, i.e., $\operatorname{supp} S \subset [0,\infty)$ and

$$\forall \varphi_n \in \mathcal{D},$$

$$\varphi_n \to 0 \Longrightarrow \|S(\varphi_n)\|_{\mathcal{L}(X,X_{\tilde{A}})} \to 0.$$

Consider the set

$$\mathcal{B} = \left\{ Sx,\; \|x\| \leq c \right\} \subset \mathcal{D}'_0(X).$$

Since for any $Sx_j \subset \mathcal{B}$ and any $\varepsilon_j \to 0$, we have $\varepsilon_j Sx_j = S(\varepsilon_j x_j) \to 0$, then $\mathcal{B}$ is bounded in $\mathcal{D}'_0(X)$. For any bounded set $\mathcal{B}$ and any $\varphi_j \to 0$ in $\mathcal{D}$

$$\exists p \in \mathbb{N},\; C > 0:\; \forall j,\; Sx \in \mathcal{B},$$

$$\|Sx(\varphi_j)\| \leq C\|\varphi_j\|_p,$$

where

$$\|\varphi\|_p = \sup_{k=0,1,\dots p,\; t\in \operatorname{supp}\varphi} \|\varphi^{(k)}(t)\|.$$

Therefore, $S(\varphi_j)x \to 0$ uniformly in x from a bounded set, or equivalently $\|S(\varphi_j)\|_{\mathcal{L}(X,X_{\tilde{A}})} \to 0$. Verification of (2.2.8), as in the nondegenerate case in Theorem 2.1.3, is based on the structure theorem. Let us prove (2.2.9). Equality (2.2.8) implies the following equalities

$$
\begin{aligned}
(P * S')x &= (P * S)'x = \delta' \otimes Bx, \quad x \in X, \\
(P * S)\tilde{A}x &= \delta \otimes Ax, \quad x \in D(\tilde{A}) = D(\mathcal{A}).
\end{aligned}
$$

By the definition of P we have

$$
(P * \delta \otimes x) = \delta' \otimes Bx - \delta \otimes Ax, \quad x \in D(\mathcal{A}).
$$

Hence

$$
P * (S'x - S\tilde{A}x - \delta \otimes x) = 0, \quad x \in D(\mathcal{A}),
$$

and since the Cauchy problem is well-posed, we have

$$
S'x - S\tilde{A}x = (S * \tilde{P})x = \delta \otimes x, \quad x \in D(\mathcal{A}).
$$

(S) $\Longrightarrow$ (W). Let

$$
S \in \mathcal{D}'_0(\mathcal{L}(X, X_{A,B})) \cap \mathcal{D}'_0(\mathcal{L}(X_{n+1}, X_{\tilde{A}})).
$$

For any $\varphi \in \mathcal{D}$ and $x \in X$ we define $U(\varphi) := S(\varphi)x = Sx(\varphi)$. Then $U \in \mathcal{D}'_0(X_{A,B})$ is a solution of (δDP):

$$
-B\langle Sx, \varphi' \rangle - A\langle Sx, \varphi \rangle = \langle \delta, \varphi \rangle Bx, \quad x \in X.
$$

For $x \in X_{n+1}$ we have

$$
-\langle Sx, \varphi' \rangle - \hat{A}\langle Sx, \varphi \rangle = \langle \delta, \varphi \rangle x,
$$

and the associativity of convolution implies the uniqueness of the solution on X_{n+1}:

$$
U = (\delta \otimes I) * U = S * \tilde{P} * U = S * (\delta \otimes x) = Sx, \quad x \in X_{n+1}.
$$

We now show that on $\ker R^{n+1}$ the degenerate solution U is also unique. Consider $x \in \ker R^{n+1}$. Since for any $\varphi \in \mathcal{D}_0$, $S(\varphi)x = 0$, then we have that $Sx = 0$ on $(0, \infty)$. Since $S \in \mathcal{D}'_0(\mathcal{L}(X, X_A))$, we have $Sx = 0$ on $(-\infty, 0)$. Thus, $\operatorname{supp} Sx = \{0\}$. By Proposition 2.1.3 we have

$$
Sx = \sum_{i=0}^{p} \delta^{(i)} z_i,
$$

where p is the local order of the distribution Sx. Since Sx is a solution of (2.2.8), we have

$$
\begin{aligned}
\delta^{(p+1)} \otimes Bz_p \;+\;& \delta^{(p)} \otimes (Bz_{p-1} - Az_p) + \ldots \\
+\;& \delta' \otimes (Bz_0 - Az_1) + \delta \otimes (-Bx - Az_0) = 0
\end{aligned}
$$

and $z_p \in \ker B$, $z_{p-1} \in K_1$, ... $z_0 \in K_p$, $-x \in K_{p+1}$. Hence, for $x \in K_i$ $(1 \leq i \leq p+1)$ and for the corresponding the sequence of A-associated vectors $\{x_0, \ldots x_{i-1}\}$, we have

$$Sx = -\sum_{j=0}^{i-1} \delta^{(j)} \otimes x_{i-1-j}. \tag{2.2.11}$$

Due to Corollary 2.2.1, $K_{n+1} = \emptyset$, and hence $p+1 \geq n$. Since A is invertible, x_k in (2.2.11) are defined uniquely. In a similar way, any degenerate on $\ker R^{n+1}$ solution U has the same form and is equal to Sx. In particular, for $x \in K_0 = \ker B$, all x_k in (2.2.11) are equal to zero and $U = 0$. Thus, $U = Sx$ for $x \in \ker R^{n+1}$ and for $x \in X_{n+1}$, hence $U = Sx$ for all $x \in X$.

Let $\|x_j\| \to 0$. Then $S(\varphi) \in \mathcal{L}(X, X_{\tilde{A}})$ for all $\varphi \in \mathcal{D}$, and

$$\|Sx_j(\varphi)\|_{\tilde{A}} \leq \|S(\varphi)x_j\| + \|S'(\varphi)x_j\| + \|\varphi(0)x_j\| \to 0,$$

i.e. $Sx_j \to 0$ in $\mathcal{D}'_0(X_{\tilde{A}})$.

(S) $\Longrightarrow$ (D). Let

$$Q \in \mathcal{D}'_0(\mathcal{L}(X, X_{\tilde{A}})) \cap \mathcal{D}'_0(\mathcal{L}(X_{n+1}, X_{\tilde{A}}))$$

be a solution of (2.2.8) and (2.2.9). For any $F \in \mathcal{D}'_+(X_{n+1})$, $U = Q * F$ is a unique solution of the equation

$$\widetilde{P} * U = F. \tag{2.2.12}$$

Using this fact, we now prove that Q satisfies properties (i) – (ii) from Definition 2.1.5, that Q is regular on X_{n+1}, degenerate on $\ker R^{n+1}$, and that ran Q is dense in X_{n+1}.

Let $\varphi, \psi \in \mathcal{D}_0$. Define $\varphi_1(t) := \varphi(-t)$, $\psi_1(t) := \psi(-t)$, $t \geq 0$. Let $u(\hat{t})$ be a solution of (2.2.12) with $F = \varphi_1 \otimes x$, where $x \in X_{n+1}$:

$$-\hat{A}u + u' = \varphi_1 \otimes x. \tag{2.2.13}$$

Let $w(\hat{t})$ and $v(\hat{t})$ be solutions of (2.2.12) with $F = (\psi_1 * \varphi_1) \otimes x$ and $F = \psi_1 \otimes u(0)$, respectively. Then $u(\cdot) = (Q * \varphi_1)x$ is an infinitely differentiable function with values in $D(\hat{A})$. The same is valid for w and v. These functions have the following initial values

$$u(0) = Q(\varphi)x, \quad w(0) = Q(\varphi * \psi)x,$$
$$v(0) = Q(\psi)u(0) = Q(\psi)Q(\varphi)x, \quad x \in X_{n+1}.$$

To prove (i) one needs to show $v(0) = w(0)$. From (2.2.13) we have

$$-\hat{A}(u * \psi_1) + (u * \psi_1)' = (\varphi_1 * \psi_1) \otimes x,$$

hence $w(t) = u * \varphi_1$ and $w(0) = u(\psi)$. Let $H(\cdot)$ be the Heaviside function, then $H(t)u(t) \in \mathcal{D}_0'(X_{\hat{A}})$. Since $\hat{A}$ does not depend on t, we have

$$-\hat{A}(Hu) + (Hu)' = H(t)(-\hat{A}u + u') + \delta \otimes u(0).$$

Since for any $\varphi \in \mathcal{D}_0$ we have $H(t)(-\hat{A}u + u') = (H\varphi_1) \otimes x = 0$, then we obtain

$$-\hat{A}(Hu) + (Hu)' = \delta \otimes u(0) \tag{2.2.14}$$

and

$$-\hat{A}((Hu) * \psi_1) + ((Hu) * \psi_1)' = \psi_1 \otimes u(0).$$

Therefore, $v = (Hu) * \psi_1$, $v(0) = \int_0^\infty u(t)\psi(t)\,dt$, and for any $\psi \in \mathcal{D}_0$ we have $v(0) = u(\psi) = w(0)$. Thus
$$\forall \varphi, \psi \in \mathcal{D}_0,$$

$$Q(\varphi * \psi)x = Q(\varphi)Q(\psi)x, \quad x \in X_{n+1}.$$

Note that $Qx = Sx$ defined for $x \in \ker R^{n+1}$ by formula (2.2.11), also satisfies relation (i). Thus, we have constructed the distribution semigroup Q degenerate on $\ker R^{n+1}$.

To show that Q is a regular distribution semigroup, we let z be a solution of (2.2.12) with $F = \delta \otimes y$, where $y \in \operatorname{ran} Q$. Since $u(0) = y$, by (2.2.14) we have

$$z(t) = Qy = H(t)u(t)$$

and Qy is a continuous function on $(0, \infty)$.

We now show that $\operatorname{ran} Q$ is dense in X_{n+1}. We have $\hat{A} \in \mathcal{L}(X_{\hat{A}}, X_{n+1})$ and $Q(\varphi) = S(\varphi) \in \mathcal{L}(X_{n+1}, X_{\hat{A}})$. Let X_{n+1}^* and $X_{\hat{A}}^*$ be the dual spaces of X_{n+1} and $X_{\hat{A}}$, then $\hat{A}^* \in \mathcal{L}(X_{n+1}^*, X_{\hat{A}}^*)$. Consider $Q(\varphi)$ on X_{n+1} and define $Q^*(\varphi) = (Q(\varphi))^*$ for $\varphi \in \mathcal{D}_0$. Then

$$Q^*(\varphi) \in \mathcal{L}(X_{\hat{A}}^*, X_{n+1}^*), \quad Q^* \in \mathcal{D}_0'(\mathcal{L}(X_{\hat{A}}^*, X_{n+1}^*)).$$

Since (2.2.8), (2.2.9) hold for Q, we have

$$Q^* * \left(\delta' \otimes I - \delta \otimes \hat{A}^*\right) = \delta \otimes I_{X_{n+1}^*},$$

and

$$\left(\delta' \otimes I - \delta \otimes \hat{A}^*\right) * Q^* = \delta \otimes I_{X_{\hat{A}}^*}.$$

Therefore, $Q^* \in \mathcal{D}_0'(\mathcal{L}(X_{\hat{A}}^*, X_{n+1}^*))$ is a distribution semigroup, that is non-degenerate on $X_{\hat{A}}^*$ as Q is nondegenerate on X_{n+1}. Hence, if $z^* \in X_{\hat{A}}^*$ is such that
$$\forall \varphi \in \mathcal{D}_0, \ x \in X_{n+1},$$

$$\langle Q(\varphi)x, z^* \rangle = \langle x, Q^*(\varphi)z^* \rangle = 0,$$

then

$$Q^*(\varphi)z^* = 0 \implies z^* = 0 \implies \overline{\{Q(\varphi)x\}} = X_{n+1}.$$

Thus, the set

$$\left\{ Q(\varphi)x \mid x \in X_{n+1} \right\}$$

is dense in X_{n+1}. Since $Q(\varphi)x = 0$ for $x \in \ker R^{n+1}$, we have $\overline{\operatorname{ran} Q(\varphi)} = X_{n+1}$.

To complete the proof we note that $\hat{A}$ generates the nondegenerate distribution semigroup Q which is densely defined on X_{n+1}. Since the domain of the generator of a degenerate distribution semigroup belongs to the complement of the set where Q is degenerate, $\hat{A}$ is the generator of the distribution semigroup Q constructed on X.

(D) $\implies$ (S). Let Q be the distribution semigroup from (D). For $\psi \in \mathcal{D}$ and the function $\psi_+ \in \mathcal{E}'_0$ introduced in (Q5) of Proposition 2.2.2, we have

$$\delta' * \psi_+ = \psi'_+ + \psi(0)\delta.$$

Hence, by (Q1) and by the definition of the generator of a distribution semigroup we have
$\forall x \in X_{n+1}, \; \varphi \in \mathcal{D}_0,$

$$
\begin{aligned}
Q(\delta' * \psi_+ * \varphi)x &= Q(\psi'_+ * \varphi)x + \psi(0)Q(\varphi)x \\
&= Q(\delta')Q(\psi_+ * \varphi)x \\
&= -\hat{A}Q(\psi_+ * \varphi)x = -\hat{A}Q(\psi_+)Q(\varphi)x \\
&= Q(\psi_+)Q(\delta' * \varphi)x = Q(\psi_+)(-\hat{A})Q(\varphi)x.
\end{aligned}
$$

Taking into account (Q5), we obtain

$$-\hat{A}Q(\psi)Q(\varphi)x = Q(\psi')Q(\varphi)x + \psi(0)Q(\varphi)x = -Q(\psi)\hat{A}Q(\varphi)x$$

or for $y = Q(\varphi)x$

$$-\hat{A}Q(\psi)y - Q(\psi')y = \psi(0)y, \tag{2.2.15}$$

$$-Q(\psi)\hat{A}y - Q(\psi')y = \psi(0)y. \tag{2.2.16}$$

Since the range of the distribution semigroup is dense in X_{n+1} and the semigroup's generator $\hat{A}$ is closed, we have that $Q(\psi)y \in D(\hat{A})$ for every $y \in X_{n+1}$ and (2.2.15) holds for every $y \in X_{n+1}$. Let $\|y_j\| \to 0$ in (2.2.15), then

$$\|\hat{A}Q(\psi)y_j\| \to 0 \implies Q(\psi) \in \mathcal{L}(X, X_{\hat{A}}).$$

If $\psi_j \to 0$ ($\psi_j \in \mathcal{D}$) in (2.2.16), then $\|Q(\psi'_j)\| \to 0$ and $\|AQ(\psi_j)\| \to 0$. Hence, since $Q \in \mathcal{D}'_0(\mathcal{L}(X, X))$, we have

$$Q \in \mathcal{D}'_0(\mathcal{L}(X_{n+1}, X_{\hat{A}})) \subset \mathcal{D}'_0(\mathcal{L}(X, X_A)).$$

Thus, $S = Q$ on X_{n+1}, and for $x \in X_{n+1}$, S satisfies (2.2.8). Furthermore, $S = Q$ defined by formula (2.2.11) satisfies (2.2.8) for $x \in \mathcal{K}_{n+1}$. Hence S satisfies (2.2.8) for all $x \in X$.

Now we show (2.2.9), i.e. we verify (2.2.16) for $y \in D(\hat{A})$. Let $x \in D(Q(-\delta'))$ and $\varphi_j \to \delta$. Then $Q(\varphi_j)x \to x$ and $Q(\delta' * \varphi_j)x \to -\hat{A}x$. From (2.2.16), for $x \in D(Q(-\delta'))$ we have

$$-Q(\psi)\hat{A}x = Q(\psi')x + \psi(0)x, \quad \psi \in \mathcal{D}. \tag{2.2.17}$$

Since
$$\forall x \in D(\hat{A}), \ \exists x_j \in D(Q(-\delta')):$$

$$x_j \to x \quad \text{and} \quad \hat{A}x_j \to \hat{A}x,$$

then (2.2.17) holds for $x \in D(\hat{A})$. Thus, considering $Sx = Qx$ on X_{n+1}, we have that equation (2.2.9) holds on $D(\hat{A})$. For S defined by (2.2.11), equation (2.2.9) holds on $D(\mathcal{A}) \cap \ker R^{n+1}$. Since $D(\hat{A}) = D(\mathcal{A}) \cap X_{n+1}$, we have (2.2.9) on $D(\mathcal{A})$.

(I) $\Longrightarrow$ (R). Let $\hat{A}$ be the generator of a local k-times integrated semigroup $\{V(t), \ 0 \le t < T\}$, then for any $\tau \in (0, T)$ the operator-function

$$R(\lambda, \tau) = \int_0^\tau \lambda^k e^{-\lambda t} V(t) dt \tag{2.2.18}$$

is defined and bounded on X. Taking into account the following equality for a local k-times integrated semigroup:

$$V(t)x = \left(\frac{t^k}{k!}\right)x + \int_0^t V(s)\hat{A}x\,ds, \qquad x \in D(\hat{A}), \ t \in (0, T),$$

we have

$$R(\lambda, \tau)(\lambda I - \hat{A})x = (I - G(\lambda))x, \quad x \in D(\hat{A}),$$

$$(\lambda I - \hat{A})R(\lambda, \tau)x = (I - G(\lambda))x, \quad x \in X_{n+1},$$

where $G(\lambda)$ is defined by the formula:

$$G(\lambda)x = \lambda^k e^{-\lambda \tau} V(\tau)x + \sum_{i=0}^{k-1} \frac{(\lambda \tau)^i}{i!} e^{-\lambda \tau} x. \tag{2.2.19}$$

It was proved in Theorem 1.2.5 that the estimates

$$\|G(\lambda)\| < \gamma, \quad \left\|(1 - G(\lambda))^{-1}\right\| < \frac{1}{1 - \gamma}$$

hold for $\lambda \in \Lambda$. Therefore, (R) holds for $(\lambda I - \hat{A})^{-1} = R_{\hat{A}}(\lambda)$ with $l = k$.

Now we prove (R) $\Longrightarrow$ (S). To construct S, we consider the operator-function:

$$P_m(t)x := \frac{1}{2\pi i} \int_{\partial \Lambda} \lambda^{-m} e^{\lambda t} R_{\hat{A}}(\lambda) x \, d\lambda, \qquad x \in X_{n+1}. \qquad (2.2.20)$$

On $\partial \Lambda$ the function under the integral sign grows not faster than $|\lambda|^{\frac{\mu}{\tau}+l-m}/\log|\lambda|$. Hence, this integral defines a continuous function on $(-\infty, \tau_m)$ with values in $\mathcal{L}(X_{n+1}, X_{\hat{A}})$, where $\tau_m = \frac{\tau}{l}(m-l-1)$ and $m > l+1$. In this case, by the abstract Cauchy theorem $P_m(t) = 0$ for $t \leq 0$. Now we define the distribution $S := P_m^{(m)}$ on $(-\infty, \tau_m)$. This definition does not depend on $m' > m$: $P_m^{(m)} = P_{m'}^{(m')}$. Furthermore, for any $\varphi \in \mathcal{D}$ with $\operatorname{supp}\varphi \in (-\infty, a)$, $\tau_m = a$ can be chosen by increasing m. In Theorem 2.1.5 it was shown that the defined operator distribution S satisfies (2.2.8) and (2.2.9) on X_{n+1} and $X_{\hat{A}}$, respectively. For $x_0 \in \ker B$ we define $P_m(t)x_0 = 0$. For $x_1 \in K_1$ using formally (2.2.20) and the equality $Bx_1 = Ax_0$, by the abstract Cauchy formula we have

$$\begin{aligned} P_m(t)x_1 &= \frac{1}{2\pi i} \int_{\partial \Lambda} \lambda^{-m} e^{\lambda t} (\lambda B - A)^{-1} B x_1 \, d\lambda \\ &= -\frac{1}{2\pi i} \int_{\partial \Lambda} \lambda^{-m} e^{\lambda t} x_0 \, d\lambda = -\frac{t^{m-1}}{(m-1)!} x_0, \end{aligned}$$

$$\cdots \qquad (2.2.21)$$

$$P_m(t)x_i = -\sum_{k=0}^{i-1} \frac{t^{m-1-k}}{(m-1-k)!} x_{i-1-k}, \qquad x_i \in K_i, \ i \leq n.$$

Then $Sx_i := P_m^{(m)}x_i$ on $\ker R^{n+1}$ coincides with S defined by (2.2.11), and satisfies (2.2.8) and (2.2.9) on $\ker R^{n+1}$ and $\ker R^{n+1} \cap D(\mathcal{A})$, respectively. Thus, S satisfies the equation (2.2.8) on X, and the equation (2.2.9) on $D(\mathcal{A})$.

(S) $\Longrightarrow$ (I). Let

$$S \in \mathcal{D}_0'(\mathcal{L}(X, X_{\hat{A}})) \cap \mathcal{D}_0'(\mathcal{L}(X_{n+1}, X_{\hat{A}}))$$

be the solution of (2.2.8) – (2.2.9) degenerate on $\ker R^{n+1}$. Then the distribution S can be extended to $\mathcal{D}^p[-1, T]$, where p is the order of S. We denote this extension by the same symbol. Let $t \in [0, T]$, consider the function $\psi_{t,p}(s) = \chi(s)\eta_p(t-s) \in \mathcal{D}^p[-1, T]$, where χ and η are the same as (2.1.19):

$$\chi(s) = \begin{cases} 0, & s \leq -1 \\ 1, & s \geq 0 \end{cases}$$

and

$$\eta_p(t) = \begin{cases} t^{p+1}/(p+1)!, & t \geq 0 \\ 0, & t < 0 \end{cases}.$$

Define $V(t) = \langle S, \psi_{t,p} \rangle$. We have that the operator-function $V(\cdot)$ is strongly continuous in t. We now show that $\{V(t),\ t \in [0,T)\}$ is a $(p+2)$-times integrated semigroup. The function $\psi_{t,p+1}$ can be approximated by $\varphi_n \in \mathcal{D}$ so that

$$\|\psi_{t,p+1} - \varphi_n\|_p \to 0 \quad \text{and} \quad \|\psi'_{t,p+1} - \varphi'_n\|_p \to 0.$$

Consider (2.2.8) and (2.2.9) with $\varphi = \varphi_n$. Taking the limit, we have

$$-B\langle S, \psi'_{t,p+1} \rangle x - A\langle S, \psi_{t,p+1} \rangle x = \psi_{t,p+1}(0)Bx$$

for $x \in X$, and

$$-\langle S, \psi'_{t,p+1} \rangle x - \langle S, \psi_{t,p+1} \rangle \tilde{A}x = \psi_{t,p+1}(0)x$$

for $x \in D(\mathcal{A})$. As in the nondegenerate case, we have

$$BV(t)x = \frac{t^{p+2}}{(p+2)!}Bx + A \int_0^t V(s)x\,ds \qquad (2.2.22)$$

for $x \in X$, and

$$V(t)x = \frac{t^{p+2}}{(p+2)!}x + \int_0^t V(s)\tilde{A}x\,ds \qquad (2.2.23)$$

for $x \in D(\mathcal{A})$. By definition of the operators $V(t)$ via S, we have that $V(t)x \in X_{n+1}$ for $x \in X_{n+1}$. Then equations (2.2.22) and (2.2.23) hold on X_{n+1} and $D(\hat{A})$, respectively. Hence, as it was proved in Theorem 2.1.6, V is a nondegenerate $(p+2)$-times integrated semigroup on X_{n+1} with the generator $\hat{A}$.

(I) $\Longleftrightarrow$ (Wk). Let $\hat{A}$ be the generator of a nondegenerate local k-times $(k \geq n)$ integrated semigroup $\{V(t),\ t \in [0,T)\}$ on X_{n+1}. Since the domain $D(\hat{A})$ of its generator is equal to D_{n+1} and is dense in X_{n+1}, then by Proposition 1.2.4, for any $x \in R^{k+1}(\lambda)X_{n+1}$ there exists the unique solution $u(t) = V^{(k)}(t)x$, $t \in [0,T)$ of the local Cauchy problem (CP). Due to the decomposition (2.2.2), $R^{k+1}(\lambda)X_{n+1}$ is equal to $R^{k+1}(\lambda)X = D_{k+1}$.

If for any $x \in D_{k+1}$, $u(\cdot)$ is a unique solution of the local Cauchy problem (CP), then for any $x \in D_{k+1}$, $u(\cdot)$ is the unique solution of (DP). Since $\rho_B(A) \neq \emptyset$, and hence $\rho(\hat{A}) \neq \emptyset$, then by Lemma 1.2.1 we have the stability of the solution:

$$\|u(t)\| \leq C\|x\|_{R^k}, \quad \text{where} \quad \|x\|_{R^k} := \inf_{y: R_{\hat{A}}^k(\lambda)y = x} \|y\|.$$

As in Theorem 1.6.4 we define operators $U(t)x := u(t)$, $t \in [0,T)$ for $x \in D_{k+1}$. We extend them to $[D_{k+1}]_k$, the closure of D_{k+1} in the norm $\|\cdot\|_{R^k}$. Then $U_1(t)$ are defined on $[D_k]_{k-1}, \ldots, U_i(t)$ on $[\mathcal{D}_{k-(i-1)}]_{k-i}$. In particular $V(t) := U_k(t)$ are defined and bounded on $\overline{D(\hat{A})} = X_{n+1}$.

By construction, $V(t)x$ is continuous in $t \in [0, T)$ for any $x \in X_{n+1}$, and satisfies equations (2.2.22) and (2.2.23) (with the semigroup parameter k) for $x \in X_{n+1}$ and $D(\hat{A})$, respectively. It follows from Theorem 1.2.5 that V is a local nondegenerate $(k+1)$-times integrated semigroup on X_{n+1} with the generator $\hat{A}$, which is densely defined in X_{n+1}.

Remark 2.2.1 We note that the condition that A is invertible implies the uniqueness of the solution on $\ker R^{n+1}$. If $\ker A \neq \{0\}$, then x_i in (2.2.11) are defined up to $\ker A$.

Summarizing the discussion of this subsection we note that assuming that X admits the decomposition (2.2.5), we have that the condition (R) is the criterion for the well-posedness of (δDP). Here we used the technique of integrated and distribution semigroups with a generator equal to the single-valued restriction of $\mathcal{A}$ on X_{n+1}.

Now, using the technique of degenerate semigroups with multivalued generators and not considering the detailed structure of semigroups on K_i, we obtain the well-posedness criterion for (δDP) and (δIP) under a less restrictive assumption.

We say that a closed linear multivalued operator $\mathcal{A}$ is the generator of a degenerate local k-times integrated semigroup $\{V(t), \ t \in [0, T)\}$ if

$$V(t)x + \frac{t^k}{k!}x = \int_0^t V(s)\mathcal{A}x\,ds, \quad x \in D(\mathcal{A}) \qquad (2.2.24)$$

and

$$V(t)x + \frac{t^k}{k!}x \in \mathcal{A}\int_0^t V(s)x\,ds, \quad x \in X. \qquad (2.2.25)$$

Theorem 2.2.2 *Let $\mathcal{A}$ be a linear multivalued operator with $\rho(\mathcal{A}) \neq \emptyset$. Suppose that the decomposition*

$$X = X_{n+1} \oplus \mathcal{A}^{n+1}0 \qquad (2.2.26)$$

holds for some $n \in \mathbb{N}$, where $X_{n+1} := \overline{D(\mathcal{A}^{n+1})}$. Then the following statements are equivalent:

($\mathcal{W}$) *the Cauchy problem (δIP) is well-posed, and its solution $U \in \mathcal{D}_0'(X_A)$ is degenerate for $x \in \mathcal{A}^{n+1}0$;*

($\mathcal{R}$) *for the resolvent of $\mathcal{A}$, the polynomial estimate (2.2.10):*

$$\|R_{\mathcal{A}}(\lambda)\| \leq \frac{M\,|\lambda|^l}{\log(1 + |\lambda|)} \leq M_1 |\lambda|^l$$

holds for some $M > 0$, $l \geq 0$, and any

$$\lambda \in \Lambda = \left\{\lambda \in \mathbb{C} \ \middle| \ \operatorname{Re}\lambda > \frac{l}{\tau}\log(1 + |\lambda|) + \frac{1}{\tau}\log\frac{C}{\gamma}\right\},$$

where

$$C > 0, \ 0 < \gamma < 1, \ \tau > 0;$$

(S) *there exists a solution operator distribution* $S \in \mathcal{D}'_0(\mathcal{L}(X, X_A))$ *such that*

$$(\mathcal{P} * S)x \ni \delta \otimes x, \quad x \in X,$$

$$(S * \mathcal{P})x = \delta \otimes x, \quad x \in D(\mathcal{A}),$$

(2.2.27)

where $\mathcal{P} := \delta' \otimes I - \delta \otimes \mathcal{A}$. *In this case*

$$U = Sx, \quad x \in X$$

and S *is degenerate on* $\mathcal{A}^{n+1}0$;

(I) *for any* $T > 0$, *the operator* $\mathcal{A}$ *is the generator of a local* k-*times integrated semigroup* $\{V(t), \ t \in [0, T)\}$ *for some* $k = k(T) \geq n$, *which is degenerate on* $\mathcal{A}^{n+1}0$;

(Wk) *for any* $T > 0$, *the local Cauchy problem* (IP) *with operator* $\mathcal{A}$ *having nonempty resolvent set is* k-*well-posed on* $D(\mathcal{A}^{k+1})$ *for some* $k = k(T) \geq n$.

Proof $(\mathcal{W}) \iff (\mathcal{S})$. As in the nondegenerate case in Section 2.1, this proof is based on the equality that connects $U \in \mathcal{D}'_0(X_A)$, the solution of ($\delta$IP), with the solution operator distribution $S \in \mathcal{D}'_0(\mathcal{L}(X, X_A))$:

$$\langle U, \varphi \rangle = \langle S, \varphi \rangle x = \langle Sx, \varphi \rangle, \quad \varphi \in \mathcal{D}, \ x \in X.$$

$(\mathcal{S}) \iff (\mathcal{I})$. To prove $(\mathcal{I}) \implies (\mathcal{S})$ we define $S := V^{(k)}$. Conversely, as in Theorem 2.2.1, we define

$$V(t)x := \langle S, \psi_{t,p} \rangle x,$$

for $x \in X_{n+1}$. This operator-function $V(\cdot)$ satisfies (2.2.24) and (2.2.25) with $k = p+2$ on X_{n+1} and $D(\mathcal{A})$, respectively. On $\mathcal{A}^{n+1}0$, $V(\cdot)x$ is defined by formula (1.6.21).

$(\mathcal{W}k) \implies (\mathcal{R}) \implies (\mathcal{I}) \implies (\mathcal{W}k)$. Let the local (IP) be k-well-posed on $D(\mathcal{A}^{k+1})$. Then one can verify the operator family

$$\{V(t) := U_k(t), \ t \in [0, T)\}$$

that was constructed in Theorem 1.6.6, satisfies the inclusion (2.2.25) with the semigroup parameter k, for $x \in X$. Hence, due to the assumption

$\rho(\mathcal{A}) \neq \emptyset$, for operators $R(\lambda, \tau)$ and $G(\lambda)$ defined by formulae (2.2.18) and (2.2.19), we have

$$R_{\mathcal{A}}(\lambda) := R(\lambda, \tau)(I - G)^{-1}$$

and the condition $(\mathcal{R})$ holds with $l = k$.

Suppose $(\mathcal{R})$ is fulfilled. Then the operator-function $V_p(\cdot)$ defined by (2.2.20):

$$V_p(t) := \frac{1}{2\pi i} \int_{\partial \Lambda} \lambda^{-p} e^{\lambda t} R_{\mathcal{A}}(\lambda) d\lambda,$$

is continuous in $t \in (-\infty, \tau_l)$, there $\tau_l := \frac{\tau}{l}(p - l - 1)$ and $p > l + 1$. Using the abstract Cauchy theorem we show that V_p satisfies (2.2.24) with $k = p$ on $D(\mathcal{A})$:

$$\int_0^t V_p(s)(\mathcal{A} \pm \lambda I)x \, ds$$

$$= \int_0^t \left(\frac{1}{2\pi i} \int_{\Gamma} \left(\lambda^{-p+1} e^{\lambda s} R_{\mathcal{A}}(\lambda)x - \lambda^{-p} e^{\lambda s} x \right) d\lambda \right) ds$$

$$= V_p(t)x - \frac{t^p}{p!}x, \quad t \in [0, \tau_l)\}, \quad x \in D(\mathcal{A}).$$

Next, applying $\mathcal{A} \pm I$ to $\int_0^t V_p(s)x ds$, $x \in X$, we obtain (2.2.25) with $k = p$. Hence, $\{V_p(t), \ t \in [0, \tau_l)\}$ is a local p-times integrated semigroup on X $(p > l + 1)$.

By Theorem 1.6.6, the Cauchy problem (IP) is p-well-posed on $R^{p+1}X_1$, which is equal to $R^{p+1}X$ since $p \geq n$. $\square$

2.2.3 Well-posedness in the space of exponential distributions

From Theorems 2.2.1 and 2.1.6 we obtain the following necessary and sufficient conditions for the well-posedness of (δDP) in the space $\mathcal{S}'_\omega(X)$.

Theorem 2.2.3 *Let $A, B : X \to Y$ be linear operators, A be closed and invertible, B be bounded. Suppose that the decomposition (2.2.5) holds for some $n \in \mathbb{N}$. Then the following statements are equivalent:*

(W_{exp}) *(δDP) is ω-well-posed. Its solution belongs to $\mathcal{D}'_0(X_{\tilde{A}})$ for $x \in X_{n+1}$ and is degenerate for $x \in \ker R^{n+1}$;*

(S_{exp}) *there exists a solution operator distribution*

$$S \in \mathcal{S}'_\omega(\mathcal{L}(X, X_{\tilde{A}})) \cap \mathcal{S}'_\omega(\mathcal{L}(X_{n+1}, X_{\tilde{A}}))$$

such that

$$(P * S)x = \delta \otimes Bx, \quad x \in X,$$

$$(S * \widetilde{P})x = \delta \otimes x, \quad x \in D(\mathcal{A}),$$

In this case,

$$U = Sx, \quad x \in X$$

and S is degenerate on $\ker R^{n+1}$;

($D_{\exp}$) *there exists a regular exponentially bounded distribution semigroup Q with* $\operatorname{ran} Q$ *dense in X_{n+1}, and degenerate on $\ker R^{n+1}$. Its generator is equal to $\hat{A}$. In this case $S = Q$;*

($I_{\exp}$) *$\hat{A}$ is the generator of an exponentially bounded n-times integrated semigroup $\{V(t),\ t \geq 0\}$ on X_{n+1};*

($R_{\exp}$) *the estimates (2.1.28) hold for the resolvent of $\hat{A}$;*

($E_{\exp}$) *for any $x \in D_{n+1}$ there exists a unique solution of local (CP).*

2.3 Ultradistributions and new distributions

In Section 2.1 we studied the Cauchy problem

$$u'(t) = Au(t), \quad t \geq 0,\ u(0) = x, \tag{CP}$$

where $A : D(A) \subset X \to X$, in spaces of abstract distributions. Recall that the polynomial estimates for the resolvent of A in a certain logarithmic region Λ of the complex plane is the criterion for the well-posedness of (CP) in the sense of distributions. This region and the resolvent estimates are connected with the order of the corresponding distribution semigroup and with the order of the solution operator distribution. Now, we consider the Cauchy problem (CP) with an operator A having the resolvent in a certain region Λ smaller than a logarithmic region from Section 2.1. We investigate the well-posedness of such a (CP) in spaces of ultradistributions. The spaces of ultradistributions are the dual of spaces of infinitely differentiable functions with a locally convex topology. These spaces are more general than the space of Schwartz distributions. We show that the existence of a solution operator ultradistribution is equivalent to exponential estimates for the resolvent of A in Λ. As it was proven in Theorem 1.3.3 the latter fact is equivalent to the existence of a unique solution of the convoluted equation

$$v(t)' = Av(t) + \Theta(t)x, \quad v(0) = 0, \tag{2.3.1}$$

where function Θ is defined by the resolvent of A.

Spaces of test functions for ultradistributions consist of infinitely differentiable functions, and they are defined in terms of estimates on derivatives of these functions. These estimates depend on some numerical sequence M_n. The same sequence allows one to estimate coefficients of differential operators of infinite order, which are called the ultradifferential operators. The spaces of ultradistributions are invariant under ultradifferential operators. Some properties of ultradistributions are considered in the first subsection. In the second subsection, the well-posedness of the (CP) in the spaces of ultradistributions is investigated. The limit case for the (CP) well-posed in the spaces of ultradistributions is the case of the Cauchy problem with an operator A having no regular points in a right semiplane (e.g. the reversed Cauchy problem for the heat equation). It is proved in the third subsection that such a (CP) is well-posed in spaces of new distributions. Spaces of test elements $x \in X$ in the construction of spaces of new distributions are defined in terms of the behaviour of $A^n x$ for any n.

2.3.1   Abstract ultradistributions

Let M_n be a sequence of positive real numbers, satisfying the following conditions:

(M.1) $M_n^2 \le M_{n-1} M_{n+1}, \quad n = 1, 2, \ldots$;

(M.2) $\exists\, \alpha, \beta \in \mathbb{R}$:

$$M_n \le \alpha \beta^n \min_{0 \le s \le n} M_s M_{n-s}, \qquad n = 0, 1, \ldots ;$$

(M.3) $\sum\limits_{s=1}^{\infty} \frac{M_{s-1}}{M_s} \le \infty$.

Let K be a compact set in $\mathbb{R}$ and $h > 0$. We consider a space of functions $\varphi \in C^\infty(\mathbb{R})$ with support K and with the estimates

$$\|\varphi^{(n)}\|_{C(K)} \le C M_n h^n, \qquad n \in \mathbb{N}, \tag{2.3.2}$$

for some constant C. The space $\mathcal{D}_K^{M_n, h}$ of such functions with the norm

$$\|\varphi\|_{M_n, h, K} := \sup_{n \in \mathbb{N}} \frac{\|\varphi^{(n)}\|_{C(K)}}{M_n h^n}$$

is a Banach space.

Definition 2.3.1 *Let $\Omega \subseteq \mathbb{R}$ be an open set. The space*

$$\mathcal{D}^{(M_n)}(\Omega) = \text{ind}\lim_{K \subset \Omega} \text{proj}\lim_{h \to 0} \mathcal{D}_K^{M_n, h}$$

with the topology of inductive limit is called the space of ultradifferentiable
functions with supports in Ω of Beurling type or of class (M_n). The dual
space $\mathcal{D}^{(M_n)'}(\Omega)$ is the space of ultradistributions of class (M_n). If $\Omega = \mathbb{R}$,
then we simply write $\mathcal{D}^{(M_n)}$ and $\mathcal{D}^{(M_n)'}$.

For more details see [135] and [153], where it is also shown that spaces
of Schwartz distributions are subspaces of spaces of ultradistributions.

Property (M.1) is called the logarithmic convexity. Property (M.2) guarantees the invariance of $\mathcal{D}^{(M_n)}(\Omega)$ under differential operators of infinite
order:

$$P(D) = \sum_{n=0}^{\infty} a_n D^n \qquad (2.3.3)$$

with coefficients satisfying the estimates

$$|a_n| \le \frac{C L^n}{M_n}$$

for some $L > 0$ and some $C > 0$. Such operators $P(D)$ are called the
ultradifferential operators of class (M_n). Property (M.3) is called the non-
quasi-analyticity. As in the case of distributions, multiplication by any
function with estimates (2.3.2) can be defined for ultradistributions. We
also note that any ultradifferential operator $P(D)$ is continuous in the space
of ultradistributions.

For every sequence M_n we define the associated function

$$M(\lambda) := \sup_n \ln \frac{\lambda^n M_0}{M_n},$$

which is closely related to Laplace transforms of ultradistributions.

For M_n satisfying (M.1) each of the conditions

$$\int_0^{\infty} \frac{M(\lambda)}{\lambda^2} d\lambda < \infty, \qquad \sum_{n=1}^{\infty} \frac{1}{M_n^{1/n}} < \infty$$

is equivalent to (M.3).

Examples of such sequences are the following Gevrey sequences: $(n!)^s$,
n^{ns} and $\Gamma(1+ns)$ for $s > 1$, where Γ is the Euler function. For such M_n the
corresponding ultradifferential polynomial $P(z)$ has the following estimates

$$e^{(lz)^{1/s}} \le P(z) \le e^{(Lz)^{1/s}}, \quad z \in \mathbb{C},$$

for some $l, L > 0$. The typical example of the ultradifferential polynomial
of class $(n!)^s$ is

$$P(z) := \prod_{n=1}^{\infty} \left(1 + \frac{lz}{n^s} \right),$$

it satisfies the above estimates with $L = 2l$. Owing to the Denjoy-Carleman-Mandelbrojt theorem [135] there are sufficiently many functions in $\mathcal{D}^{(M_n)'}(\Omega)$.

The structure theorem for ultradistributions states that

$$f \in \mathcal{D}^{(M_n)'}(\Omega)$$

if and only if on every relatively compact open set $G \subset \Omega$, f can be written in the form

$$f = \sum_{n=0}^{\infty} D^n f_n$$

with measures f_n on G satisfying the estimate

$$\|f_n\|_{C'(\overline{G})} \le \frac{C L^n}{M_n}, \qquad n \in \mathbb{N},$$

for some constants L and C. Let K be a compact subset of $\mathbb{R}$. We remind here that $C(K)$ is the space of all continuous functions in K equipped with the supremum norm

$$\|u\|_{C(K)} = \sup_{t \in K} |u(t)|.$$

Then $C'(K)$, the dual to $C(K)$, is the space of all finite Borel measures defined on K. For $f \in C'(K)$ define the total variation $|f|$ by

$$|f|(G) = \sup \sum_{j} |f(G_j)|,$$

where G is any f-measurable set and supremum is taken over all possible decompositions of G into a disjoint union of finite number of f-measurable sets G_j. Then $|f| \in C'(K)$ and we can define a norm in $C'(K)$ by

$$\|f\| = |f|(K).$$

The space $C'(K)$ equipped with this norm is a Banach space.

Definition 2.3.2 *Let X be a Banach space. The space $\mathcal{D}_0^{(M_n)'}(X)$ of abstract ultradistributions of class (M_n) is the space $\mathcal{L}(\mathcal{D}^{(M_n)}, X)$ of bounded linear operators with supports in $[0, \infty)$, equipped with the topology of uniform convergence on bounded sets from $\mathcal{D}^{(M_n)}$.*

We now define the convolution of abstract ultradistributions. Let X_1, X_2 and X_3 be Banach spaces. Suppose that multiplication

$$(x_1, x_2) \to x_1 \cdot x_2 \; : \; X_1 \times X_2 \to X_3, \; x_i \in X_i$$

is defined. Consider $T \in \mathcal{D}_0^{(M_n)'}(X_1)$ and $S \in \mathcal{D}_0^{(M_n)'}(X_2)$. Then their convolution

$$(T, S) \to T * S : \; \mathcal{D}_0^{(M_n)'}(X_1) \times \mathcal{D}_0^{(M_n)'}(X_2) \to \mathcal{D}_0^{(M_n)'}(X_3),$$

is defined by

$$\langle T * S, \phi \rangle := \langle S, \widehat{T * \phi} \rangle,$$

where $\phi \in \mathcal{D}^{(M_n)}$ and $(\widehat{T * \phi})(\tau) := (T * \phi)(-\tau)$. This mapping is bilinear and is bounded [135]. In the particular case when $T, S \in \mathcal{D}_0^{(M_n)'}(\mathbb{R})$ and $x_i \in X_i$, the convolution of ultradistributions is defined by

$$(T \otimes x_1) * (S \otimes x_2) = (T * S) \otimes (x_1 \cdot x_2).$$

2.3.2 The Cauchy problem in spaces of abstract ultradistributions

Consider the Cauchy problem

$$u'(t) = Au(t), \quad t \ge 0, \; u(0) = x, \tag{CP}$$

where $A : D(A) \subset X \to X$. Denote

$$Y = [D(A)] := \Big\{ D(A), \; \|x\|_A = \|x\| + \|Ax\| \Big\}.$$

Recall that Y is a Banach space. Furthermore, the operator $A : Y \to X$ belongs to $\mathcal{L}(Y, X)$, the space of bounded linear operators from Y to X. Introduce

$$P := \delta' \otimes I_Y - \delta \otimes A, \tag{2.3.4}$$

then

$$P \in \mathcal{D}_0^{(M_n)'}(\mathcal{L}(Y, X)).$$

Thus, in the space of ultradistributions $\mathcal{D}_0^{(M_n)'}(X)$, the Cauchy problem (CP) can be written in the form

$$P * U = \delta \otimes x,$$

where $x \in X$, $U \in \mathcal{D}_0^{(M_n)'}(Y)$ and $Y = [D(A)]$.

In general, Y does not have to be equal to $[D(A)]$. Let X and Y be Banach spaces such that $Y \subseteq X$ and $A \in \mathcal{L}(Y, X)$. Consider the Cauchy problem

$$P * U = \delta \otimes x, \qquad x \in X, \tag{uδP}$$

where $U \in \mathcal{D}_0^{(M_n)'}(Y)$ and $P \in \mathcal{D}_0^{(M_n)'}(\mathcal{L}(Y, X))$ is defined by (2.3.4). This problem is referred to as the Cauchy problem (CP) in the space of ultradistributions $\mathcal{D}_0^{(M_n)'}(X)$.

Definition 2.3.3 *The Cauchy problem (CP) is said to be well-posed in the sense of ultradistributions (or (uδP) is well-posed) if for any $x \in X$, there exists a unique solution $U \in \mathcal{D}_0^{(M_n)'}(Y)$ such that for any $x_n \to 0$ in X, corresponding solutions of (uδP) $U_n \to 0$ in $\mathcal{D}_0^{(M_n)'}(Y)$.*

The next theorem generalizes Theorems 2.1.3 – 2.1.5.

Theorem 2.3.1 *Let X and Y be Banach spaces, $Y \subseteq X$ and $A \in \mathcal{L}(Y, X)$. Then the following statements are equivalent:*

(I) *the Cauchy problem (CP) is well–posed in the sense of ultradistributions;*

(II) *there exists a solution operator ultradistribution $E \in \mathcal{D}_0^{(M_n)'}(\mathcal{L}(X, Y))$ satisfying*

$$P * E = \delta \otimes I_X, \quad E * P = \delta \otimes I_Y; \qquad (2.3.5)$$

(III) *for any $\gamma > 0$, there exists region*

$$\Lambda := \left\{ \lambda \in \mathbb{C} \mid \operatorname{Re} \lambda \geq \gamma M(\alpha \lambda) + \beta \right\}$$

such that for any $\lambda \in \Lambda$,

$$\|R(\lambda)\|_{\mathcal{L}(X,Y)} \leq C e^{M(\alpha \lambda)}. \qquad (2.3.6)$$

for some constant $C > 0$.

Proof (I) $\implies$ (II). We have that for any $x \in X$ there exists a unique solution $U \in \mathcal{D}_0^{(M_n)'}(Y)$. Define the operator $E(\varphi)$ by

$$E(\varphi)x := U(\varphi), \quad \varphi \in \mathcal{D}^{(M_n)}, \ x \in X.$$

Then $Ex \in \mathcal{D}_0^{(M_n)'}(Y)$, where $Ex(\varphi) = E(\varphi)x$. It follows from the well-posedness of the Cauchy problem that for any sequence of initial values $\{x_n\} \subset X$ such that $x_n \to 0$, the corresponding sequence $\|E(\varphi)x_n\|_Y = \|U_n(\varphi)\|_Y$ converges to 0 uniformly in φ from a bounded set. Therefore, $E(\varphi) \in \mathcal{L}(X, Y)$.

We now show that $E \in \mathcal{D}_0^{(M_n)'}(\mathcal{L}(X, Y))$, i.e., E is a continuous linear operator from $\mathcal{D}^{(M_n)}$ to $\mathcal{L}(X, Y)$:

$\forall \varphi_j \in \mathcal{D}^{(M_n)}$,

$$\varphi_j \to 0 \implies \|E(\varphi_j)\|_{\mathcal{L}(X,Y)} \to 0,$$

and supp $E \subset [0, \infty)$. Consider the set

$$\mathcal{B} = \left\{ Ex \mid \|x\|_X \leq C \right\} \subset \mathcal{D}_0^{(M_n)'}(Y).$$

From the stability of the solution we have

$$\sup_{Ex \in B} \|E(\varphi)x\|_Y = \sup_{\|x\|_X \le C} \|U(\varphi)\|_Y < \infty,$$

therefore B is a bounded set. As in the scalar case, the set B is bounded in $\mathcal{D}_0^{(M_n)'}(Y)$ if and only if for any relatively compact open set $G \subset \Omega$ there exist constants C_1, L such that for any element $Ex \in B$ there exist measures f_n satisfying such that

$$Ex\big|_G = \sum_{n=0}^{\infty} D^n f_n$$

and

$$\|f_n\|_{C'\{\overline{G}, Y\}} \le \frac{C_1 L^n}{M_n},$$

where $C'\{\overline{G}, Y\}$ is the space of functions continuous on $\overline{G}$ with values in Y. Hence, for $Ex \in B$ and any $\varphi_j \to 0$, we have

$$\|E(\varphi_j)x\|_Y$$

$$= \left\| \sum_{n=0}^{\infty} \langle D^n f_n, \varphi_j \rangle \right\|_Y = \left\| \sum_{n=0}^{\infty} \langle f_n, (-1)^n \varphi_j^{(n)} \rangle \right\|_Y$$

$$\le \sum_{n=0}^{\infty} \|f_n\|_{C'\{\overline{G}, Y\}} \|\varphi_j^{(n)}\|_{C(\overline{G})} \le C_1 \sum_{n=0}^{\infty} \frac{L^n h^n \|\varphi_j^{(n)}\|_{C(\overline{G})}}{M_n h^n}.$$

Since

$$\sup_n \frac{\|\varphi_j^{(n)}\|_{C(\overline{G})}}{M_n h^n} = \|\varphi_j\|_{M_n, h, \overline{G}},$$

we obtain

$$\|E(\varphi_j)x\|_Y \le C_1 \|\varphi_j\|_{M_n, h, \overline{G}} \sum_{n=0}^{\infty} L^n h^n.$$

This estimate holds for any $h > 0$, in particular for $h = \frac{1}{2L}$ we have

$$\|E(\varphi_j)x\|_Y \le 2C_1 \|\varphi_j\|_{M_n, \frac{1}{2L}, \overline{G}} \longrightarrow 0 \quad \text{as} \quad \varphi_j \to 0.$$

Thus, for $\varphi_j \to 0$, $E(\varphi_j)x$ tends to zero in Y uniformly in x from a bounded set in X. Hence $E(\varphi) \to 0$ in the norm of the space $\mathcal{L}(X, Y)$.

Now we show that E satisfies equations (2.3.5). We have

$$\langle P * E, \varphi \rangle x = \langle (\delta' \otimes I_Y - \delta \otimes A) * E, \varphi \rangle x$$
$$= \langle E' - AE, \varphi \rangle x = \langle (E' - AE)x, \varphi \rangle.$$

For $E \in \mathcal{D}_0^{(M_n)'}(\mathcal{L}(X,Y))$ we have the equality $E'x = (Ex)'$, therefore

$$
\begin{aligned}
\langle (Ex)' - AEx, \varphi \rangle &= \langle U' - AU, \varphi \rangle = \langle P * U, \varphi \rangle \\
&= \langle \delta \otimes x, \varphi \rangle = \langle \delta \otimes I_X, \varphi \rangle x.
\end{aligned}
$$

The second equation:

$$
\begin{aligned}
(P * E')x &= P * (E'x) = P * U' = U'' \otimes I_X - U' \otimes A \\
&= (P * Ex)' = \delta' \otimes x, \qquad x \in X, \\
(P * E)Ax &= \delta \otimes Ax, \qquad x \in Y, \\
P * (\delta \otimes x) &= \delta' \otimes x - \delta \otimes Ax, \qquad x \in Y.
\end{aligned}
$$

For $x \in Y$ we have the equality $P * (E'x - EAx - \delta \otimes x) = 0$. Hence, $E' - EA = \delta \otimes I_Y$, that is $E * P = \delta \otimes I_Y$.

(II) $\Longrightarrow$ (I). For $E \in \mathcal{D}_0^{(M_n)'}(\mathcal{L}(X,Y))$ we consider $U := Ex$, $x \in X$. Then U belongs to $\mathcal{D}_0^{(M_n)'}(Y)$ and satisfies the equation:

$$
P * U = P * Ex = \delta \otimes x.
$$

Let $\{x_j\} \subset X$ be a sequence of initial data converging to zero. We now show that the corresponding sequence $U_j = Ex_j$ tends to zero in $\mathcal{D}_0^{(M_n)'}(Y)$, i.e., converges uniformly in φ from any bounded set $\mathcal{K}$ in $\mathcal{D}^{(M_n)}$. Let

$$
\mathcal{K} = \left\{ \varphi \in \mathcal{D}^{(M_n)} \;\middle|\; \operatorname{supp} \varphi \in K, \quad \|\varphi\|_{M_n, h, K} < C \right\}
$$

for some compact set $K \subset \mathbb{R}$. Since $E(\varphi)$ is a bounded linear operator, we have

$$
\|U_j(\varphi)\|_Y = \|Ex_j(\varphi)\|_Y \leq \|E(\varphi)\|_{\mathcal{L}(X,Y)} \|x_j\|_X.
$$

By the structure theorem, for any $E \in \mathcal{D}_0^{(M_n)'}(\mathcal{L}(X,Y))$ and any relatively compact open set $G \in \mathbb{R}$ with $\overline{G} = K$, there exist measures $f_n \in C'\{K, \mathcal{L}(X,Y)\}$ such that for some constants $L, C_1 > 0$

$$
E\big|_G = \sum_{n=0}^{\infty} D^n f_n,
$$

where

$$
\|f_n\|_{C'\{K, \mathcal{L}(X,Y)\}} \leq C_1 \frac{L^n}{M_n}.
$$

Using the boundedness of test functions we have

$$
\|E(\varphi)\|_{\mathcal{L}(X,Y)}
$$

$$= \left\| \sum_{n=0}^{\infty} \langle D^n f_n, \varphi \rangle \right\|_{\mathcal{L}(X,Y)} = \left\| \sum_{n=0}^{\infty} \langle f_n, \varphi^{(n)} \rangle \right\|_{\mathcal{L}(X,Y)}$$

$$\leq \sum_{n=0}^{\infty} \left\| \langle f_n, \varphi^{(n)} \rangle \right\|_{\mathcal{L}(X,Y)} \leq \sum_{n=0}^{\infty} \|f_n\|_{C'\{K, \mathcal{L}(X,Y)\}} \cdot \|\varphi^{(n)}\|_{C(K)}$$

$$\leq C_1 \sum_{n=0}^{\infty} \frac{L^n}{M_n} \|\varphi^{(n)}\|_{C(K)} \leq C_1 C \sum_{n=0}^{\infty} L^n h^n.$$

Let $C_1 C \sum_{n=0}^{\infty} L^n h^n = C_2$ for $h = \frac{1}{2L}$, then

$$\|U_j(\varphi)\|_Y \leq \|E(\varphi)\|_{\mathcal{L}(X,Y)} \|x_j\|_X \leq C_2 \|x_j\|_X.$$

Hence $\|U_j(\varphi)\|_Y$ tends to zero uniformly with respect to $\varphi \in \mathcal{K}$.

Now we show the uniqueness of the solution. Let $P * V = \delta \otimes x$. Then

$$E * P * V = E * (\delta \otimes x) = Ex = U \in \mathcal{D}_0^{(M_n)'}(Y).$$

Using the second equation in (2.3.5) we obtain

$$U = E * P * V = (\delta \otimes I_Y) * V = V \in \mathcal{D}_0^{(M_n)'}(Y).$$

(II) $\Longrightarrow$ (III). For $0 < c < c'$ we consider the 'local unity' function ψ in the space $\mathcal{D}^{(M_n)}$:

$$\psi(t) = \begin{cases} 0, & t < -1 \\ 1, & 0 \leq t \leq c \\ 0, & t > c' \end{cases}.$$

For arbitrary $\lambda \in \mathbb{C}$ we denote $\psi_\lambda(t) = e^{-\lambda t} \psi(t)$ and consider the first equation in (2.3.5) for this test function

$$(P * E)(\psi_\lambda) = E'(\psi_\lambda) - AE(\psi_\lambda) = I_X \psi_\lambda(0) = I_X. \tag{2.3.7}$$

From the linearity of $E(\varphi)$ we have

$$E'(\psi_\lambda) = -E([\psi_\lambda(t)]') = -E(-\lambda e^{-\lambda t} \psi(t) + e^{-\lambda t} \psi'(t))$$
$$= \lambda E(\psi_\lambda) - E(e^{-\lambda t} \psi'(t)).$$

Hence, from (12.3.7) we obtain

$$(\lambda I - A)E(\psi_\lambda) = I_X + E(e^{-\lambda t} \psi'(t)). \tag{2.3.8}$$

To show the existence of a resolvent, we estimate the function $e^{-\lambda t} \varphi(t)$ for any test function $\varphi \in \mathcal{D}^{(M_n)}$ and then we estimate the operator-functions $E(e^{-\lambda t} \psi(t))$ and $E(e^{-\lambda t} \psi'(t))$. Let $\operatorname{supp} \varphi \cap \mathbb{R}_+ \subset K = [a, b]$, then

$$D^p(e^{-\lambda t} \varphi(t)) = e^{-\lambda t} \sum_{j=0}^{p} C_j^p (-\lambda)^{p-j} \varphi^{(j)}(t),$$

and

$$\sup_{t \in K} |D^p(e^{-\lambda t}\varphi(t))|$$

$$\leq e^{-a\operatorname{Re}\lambda} \sum_{j=0}^{p} \sup_{t \in K} \frac{|\varphi^{(j)}(t)|}{M_j L^j} C_j^p \frac{|\lambda|^{p-j}}{L^{p-j}} M_j L^p$$

$$\leq \|\varphi\|_{M_n,L,K}\, e^{-a\operatorname{Re}\lambda} \sum_{j=0}^{p} \sup_{p-j} \frac{|\lambda|^{p-j}}{L^{p-j} M_{p-j}} L^p C_j^p M_j M_{p-j}.$$

From the equality $e^{M(\lambda)} = \sup_n \frac{|\lambda|^n}{M_n}$ and the property of logarithmic convexity: $M_j M_{p-j} \leq M_p$, we have

$$\sup_{t \in K} |D^p(e^{-\lambda t}\varphi(t))| \;\leq\; \|\varphi\|_{M_n,L,K}\, e^{M(\lambda/L)-a\operatorname{Re}\lambda} L^p \sum_{j=0}^{p} C_j^p M_j M_{p-j}$$

$$\leq\; \|\varphi\|_{M_n,L,K}\, e^{M(\lambda/L)-a\operatorname{Re}\lambda} L^p 2^p M_p.$$

Hence for any $\varphi \in \mathcal{D}^{(M_n)}$

$$\|e^{-\lambda t}\varphi(t)\|_{M_n,2L,K} \leq \|\varphi\|_{M_n,L,K}\, e^{M(\lambda/L)-a\operatorname{Re}\lambda}. \qquad (2.3.9)$$

From the structure theorem, for $E \in \mathcal{D}_0^{(M_n)'}(\mathcal{L}(X,Y))$ and any relatively compact set $G \subset [0,\infty)$ there exist $f_n \in C'\{\overline{G}, \mathcal{L}(X,Y)\}$ and constants $\varpi, C > 0$ such that

$$E\big|_G = \sum_{n=0}^{\infty} D^n f_n,$$

where

$$\|f_n\|_{C'\{\overline{G}, \mathcal{L}(X,Y)\}} \leq C \frac{\varpi^n}{M_n} \quad \text{and} \quad K = \overline{G}.$$

Hence

$$\left\| E(e^{-\lambda t}\psi(t)) \right\|_{\mathcal{L}(X,Y)} \;=\; \left\| \sum_{n=0}^{\infty} \langle D^n f_n, e^{-\lambda t}\psi(t)\rangle \right\|_{\mathcal{L}(X,Y)}$$

$$\leq\; \sum_{n=0}^{\infty} \|f_n\|_{C'\{\overline{G}, \mathcal{L}(X,Y)\}} \left\| D^n(e^{-\lambda t}\psi(t)) \right\|_{C(K)}$$

$$\leq\; C \sum_{n=0}^{\infty} \frac{\varpi^n \|D^n e^{-\lambda t}\psi(t)\|_{C(K)} L^n}{M_n L^n}$$

$$\leq\; 2C \|e^{-\lambda t}\psi'(t)\|_{M_n,L,K} \sum_{n=0}^{\infty} L^n \varpi^n.$$

Since $K = [0, c']$ for ψ, and $K = [c, c']$ for ψ', from (2.3.9) we obtain for $L = \frac{1}{2\varpi}$:

$$\|E(e^{-\lambda t}\psi'(t))\|_{\mathcal{L}(X,Y)} \leq 2Ce^{M(2\varpi\lambda) - c\operatorname{Re}\lambda}\|\psi'\|_{M_n, \frac{1}{4\varpi}, K},$$

$$\|E(e^{-\lambda t}\psi(t))\|_{\mathcal{L}(X,Y)} \leq 2Ce^{M(2\varpi\lambda)}\|\psi\|_{M_n, \frac{1}{4\varpi}, K}.$$

From (2.3.8) and the equality $2Ce^{M(2\varpi\lambda) - c\operatorname{Re}\lambda} = \delta < 1$ we obtain that the operator $\lambda I - A$ has the inverse on

$$\Lambda := \left\{ \lambda \in \mathbb{C} \mid \operatorname{Re}\lambda \geq 1/c[M(2\varpi\lambda) - \ln(\delta/2C)] =: \gamma M(\alpha\lambda) + \beta \right\},$$

and the resolvent estimate there:

$$\|R_A(\lambda)\| \leq \frac{2C}{1-\delta} e^{M(\alpha\lambda)}\|\psi\|_{M_n, \frac{1}{2\varpi}, K}.$$

Thus
$$\forall \gamma > 0 \; \exists \; \alpha, \beta, C > 0 \text{ such that}$$

$$\forall \lambda \in \Lambda, \qquad \|R_\lambda\| \leq Ce^{M(\alpha\lambda)}.$$

(III) $\implies$ (II). For $\varphi \in \mathcal{D}^{(M_n)}$ we consider the operator

$$E(\varphi) := \frac{1}{2\pi i} \int_{\partial\Lambda} R_A(\lambda)\theta(\lambda)\, d\lambda,$$

where $\theta(\lambda) = \int_{\mathbb{R}} e^{\lambda t}\varphi(t)\, dt$. From (2.3.6) we have

$$\|R_A(\lambda)\| \leq Ce^{\frac{1}{\gamma}(\operatorname{Re}\lambda - \beta)}, \qquad \lambda \in \partial\Lambda. \tag{2.3.10}$$

Let $\operatorname{supp}\varphi \in K = [a, b]$, then

$$
\begin{aligned}
|\theta(\lambda)| &= \left| \frac{(-1)^n}{\lambda^n} \int_{K=[a,b]} e^{(\lambda t)}\varphi^{(n)}(t)\, dt \right| \\
&\leq \sup_n \frac{\|\varphi^{(n)}\|_{C_K}}{M_n h^n} \cdot \frac{M_n h^n}{|\lambda|^n} \int_K e^{(\operatorname{Re}\lambda t)}\, dt \\
&= \|\varphi\|_{M_n, h, K} \frac{M_n h^n}{|\lambda|^n} \frac{e^{(b\operatorname{Re}\lambda)} - e^{(a\operatorname{Re}\lambda)}}{\operatorname{Re}\lambda}, \qquad \forall h > 0.
\end{aligned}
$$

Since this inequality holds for any n, it is valid for

$$\inf_n \frac{M_n h^n}{|\lambda|^n} = e^{-M(\lambda/h)},$$

therefore

$$|\theta(\lambda)| \leq C_1 \|\varphi\|_{M_n,h,K} e^{b\operatorname{Re}\lambda - M(\lambda/h)}. \tag{2.3.11}$$

Since

$$M(\lambda/h) = \frac{\operatorname{Re}\lambda}{\gamma}\left(\frac{1}{\alpha h} - \beta\right)$$

on $\partial\Lambda$, from (2.3.10), (2.3.11) we have

$$\|E(\varphi)\|_{\mathcal{L}(X,Y)}$$
$$\leq \frac{1}{2\pi}\int_{\partial\Lambda} \|R_A(\lambda)\|_{\mathcal{L}(X,Y)}|\theta(\lambda)||d\lambda|$$
$$\leq \frac{1}{2\pi}\int_{\partial\Lambda} CC_1 e^{1/\gamma(\operatorname{Re}\lambda-\beta)+b\operatorname{Re}\lambda-M(\lambda/h)}\|\varphi\|_{M_n,h,K}|d\lambda|$$
$$= \int_{\partial\Lambda} C_2 e^{(\frac{1}{\gamma}+b-\frac{1}{\gamma\alpha h})\operatorname{Re}\lambda}\|\varphi\|_{M_n,h,K}|d\lambda|.$$

For any γ, α, b we can take h such that $\frac{1}{\gamma}+b-\frac{1}{\gamma\alpha h} < 0$, hence, the operator $E(\varphi)$ is defined. We now show that it satisfies (2.3.5). We have

$$\begin{aligned}\langle P * E, \varphi\rangle &= \langle(\delta' \otimes I_Y - \delta \otimes A) * E, \varphi\rangle = \langle E' - AE, \varphi\rangle \\ &= -I_Y\langle E, \varphi'\rangle - A\langle E, \varphi\rangle.\end{aligned}$$

Integrating by parts and taking into account that $R_A(\lambda)$ is the resolvent of A we obtain

$$\begin{aligned}\langle P * E, \varphi\rangle &= -\frac{1}{2\pi i}\int_{\partial\Lambda} R_A(\lambda)\int_{\mathbb{R}} e^{\lambda t}\varphi'(t)\,dt\,d\lambda - AE(\varphi) \\ &= I_Y \frac{1}{2\pi i}\int_{\partial\Lambda}\int_K e^{\lambda t}\varphi(t)\,dt\,d\lambda,\end{aligned}$$

$\operatorname{supp}\psi \subset K$. It follows from (2.3.10) that we can continuously deform contour $\partial\Lambda$ into the imaginary axis

$$\begin{aligned}\langle P * E, \varphi\rangle &= I_Y \frac{1}{2\pi i}\int_{-i\infty}^{+i\infty}\int_K e^{\lambda t}\varphi(t)\,dt\,d\lambda \\ &= I_Y \frac{1}{2\pi i}\int_{-i\infty}^{+i\infty}\int_K \frac{e^{\lambda t}}{\lambda^2}\varphi''(t)\,dt\,d\lambda.\end{aligned}$$

Using the inverse Laplace transform of the identity in the space of ultra-distributions

$$\langle \mathcal{L}^{-1}(1), \varphi\rangle = \left\langle \frac{1}{2\pi i}\int_{-i\infty}^{+i\infty}\frac{e^{\lambda t}}{\lambda^2}\,d\lambda, \varphi''\right\rangle = \langle\delta, \varphi\rangle,$$

we have

$$\langle P * E, \varphi \rangle = \langle \delta \otimes I_Y, \varphi \rangle.$$

Corollary 2.3.1 By Theorem 1.3.3, each statement (I), (II) and (III) of Theorem 2.3.1 is equivalent to the existence of κ-convoluted semigroup, where $\tilde{\kappa}(\lambda) =_{|\lambda| \to \infty} \mathcal{O}(e^{-M(\alpha\lambda)})$.

2.3.3   The Cauchy problem in spaces of new distributions

We continue the study of the generalized well-posedness of the Cauchy problem (CP):

$$u'(t) = Au(t), \quad t \geq 0, \ u(0) = x. \tag{CP}$$

In Section 2.1 we considered the (CP) with an operator A generating an n-times integrated semigroup. Such a (CP) is well-posed in the space of Schwartz distributions and can be referred to as 'slightly' ill-posed problem. In this subsection, we study the generalized well-posedness of 'essentially' ill-posed Cauchy problems, namely problems with operators A generating C-regularized semigroups. We note that Cauchy problems that are well-posed in the space of ultradistributions, i.e., problems with operators A generating κ-convoluted semigroups, can be referred to as intermediate case between 'slightly' and 'essentially' ill-posed Cauchy problems.

In this subsection, we use spaces of new distributions (or new generalized functions) introduced by V.K. Ivanov. The construction of spaces of new distributions is based on the development of Sobolev's and Schwartz's concepts: instead of differentiation operators applied to test functions we take an unbounded operator $\Phi : X \to X$ giving a solution of an ill-posed problem. The test functions $\varphi \in X$ are constructed in terms of their Fourier coefficients in the case Hilbert space X, and as functions from $D(\Phi^k)$ or $\bigcap_{k=1}^{\infty} D(\Phi^k)$ in the general case.

In a Hilbert space H we consider a self-adjoint operator Φ generating an orthonormal basis of eigenvectors $\{\alpha_k\}$ corresponding to the eigenvalues $|\mu_1| \leq |\mu_2| \leq \cdots$. We denote by $\mathcal{H}$ the class of such operators. The domain $D(\Phi)$ of such an operator Φ is dense in H, but generally does not coincide with H. The same holds for $D(\Phi^k)$. We introduce the normed spaces

$$\mathcal{P}_k = \left\{ u \in D(\Phi^k) \subset H, \quad \|u\|_{\Phi^k} = \sum_{i=0}^{k} \|\Phi^i u\|_H \right\}, \quad k = 1, 2, \dots$$

and the countably normed space

$$\mathcal{P}_\infty = \left\{ u \in \bigcap_{k=1}^{\infty} D(\Phi^k) \subset H, \quad \|u\|_{\Phi^k}, \ k \geq 1 \right\}.$$

The convergence of a sequence $u_n \to u$ in $\mathcal{P}_\infty$ means that $\Phi^k u_n \to \Phi^k u$ in H for all $k \geq 1$. Note that $\mathcal{P}_\infty$ is not empty since it contains all eigenvectors of operator Φ.

Spaces of new distributions are defined as the dual spaces $\mathcal{P}'_k$ ($\mathcal{P}'_\infty$). The space $\mathcal{P}_\infty$ is the generalization of the Schwartz space of test functions $\mathcal{E}$. The space $\mathcal{P}'_\infty$ is the generalization of the space $\mathcal{E}'$ of distributions with compact support. Spaces $\mathcal{P}_k$ generalize the Sobolev spaces. If Φ is equal to the differentiation operator defined on $H = L^2(G)$, then $\mathcal{P}_k$ coincides with $W^{k,2}(G) \equiv H^k(G)$, and $\mathcal{P}'_k$ coincides with $H^{-k}(G)$. This is the reason that we denote $\mathcal{P}'_k$ by $\mathcal{P}_{-k}$.

Later on we use the sequences of values of functionals generated by elements $\varphi \in \mathcal{P}_k$ and $f \in \mathcal{P}_{-k}$ on eigenvectors α_n: $\tilde{\varphi}_n = \langle \varphi, \alpha_n \rangle = (\varphi, \alpha_n)$ and $\tilde{f}_n = \langle f, \alpha_n \rangle$. Here $\tilde{\varphi}_n$ and $\tilde{f}_n$ are analogous to the Fourier coefficients. The inverse transforms have the form:

$$\varphi = \sum_{n=1}^{\infty} \tilde{\varphi}_n \alpha_n, \quad f = \sum_{n=1}^{\infty} \tilde{f}_n \alpha_n. \qquad (2.3.12)$$

It follows from the definition of $\mathcal{P}_k$ that the series for $\varphi \in \mathcal{P}_k$ is convergent in $\mathcal{P}_k$. Since for any $\varphi \in \mathcal{P}_k$,

$$\lim_{m \to \infty} \left\langle \sum_{n=1}^{m} \tilde{f}_n \alpha_n, \varphi \right\rangle = \lim_{m \to \infty} \left\langle \sum_{n=1}^{m} \tilde{f}_n \alpha_n, \sum_{n=1}^{\infty} \tilde{\varphi}_n \alpha_n \right\rangle$$

$$= \lim_{m \to \infty} \sum_{n=1}^{m} \tilde{f}_n, \tilde{\varphi} = \lim_{m \to \infty} \left\langle f, \sum_{n=1}^{m} \tilde{\varphi}_n \alpha_n, \right\rangle = \langle f, \varphi \rangle,$$

the series for $f \in \mathcal{P}_{-k}$ is convergent in $\mathcal{P}_{-k}$. This implies that spaces $\mathcal{P}_k$ are dense in H, and H is dense in $\mathcal{P}_{-k}$ in the sense of weak topology.

Now we construct spaces of test functions and new distributions defined in the form of Fourier series, via the behaviour of the coefficients of the series. Such spaces can coincide with those already introduced, $\mathcal{P}_k$ and $\mathcal{P}_\infty$, but both these approaches turn out to be useful for the investigation of generalized well-posedness.

Let $\{\gamma_k\}$ with $|\gamma_1| \leq |\gamma_2| \leq \ldots$, be some given numerical sequence, and let $\{\alpha_n\}$ be an orthonormal system in H. We consider the space of test functions:

$$\mathcal{J}_\infty := \left\{ \varphi \in H \ \Big|\ \forall k \geq 1, \ \sum_{n=1}^{\infty} |\tilde{\varphi}_n|^2 |\gamma_n|^{2k} < \infty \right\}$$

with the countable system of norms

$$\|\varphi\|_{\mathcal{J}_k} = \left(\sum_{n=1}^{\infty} |\tilde{\varphi}_n|^2 |\gamma_n|^{2k} \right)^{1/2}, \qquad (2.3.13)$$

and consider the corresponding space $\mathcal{J}_{-\infty} = \mathcal{J}'_{\infty}$ of new distributions f defined in the form of series (2.3.12) whose coefficients are subject to the conditions

$$\exists k_0 \geq 1 : \quad \sum_{n=1}^{\infty} |\tilde{f}_n|^2 |\gamma_n|^{-2k_0} < \infty.$$

We also define the spaces $\mathcal{J}_k$ and $\mathcal{J}_{-k}$ such that for $\varphi = \sum_{n=1}^{\infty} \tilde{\varphi}_n \alpha_n \in \mathcal{J}_k$ and $f = \sum_{n=1}^{\infty} \tilde{f}_n \alpha_n \in \mathcal{J}_{-k}$ the coefficients $\tilde{\varphi}_n$ and $\tilde{f}_n$ are subject to the conditions

$$\sum_{n=1}^{\infty} |\tilde{\varphi}_n|^2 |\gamma_n|^{2k} < \infty \quad \text{and} \quad \sum_{n=1}^{\infty} |\tilde{f}_n|^2 |\gamma_n|^{-2k} < \infty,$$

and the norm in $\mathcal{J}_k$ is defined by (2.3.13). Then

$$\mathcal{J}_{\infty} = \bigcap_{k=1}^{\infty} \mathcal{J}_k, \quad \mathcal{J}_{-\infty} = \bigcup_{k=1}^{\infty} \mathcal{J}_{-k}.$$

If we take the system of eigenvectors of an operator $\Phi \in \mathcal{H}$ as the orthonormal system α_n, and the sequence of eigenvalues $\{\mu_n\}$ as the sequence $\{\gamma_n\}$, then the spaces $\mathcal{J}_{\infty}$ and $\mathcal{J}_k$ coincide with $\mathcal{P}_{\infty}$ and $\mathcal{P}_k$, respectively. The spaces $\mathcal{J}_k$ with $\{\gamma_n\}$ chosen in this way are used for studying the generalized well-posedness of the problem of evaluating of unbounded operator $\Phi : \quad u = \Phi f$. If there exists $\Phi^{-1} = \Psi$, then this problem is equivalent to the operator equation of the first kind

$$\Psi u = f. \tag{2.3.14}$$

If $\{\gamma_n\}$ is taken to be a sequence of the form $\{F(\lambda_n)\}$, where the λ_n are the eigenvalues of an operator $A \in \mathcal{H}$ and F is an analytic function, then $\mathcal{J}_k$ coincides with the space $\mathcal{P}_k$ generated by the operator $\Phi = F(A)$. Such spaces are used in the study of generalized well-posedness of different boundary value problems for differential-operator equations: if the initial data are taken in $\mathcal{P}_{-k}$ ($\mathcal{J}_{-k}$), $k = 0, 1, \ldots$ (in particular, in $H = \mathcal{P}_0 = \mathcal{J}_0$), then the existence, the uniqueness, and the stability of a solution can be established in the spaces $\mathcal{P}_{-m}$ ($\mathcal{J}_{-m}$) for $m > k$. If initial data are taken in $\mathcal{P}_{-\infty}$ ($\mathcal{J}_{-\infty}$), then the well-posedness can be proved in the same spaces.

Let $\mathcal{H}_+$ be the subclass of operators from $\mathcal{H}$ with eigenvalues μ_k such that $\mu_1 \leq \mu_2 \leq \cdots$. We now investigate the generalized well-posedness of the following problems:

1) the ill-posed Cauchy problem (CP) with an operator $A \in \mathcal{H}_+$ and $\Phi = e^{At}$;

2) the problem of evaluating of unbounded operator Φ written in the form (2.3.14) with $\Psi = \Phi^{-1}$.

In the next chapter we show that these 'essentially' ill-posed Cauchy problems can be solved by 'differential' regularization methods. Writing them in the form (2.3.14) one can use the variational regularization methods.

Theorem 2.3.2 *Suppose that* $\Phi = \Psi^{-1} \in \mathcal{H}$. *Then for any* $f \in \mathcal{P}_{-k}$ $(\mathcal{P}_{-\infty})$, $k = 1, 2, \ldots$, *the solution of the equation*

$$\langle \Psi u, \varphi \rangle = \langle u, \Psi \varphi \rangle = \langle f, \varphi \rangle, \quad \varphi \in \mathcal{P}_k \qquad (2.3.15)$$

exists, is unique, and is stable in the space $\mathcal{P}_{-k-1}$ $(\mathcal{P}_{-\infty})$. *For this solution the following expansion holds:*

$$u = \sum_{n=1}^{\infty} \mu_n \tilde{f}_n \alpha_n. \qquad (2.3.16)$$

Proof We show that for any $f \in \mathcal{P}_{-k}$ the formal solution Φf of (2.3.15) belongs to $\mathcal{P}_{-k-1}$. According to the definition of the space $\mathcal{P}_k$, if $\varphi \in \mathcal{P}_k$, then $\Phi \varphi \in \mathcal{P}_{k-1}$. Since $\langle \Phi f, \varphi \rangle = \langle f, \Phi \varphi \rangle$ for $f \in \mathcal{P}_{-k}$, and the operator Φ maps $\mathcal{P}_{k+1}$ to $\mathcal{P}_k$ and $\mathcal{P}_{-k}$ to $\mathcal{P}_{-k-1}$ for $k = 0, 1, \ldots$, we have $\Phi f \in \mathcal{P}_{-(k+1)}$.

As $\Psi = \Phi^{-1} : \mathcal{P}_k \to \mathcal{P}_{k+1}$, for the unbounded operator Φ we have that $\Phi f =: u$ satisfies the equation
$\forall \varphi \in \mathcal{P}_k$,

$$\langle \Psi u, \varphi \rangle = \langle u, \Psi \varphi \rangle = \langle f, \varphi \rangle, \quad k = 0, 1, \ldots .$$

We now show that the expansion (2.3.16) holds. According to (2.3.15), a function u can be expanded in the series $u = \sum_{n=1}^{\infty} \tilde{u}_n \alpha_n$ in the space $\mathcal{P}_{-k-1}$, where $\tilde{u}_n$ satisfies the equalities

$$\tilde{u}_n = \langle u, \varphi_n \rangle = \langle \Phi f, \varphi_n \rangle = \langle f, \Phi \varphi_n \rangle = \mu_n \langle f, \varphi_n \rangle = \mu_n \tilde{f}_n.$$

This implies (2.3.16). It remains to verify the stability of the solution in the space $\mathcal{P}_{-k-1}$, which is equivalent to the continuity of the operator $\Phi : \mathcal{P}_{-k} \to \mathcal{P}_{-k-1}$. Let $f_n \to f$ in $\mathcal{P}_{-k}$, i.e.

$$\forall \varphi \in \mathcal{P}_k, \quad \lim_{n \to \infty} \langle f_n, \varphi \rangle = \langle f, \varphi \rangle.$$

Then for any $\varphi \in \mathcal{P}_k$,

$$\langle u_n, \varphi \rangle = \langle \Phi f_n, \varphi \rangle = \langle f_n, \Phi \varphi \rangle \to \langle f, \Phi \varphi \rangle = \langle \Phi f, \varphi \rangle = \langle u, \varphi \rangle,$$

i.e. $u_n \to u$ in $\mathcal{P}_{-k-1}$. $\square$

Consider $A \in \mathcal{H}_+$ with eigenvalues μ_n. Let $T < \infty$. By $\mathcal{J}_k^{\exp}$ we denote $\mathcal{J}_k$ with $\{\gamma_n = e^{\mu_n T}\}$. We now establish the well-posedness of the local Cauchy problem

$$u'(t) = Au(t), \quad 0 \le t < T, \ u(0) = x, \qquad (\text{LCP})$$

in these spaces of new distributions.

Theorem 2.3.3 *Suppose that $A \in \mathcal{H}_+$. Then for any $x \in \mathcal{J}_{-k}^{\exp}$ ($\mathcal{J}_{-\infty}^{\exp}$) the solution of* (LCP) *exists, is unique, and is stable in the space $\mathcal{J}_{-k-1}^{\exp}$ ($\mathcal{J}_{-\infty}^{\exp}$):*
$$\forall \varphi \in \mathcal{J}_{k+1}^{\exp},$$

$$\left\langle \frac{du(t)}{dt}, \varphi \right\rangle = \langle Au(t), \varphi \rangle, \qquad \langle u(0), \varphi \rangle = \langle x, \varphi \rangle,$$

here

$$\langle u(t), \varphi \rangle = \left\langle \sum_{n=1}^{\infty} e^{\mu_n t} x_n \alpha_n, \varphi_n \right\rangle = \sum_{n=1}^{\infty} e^{\mu_n t} x_n \varphi_n. \qquad \square$$

The operator A considered in Theorem 2.3.3 generates the local C-regularized semigroup

$$S(t)x = \sum_{n=1}^{\infty} e^{\mu_n(t-T)} x_n \alpha_n, \quad \text{with} \quad Cx := \sum_{n=1}^{\infty} e^{-\mu_n T} x_n \alpha_n,$$

and the space $\mathcal{J}_k^{\exp}$ coincides with $\mathcal{P}_k$, where $\Phi = C^{-1}$. It follows from the construction of $\mathcal{J}_k^{\exp}$ and the definition of C-semigroups that Theorem 2.3.3 can be extended in terms of C-regularized semigroups in the following way.

Theorem 2.3.4 *Suppose that an operator $A \in \mathcal{H}_+$ is the generator of a C-regularized semigroup. Then for any $x \in \mathcal{P}_{-k}$, where $\Phi = C^{-1}$, a solution of* (LCP) *exists, is unique, and is stable in the space $\mathcal{P}_{-k-1}$. In the space $\mathcal{P}_{-\infty}$ the* (LCP) *is well-posed in the Hadamard sense: for any $x \in \mathcal{P}_{-\infty}$, the solution exists, is unique, and is stable in $\mathcal{P}_{-\infty}$.*

The unbounded operators $U(t)$ giving the weak solution of the (LCP), are defined by
$$U(t)x := C^{-1} S(t)x, \quad x \in X$$

They satisfy the equations
$$\forall \varphi \in \mathcal{P}_k, \; k \geq 1,$$

$$\langle U(t+s)x, \varphi \rangle = \langle U(t)U(s)x, \varphi \rangle, \qquad \langle U(0)x, \varphi \rangle = \langle x, \varphi \rangle,$$

i.e., they form a semigroup in the space of new distributions. $\square$

Chapter 3

Regularization Methods

In the first two chapters we investigated the Cauchy problem

$$u'(t) = Au(t), \quad t \geq 0, \ u(0) = x, \qquad \text{(CP)}$$

where $A : D(A) \subset X \to X$, which is not uniformly well-posed. The technique of integrated and C-regularized semigroups presented in Chapter 1 allows one to construct a solution of (LCP) for initial values x from various subsets of $D(A)$ stable in X with respect to x in corresponding graph-norms. The technique of distributions presented in Chapter 2 allows the construction of a generalized solution for any $x \in X$ stable in a space of distributions.

In this chapter, we consider regularizing operators that allow the construction of an approximate solution stable in X for concordant parameters: δ - the error of initial data and the regularization parameter ε. In Section 3.1, we study three main 'differential' regularization methods for an ill-posed Cauchy problem: Lattes-Lions' quasi-reversibility method, the auxiliary boundary conditions (ABC) method, and the method of reduction to a Dirichlet problem (Carasso's method). In Section 3.2 we study the connections between different regularization methods and between regularization methods and the C-regularized semigroup method.

3.1   The ill-posed Cauchy problem

3.1.1   Quasi-reversibility method

Let X be a Banach space. Consider the local Cauchy problem

$$u'(t) = Au(t), \quad 0 \leq t < T, \ u(0) = x, \qquad \text{(LCP)}$$

179

with a densely defined linear operator A. Suppose that spectrum of A lies in the region

$$G_1 = \left\{ \lambda \in \mathbb{C} \ \Big| \ |\arg \lambda| < \beta < \frac{\pi}{4} \right\},$$

and

$$\exists M > 0, \ \forall \lambda \notin G_1,$$

$$\|R_A(\lambda)\| \leq \frac{M}{1 + |\lambda|}. \tag{3.1.1}$$

Denote the class of such operators by $\mathcal{A}_1$. In a general case, under the assumption (3.1.1) the problem (LCP) is ill-posed: its solution does not exist for every initial data from $D(A)$ and it is not stable in any graph-norm. We can say that such problems are 'essentially' ill-posed, in contrast to n-well-posed problems from the first two chapters, which can be referred to as 'slightly' ill-posed problems. 'Essentially' ill-posed problems can be treated with the help of regularization methods.

Suppose, as it is customary in the theory of ill-posed problems, that for some exactly given initial value x there exists a solution of (LCP) $u(\cdot)$ (it is unique due to the estimate (3.1.1)) and the initial value x_δ is given with an error $\delta : \|x_\delta - x\| \leq \delta$. We show that a regularizing operator for (LCP) with $A \in \mathcal{A}_1$ can be constructed by the quasi-reversibility method. The following definition of a regularizing operator for the ill-posed Cauchy problem is given according to the general definition of regularizing operators in the theory of ill-posed problems [129], [258].

Definition 3.1.1 *An operator $\mathbf{R}_{\varepsilon,t} : X \to C\{[0,T], X\}$ is called the regularizing operator for the Cauchy problem (LCP) if the following conditions are fulfilled:*

1) *operator $\mathbf{R}_{\varepsilon,t}$ is defined for all $x \in X$. (Usually $\mathbf{R}_{\varepsilon,t}$ supposed to be bounded).*

2) *there exists a dependence $\varepsilon = \varepsilon(\delta)$ $(\varepsilon(\delta) \longrightarrow_{\delta \to 0} 0)$ such that*

$$\|\mathbf{R}_{\varepsilon(\delta),t} x_\delta - u(t)\| \underset{\delta \to 0}{\longrightarrow} 0.$$

Consider the Cauchy problem corresponding to the quasi-reversibility method

$$u'_{\varepsilon,\delta}(t) = (A - \varepsilon A^2) u_{\varepsilon,\delta}(t) =: A_\varepsilon u_{\varepsilon,\delta}(t), \quad 0 < t \leq T, \tag{3.1.2}$$

$$u_{\varepsilon,\delta}(0) = x_\delta.$$

We now show that there exist the unique generalized solution and a dependence $\varepsilon = \varepsilon(\delta)$ $(\varepsilon(\delta) \longrightarrow_{\delta \to 0} 0)$ such that for any $x_\delta \in X$ the generalized solution of the Cauchy problem (3.1.2) converges to the solution of (LCP) when $\delta \to 0$:

$$\|u_{\varepsilon(\delta),\delta}(t) - u(t)\| \xrightarrow[\delta \to 0]{} 0, \quad 0 < t \leq T.$$

First we prove the well-posedness of the Cauchy problem (3.1.2) and then find its solution.

Theorem 3.1.1 *Let $A \in \mathcal{A}_1$, then the Cauchy problem (3.1.2) is well-posed on $D(A^2)$.*

Proof Consider the identity

$$A_\varepsilon - \lambda I = -\varepsilon \left(A - \frac{1 + \sqrt{1 - 4\lambda\varepsilon}}{2\varepsilon} I \right) \left(A - \frac{1 - \sqrt{1 - 4\lambda\varepsilon}}{2\varepsilon} I \right)$$

$$=: -\varepsilon (A - \mu_1 I)(A - \mu_2 I).$$

From the estimate (3.1.1), for $R_A(\mu_j)$, $j = 1, 2$, we obtain the corresponding estimate for $R_{A_\varepsilon}(\lambda)$ at points $\lambda = \mu_j - \varepsilon\mu_j^2$, $j = 1, 2,$:

$$\|R_{A_\varepsilon}(\lambda)\| \tag{3.1.3}$$

$$\leq \frac{1}{\varepsilon} \|R_A(\mu_1)\| \|R_A(\mu_2)\| \leq \frac{1}{\varepsilon} M^2 (1 + |\mu_1|)^{-1} (1 + |\mu_2|)^{-1}$$

$$= \frac{1}{\varepsilon} M^2 \left(1 + \left| \frac{1 + \sqrt{1 - 4\lambda\varepsilon}}{2\varepsilon} \right| \right)^{-1} \left(1 + \left| \frac{1 - \sqrt{1 - 4\lambda\varepsilon}}{2\varepsilon} \right| \right)^{-1}$$

$$\leq \frac{1}{\varepsilon} M^2 \left(1 + \frac{|\lambda|}{\varepsilon} \right)^{-1} \leq M^2 |\lambda|^{-1}.$$

If μ_1 and μ_2 are lying on the left of the curve

$$\gamma = \left\{ \mu \in \mathbb{C} \mid \mu = t(1 \pm i \tan \beta), \ 0 \leq t < \infty \right\},$$

then the estimate (3.1.3) for $R_{A_\varepsilon}(\lambda)$ holds for λ lying on the right of the curve

$$\gamma_1 = \left\{ \lambda = \mu - \varepsilon\mu^2, \ \mu \in \gamma \mid \right.$$
$$\left. \lambda = t - \varepsilon(1 - \tan^2 \beta)t^2 + i(\pm t \tan \beta \mp 2\varepsilon t^2 \tan \beta) \right\}.$$

In particular, for

$$\operatorname{Re} \lambda > C(\varepsilon) = \max_{0 \leq t < \infty} \left\{ t - \varepsilon(1 - \tan^2 \beta) t^2 \right\} = \frac{1}{4\varepsilon(1 - \tan^2 \beta)}$$

the following condition is fulfilled

$\exists L = M^2 > 0:$

$$\forall\lambda,\quad \operatorname{Re}\lambda > C(\varepsilon) \Longrightarrow \|R(\lambda, A_e)\| \le L|\lambda|^{-1}. \tag{3.1.4}$$

As $C(\varepsilon) \ge 0$, for $\operatorname{Re}\lambda > C(\varepsilon)$ we have $|\lambda - C(\varepsilon)| < |\lambda|$ and

$$\|R_{A_\varepsilon}(\lambda)\| \le L|\lambda - C(\varepsilon)|^{-1}.$$

Therefore, the operator A_ε generates an analytic C_0-semigroup $V_{A_\varepsilon}(\cdot)$, and the Cauchy problem (3.1.2) is well-posed on $D(A_\varepsilon) = D(A^2)$. □

According to Theorem 1.1.2, for $x_\delta \in D(A_\varepsilon^2) = D(A^4)$ the solution of (3.1.2) equal to $V_{A_\varepsilon} x_\delta$ can be written in the integral form:

$$u_{\varepsilon,\delta}(t) \;=\; \text{v.p.} -\frac{1}{2\pi i} \int_{\sigma-i\infty}^{\sigma+i\infty} e^{\lambda t} R_{A_\varepsilon}(\lambda) x_\delta d\lambda, \tag{3.1.5}$$

$$t > 0,\quad \sigma > C(\varepsilon).$$

We substitude the resolvent $R_{A_\varepsilon}(\lambda)$ in the form of the Cauchy-Dunford integral:

$$R_{A_\varepsilon}(\lambda) = \frac{1}{2\pi i} \int_\gamma \frac{R_A(\mu)}{\lambda - (\mu - \varepsilon\mu^2)} d\mu, \quad \gamma = \partial G_1$$

into (3.1.5) with $t = T$. For $x_\delta \in D(A^4)$ we obtain the solution of (3.1.2)

$$u_{\varepsilon,\delta}(T) = -\frac{1}{2\pi i} \int_\gamma e^{(\mu-\varepsilon\mu^2)T} R_A(\mu) x_\delta d\mu. \tag{3.1.6}$$

The integral (3.1.6), giving for $x_\delta \in D(A^4)$ the classical solution of problem (3.1.2) at $t = T$, converges for all $x_\delta \in X$. Note that $\overline{D(A^4)} = X$. Extending (3.1.6) by continuity from $D(A^4)$ to X, we obtain the generalized solution of problem (3.1.2) equal to $V_{A_\varepsilon}(t)x_\delta$. We define the operator $\mathbf{R}_{\varepsilon,T}$ as the operator giving the generalized solution of (3.1.2) at $t = T$:

$$\mathbf{R}_{\varepsilon,T} x_\delta := \; V_{A_\varepsilon}(T) x_\delta = -\frac{1}{2\pi i} \int_\gamma e^{(\mu-\varepsilon\mu^2)T} R_A(\mu) x_\delta d\mu.$$

We now show that for $\mathbf{R}_{\varepsilon,T}$ defined on the whole space X, there exists a dependence

$$\varepsilon = \varepsilon(\delta), \qquad \text{with} \qquad \varepsilon(\delta) \xrightarrow[\delta\to0]{} 0$$

such that

$$\mathbf{R}_{\varepsilon,T} x_\delta \xrightarrow[\delta\to0]{} u(T),$$

i.e. $\mathbf{R}_{\varepsilon,T}$ is the regularizing operator for (LCP) at $t = T$.

Theorem 3.1.2 Let $A \in \mathcal{A}_1$. Suppose that for some x there exists a solution of the Cauchy problem (LCP). Then $\mathbf{R}_{\varepsilon,T}$ is the regularizing operator for (LCP) at $t = T$.

Proof Consider supremum of the difference $\|\mathbf{R}_{\varepsilon,T}x_\delta - u(T)\|$ for all x_δ from δ-neighbourhood of x:

$$
\begin{aligned}
\Delta_Q(A, & u(T), \varepsilon, \delta) \\
&= \Delta = \sup_{\|x - x_\delta\| \le \delta} \|\mathbf{R}_{\varepsilon,T}x_\delta - u(T)\| \\
&\le \|u_\varepsilon(T) - u(T)\| + \sup_{\|x - x_\delta\| \le \delta} \|\mathbf{R}_{\varepsilon,T}x_\delta - u_\varepsilon(T)\| \\
&=: \Delta_1 + \Delta_2,
\end{aligned}
$$

here $u_\varepsilon(T)$ is the generalized solution of problem (3.1.2) with the initial data x: $u_\varepsilon(T) = \mathbf{R}_{\varepsilon,T}x = V_{A_\varepsilon}(T)x$. Since $V_{A_\varepsilon}(t)$ is a C_0-semigroup, then

$$
\|V_{A_\varepsilon}(t)\| \le L e^{C(\varepsilon)t}, \quad t \ge 0. \tag{3.1.7}
$$

Hence we obtain

$$
\begin{aligned}
\Delta_2 &= \sup_{\|x - x_\delta\| \le \delta} \|V_{A_\varepsilon}(T)(x - x_\delta)\| \tag{3.1.8} \\
&\le L\delta e^{C(\varepsilon)T} = L\delta e^{T/4\varepsilon(1 - \tan^2 \beta)}.
\end{aligned}
$$

Now we estimate $\Delta_1 = \|u_\varepsilon(T) - u(T)\|$. First we show that $u_\varepsilon(T) = V_{-A^2}(\varepsilon T)u(T)$. Due to condition (3.1.1) the Cauchy problem

$$
w'(t) = -Aw(t), \quad w(0) = w_0, \tag{3.1.9}
$$

is well-posed. Its solution at $t = T$ for $w = u(T)$ can be written in the form of the contour integral:

$$
x = V_{-A}(T)u(T) = \int_{\gamma'} e^{\lambda T} R_{-A}(\lambda)u(T)d\lambda, \tag{3.1.10}
$$

where contour γ' is lying on the right of contour $-\gamma$ and for all $\lambda \in \gamma'$ with sufficiently large absolute values, $|\arg(-\lambda)| < \pi/4$. This choice of γ' guarantees the convergence of integral in (3.1.10). From equality (3.1.10) and estimate (3.1.3) we have

$$
\begin{aligned}
u_\varepsilon(T) &= V_{A_\varepsilon}(T)x \\
&= -\frac{1}{2\pi i} \int_\gamma e^{(\mu - \varepsilon\mu^2)T} R_A(\mu)x\,d\mu \\
&= -\frac{1}{2\pi i} \int_{-\gamma} e^{-(\mu + \varepsilon\mu^2)T} R_{-A}(\mu)x\,d\mu \\
&= -\frac{1}{2\pi i} \int_{-\gamma} e^{-(\mu + \varepsilon\mu^2)T} R_{-A}(\mu) \\
&\qquad\qquad \times \left[-\frac{1}{2\pi i} \int_{\gamma'} e^{\lambda T} R_{-A}(\lambda)u(T)d\lambda \right] d\mu.
\end{aligned}
$$

Using the resolvent identity, the Cauchy formula, and the analyticity of functions $e^{\lambda T}$, $e^{-(\mu+\varepsilon\mu^2)T}$ in the region $\mathbb{C}\setminus G_1$, we obtain

$$
\begin{aligned}
u_\varepsilon(T) &= -\frac{1}{2\pi i}\int_{-\gamma} e^{-(\mu+\varepsilon\mu^2)T} \\
&\qquad \times \left[-\frac{1}{2\pi i}\int_{\gamma'} e^{\lambda T}\frac{R_{-A}(\mu)-R_{-A}(\lambda)}{\mu-\lambda}u(T)d\lambda\right]d\mu \\
&= -\frac{1}{2\pi i}\int_\gamma e^{-\varepsilon\mu^2 T}R_A(\mu)u(T)d\mu \\
&= -\frac{1}{2\pi i}\int_\gamma e^{-\varepsilon\mu^2 T}[R_A(-\mu)+2\mu R_{A^2}(\mu^2)]u(T)d\mu \\
&= -\frac{1}{2\pi i}\int_\gamma e^{-\varepsilon\mu^2 T}R_{A^2}(\mu^2)u(T)2\mu d\mu \\
&= -\frac{1}{2\pi i}\int_{-\gamma^2} e^{\varepsilon\lambda T}R_{-A^2}(\lambda)u(T)d\lambda,
\end{aligned}
$$

where

$$
-\gamma^2 = \left\{\lambda\in\mathbb{C}\ \middle|\ \lambda=-\mu^2,\ \mu\in\gamma\right\}.
$$

Therefore

$$
u_\varepsilon(T) = V_{-A^2}(\varepsilon T)u(T),
$$

where V_{-A^2} is the analytic C_0-semigroup generated by $-A^2$. The analyticity of V_{-A^2} may be proved similarly to the analyticity of the semigroup V_{A_ε}. Hence, using the properties of integrals of a C_0-semigroup, for $u(T)\in D(A^4)$ we obtain

$$
\begin{aligned}
\Delta_1 &= \|V_{-A^2}(\varepsilon T)u(T)-u(T)\| \\
&= \left\|\int_0^{\varepsilon T} V_{-A^2}(s)A^2 u(T)ds\right\| \le L_1\varepsilon, \qquad (3.1.11)
\end{aligned}
$$

and

$$
\Delta \le L\delta e^{T/4\varepsilon(1-\tan^2\beta)} + L_1\varepsilon. \qquad (3.1.12)
$$

It follows from (3.1.12) that we can choose $\varepsilon = \varepsilon(\delta)$ so that $\Delta\to 0$ as $\delta\to 0$. Hence, if we eliminate the additional condition $u(T)\in D(A^4)$, it will be proved that $\mathbf{R}_{\varepsilon,T}$ is the regularizing operator for problem (LCP) at $t=T$. Note that, in general, the condition $u(T)\in D(A^4)$ used in the estimate (3.1.11) is not fulfilled. If $u(T)\notin D(A^4)$, then due to the density of $D(A^4)$ and the well-posedness of the Cauchy problem (3.1.9), there exists $w_0\in D(A^4)$ in a neighbourhood of $u(T)$ such that for the solution of (3.1.9) with $w_0 = w_1$, the estimate $\|x-w(T)\|\le\delta$ holds. Now if we consider (LCP) with the initial data $w(T)$, then

$$
u(T) = w_0\in D(A^4),\quad \|w(T)-x_\delta\|\le\|w(T)-x\|+\|x-x_\delta\|\le 2\delta,
$$

and the order of the obtained estimate (3.1.12) is unchanged.

3.1.2   Auxiliary bounded conditions (ABC) method

Consider the ill-posed Cauchy problem (LCP) with a densely defined linear operator A such that the spectrum of A lies in the half-strip

$$G_2 = \left\{ \lambda \in \mathbb{C} \ \middle| \ |\operatorname{Im} \lambda| < \alpha < \frac{\pi}{T+\tau}, \quad \operatorname{Re} \lambda > \omega, \ \omega \in \mathbb{R}, \ \tau > 0 \right\},$$

and suppose that for $\lambda \notin G_2$ the resolvent of A is bounded:

$$\exists M > 0: \ \forall \lambda \notin G_2, \qquad \|R_A(\lambda)\| \le M. \tag{3.1.13}$$

Denote the class of such operators by $\mathcal{A}_2$. Suppose for some x there exists a solution of (LCP) $u(\cdot)$. We show that (LCP) with $A \in \mathcal{A}_2$ can be regularized by the ABC method: i.e. there exists a dependence $\varepsilon = \varepsilon(\delta)$ ($\varepsilon(\delta) \to_{\delta \to 0} 0$) such that the generalized solution of the boundary problem

$$\tilde{u}'_{\varepsilon,\delta}(t) = A\tilde{u}_{\varepsilon,\delta}(t), \quad 0 < t < T+\tau, \quad \tau > 0, \tag{3.1.14}$$

$$\tilde{u}_{\varepsilon,\delta}(0) + \varepsilon\tilde{u}_{\varepsilon,\delta}(T+\tau) = x^\delta, \quad \varepsilon > 0,$$

converges to the solution of (LCP):

$$\|\tilde{u}_{\varepsilon,\delta}(t) - u(t)\| \xrightarrow[\delta \to 0]{} 0, \quad 0 \le t < T+\tau.$$

In [128] we investigated the well-posedness of the boundary value problem

$$u'(t) = Au(t), \quad 0 < t < T, \quad \mu u(0) - u(T) = x, \tag{3.1.15}$$

in a Banach space X. It was proved that if a densely defined operator A satisfies the condition
$\exists K > 0$, a region $G \in \mathbb{C}$, and an integer $k \ge -1$:

$$\forall \lambda \in G, \qquad \|R_A(\lambda)\| \le K(1 + |\lambda|)^k,$$

and $\mu \notin \exp(T\overline{G})$, then there exists a unique solution of (3.1.15). The assumptions of this theorem are fulfilled for $A \in \mathcal{A}_2$, $x_\delta \in D(A^2)$, and any positive ε. Therefore, the problem (3.1.14) for $x_\delta \in D(A^2)$ has a unique solution that can be written in the integral form:

$$\tilde{u}(t) = -\frac{1}{2\pi i} \int_\gamma \frac{e^{\lambda t}}{1 + \varepsilon e^{\lambda(T+\tau)}} R_A(\lambda) x_\delta d\lambda, \quad \gamma = \partial G_2. \tag{3.1.16}$$

The integral in (3.1.16) converges for all $x_\delta \in X$ and $0 \le t < T + \tau$. Extending by continuity the equality (3.1.16) to the whole space X, we define a generalized solution of (3.1.14)

$$\widetilde{\mathbf{R}}_{\varepsilon,T} x_\delta := -\frac{1}{2\pi i} \int_\gamma \frac{e^{\lambda T}}{1 + \varepsilon e^{\lambda(T+\tau)}} R_A(\lambda) x_\delta d\lambda, \quad x_\delta \in X,$$

and show that $\widetilde{\mathbf{R}}_{\varepsilon,T}$ is a regularizing operator for (LCP) at $t = T$.

Theorem 3.1.3 *Let $A \in \mathcal{A}_2$. Suppose that for some x there exists a solution of* (LCP) *for $0 \le t < T_1$, where $T_1 > T$. Then $\widetilde{\mathbf{R}}_{\varepsilon,T}$ is the regularizing operator for* (LCP) *at $t = T$.*

Proof Consider $\|\widetilde{\mathbf{R}}_{\varepsilon,T} x_\delta - u(T)\|$. Let

$$\Delta_A(A, u(T), \varepsilon, \delta)$$
$$= \Delta = \sup_{\|x - x_\delta\| \le \delta} \|\widetilde{\mathbf{R}}_{\varepsilon,T} x_\delta - u(T)\|$$
$$\le \|\widetilde{\mathbf{R}}_{\varepsilon,T} x_\delta - u(T)\| + \sup_{\|x - x_\delta\| \le \delta} \|\widetilde{\mathbf{R}}_{\varepsilon,T} u_0 - \widetilde{\mathbf{R}}_{\varepsilon,T} x_\delta\|$$
$$=: \Delta_1 + \Delta_2.$$

Here $\widetilde{\mathbf{R}}_{\varepsilon,T} x = u_\varepsilon(T)$ is a solution of (3.1.14) at $t = T$ with the boundary condition $u_\varepsilon(0) + \varepsilon u_\varepsilon(T + \tau) = x$, $\tau < T_1 - T$. We have

$$\Delta_2 = \Delta_2(T)$$
$$= \sup_{\|x - x_\delta\| \le \delta} \left\| \frac{1}{2\pi i} \int_\gamma \frac{e^{\lambda T}}{1 + \varepsilon e^{\lambda(T+\tau)}} R_A(\lambda)(x - x_\delta) d\lambda \right\|$$
$$\le \frac{\delta}{2\pi} \left(\int_{\gamma_+} \left| \frac{e^{\lambda T}}{1 + \varepsilon e^{\lambda(T+\tau)}} \right| \|R_A(\lambda)\| |d\lambda| \right.$$
$$\left. + \int_{\gamma_-} |e^{\lambda T}| \|R_A(\lambda)\| |d\lambda| \right)$$
$$\le \frac{\delta}{\varepsilon} C_1(T),$$

where

$$\gamma_+ = \{\lambda \in \gamma \mid \operatorname{Re} \lambda > 0\} \quad \text{and} \quad \gamma_- = \{\lambda \in \gamma \mid \operatorname{Re} \lambda \le 0\}.$$

To estimate $\Delta_1 = \Delta_1(T)$ we introduce the function $v_\varepsilon(\cdot) := u_\varepsilon(\cdot) - u(\cdot)$. Then $v_\varepsilon(\cdot)$ is a solution of problem (3.1.14) with the boundary condition

$$v_\varepsilon(0) + \varepsilon v_\varepsilon(T + \tau) = -\varepsilon u(T + \tau).$$

As was shown in Theorem 3.1.2, we can assume without loss of generality that $u(T + \tau) \in D(A^2)$. Therefore,

$$
\begin{aligned}
\Delta_1(T) &= \|v_\varepsilon(T)\| && (3.1.17) \\
&= \left\| \frac{1}{2\pi i} \int_\gamma \frac{\varepsilon e^{\lambda T}}{1 + \varepsilon e^{\lambda(T+\tau)}} R_A(\lambda) u(T + \tau) d\lambda \right\|.
\end{aligned}
$$

We divide γ into three parts:

$$
\begin{aligned}
\gamma_- &= \{\lambda \in \gamma \mid \operatorname{Re}\lambda < 0\}, \\
\gamma_N &= \{\lambda \in \gamma \mid \operatorname{Re}\lambda > N\}, \\
\gamma_0 &= \gamma \setminus (\gamma_- \cup \gamma_N),
\end{aligned}
$$

and obtain estimates of integral (3.1.17) along each part of γ:

$$
\left\| \frac{1}{2\pi i} \int_{\gamma_-} \frac{\varepsilon e^{\lambda T}}{1 + \varepsilon e^{\lambda(T+\tau)}} R_A(\lambda) u(T + \tau) d\lambda \right\| \tag{3.1.18}
$$

$$
\leq \frac{\varepsilon}{2\pi} \int_{\gamma_-} |e^{\lambda T}| \, \|R_A(\lambda)\| \, \|u(T + \tau)\| \, |d\lambda| \leq C_2\varepsilon,
$$

$$
\left\| \frac{1}{2\pi i} \int_{\gamma_N} \frac{\varepsilon e^{\lambda T}}{1 + \varepsilon e^{\lambda(T+\tau)}} R_A(\lambda) u(T + \tau) d\lambda \right\| \tag{3.1.19}
$$

$$
\leq \frac{1}{2\pi} \int_{\gamma_N} |e^{-\lambda\tau}| \, \|R_A(\lambda)\| \, \|u(T + \tau)\| \, |d\lambda| \leq C_3 e^{-N\tau}.
$$

Taking into account that

$$
\max_{0 < \operatorname{Re}\lambda \leq N} \left| \frac{\varepsilon e^{\lambda T}}{1 + \varepsilon e^{\lambda(T+\tau)}} \right| = \frac{T^{\frac{T}{T+\tau}} (\tau\varepsilon)^{\frac{\tau}{T+\tau}}}{T + \tau},
$$

for the integral along γ_0 we obtain

$$
\left\| \frac{1}{2\pi i} \int_{\gamma_0} \frac{\varepsilon e^{\lambda T}}{1 + \varepsilon e^{\lambda(T+\tau)}} R_A(\lambda) u(T + \tau) d\lambda \right\| \tag{3.1.20}
$$

$$
\leq \frac{1}{2\pi} \int_{\gamma_0} \frac{T^{\frac{T}{T+\tau}} (\tau\varepsilon)^{\frac{\tau}{T+\tau}}}{T} \|R_A(\lambda)\| \, \|u(T + \tau)\| \, |d\lambda|
$$

$$
\leq C_4 N \varepsilon^{\frac{\tau}{T+\tau}}.
$$

From (3.1.18) – (3.1.20) we obtain the estimate for $\Delta_1(T)$:

$$
\Delta_1(T) \leq C_2\varepsilon + C_3 e^{-N\tau} + C_4 N \varepsilon^{\frac{\tau}{T+\tau}}.
$$

Choosing $N = \frac{1}{T+\tau} \ln \frac{1}{\varepsilon}$, we have

$$\Delta_1(T) \leq \left(C_2 \varepsilon^{\frac{T}{T+\tau}} + C_3 + C_4 \frac{1}{T+\tau} \ln \frac{1}{\varepsilon} \right) \varepsilon^{\frac{\tau}{T+\tau}}.$$

Finally, we obtain

$$\Delta(T) \leq \left(C_2 \varepsilon^{\frac{T}{T+\tau}} + C_3 + C_4 \frac{1}{T+\tau} \ln \frac{1}{\varepsilon} \right) \varepsilon^{\frac{\tau}{T+\tau}} + C_1(T) \frac{\delta}{\varepsilon}. \qquad (3.1.21)$$

Therefore, we can choose $\varepsilon = \varepsilon(\delta)$ such that $\Delta(T) \to 0$ as $\delta \to 0$, i.e. $\widetilde{\mathbf{R}}_{\varepsilon,T}$ is the regularizing operator of (LCP) at $t = T$.

3.1.3   Carasso's method

Consider the ill-posed Cauchy problem (LCP) with a densely defined linear operator A such that the spectrum of A lies in a half-strip

$$G_3 = \left\{ \lambda \in \mathbb{C} \mid |\operatorname{Im} \lambda| < b, \ \operatorname{Re} \lambda > a \right\},$$

and suppose that for $\lambda \notin G_3$ the resolvent of A is bounded:

$$\exists M > 0 : \ \forall \lambda \notin G_3, \qquad \|R_A(\lambda)\| \leq M.$$

Denote the class of such operators by $\mathcal{A}_3$. Suppose that for some x there exists a solution of (LCP) such that $\|u(T)\| \leq K$ for some $K > 0$, and the initial value x_δ is given with an error δ. We show that (LCP) with $A \in \mathcal{A}_3$ can be regularized by reduction to a Dirichlet problem.

Using the substitution $w(t) = e^{k(T-t)}u(t)$ for some $k > 0$ and differentiating both sides of the equation in (LCP), we obtain the Dirichlet problem

$$w''(t) = (A - kI)^2 w(t), \quad 0 < t < T, \qquad (3.1.22)$$

$$w(0) = e^{kT}x, \quad \|w(T)\| \leq K.$$

We show that if w is the solution of (3.1.22) with boundary conditions

$$w(0) = e^{kT}x_\delta, \quad w(T) = 0,$$

then $w(t)e^{-k(T-t)}$ converges to $u(t)$ as $\delta \to 0$, where $k = k(\delta) = \frac{1}{T} \ln (K/\delta)$. This method of construction of regularizing operator $\mathbf{R}_{k,t}$ by reduction to the Dirichlet problem

$$w_k''(t) = (A - kI)^2 w_k(t), \quad 0 < t < T, \qquad (3.1.23)$$
$$w_k(0) = e^{kT}x_\delta, \quad w_k(T) = 0,$$

we call Carasso's method. Here $\mathbf{R}_{k,t}x_\delta := w_k(t)e^{-k(T-t)}$.

Using the technique of Cauchy-Dunford integrals we obtain a solution of (3.1.23).

Theorem 3.1.4 *Let $A \in \mathcal{A}_3$, then*

$$J_k(t) = -\frac{e^{k(T-t)}}{2\pi i} \int_\gamma \frac{e^{\lambda t}(1 - \varepsilon^{\frac{T-t}{T}} e^{2\lambda(T-t)})}{1 - \varepsilon e^{2\lambda T}} R_A(\lambda)\, x_\delta\, d\lambda, \quad \gamma = \partial G_3,$$

is a twice continuously differentiable solution of the Dirichlet problem (3.1.23). Here $\varepsilon = e^{-2kT}$ and ∂G_3 (that is b and a) are chosen so that $k = \frac{1}{T}\ln(K/\delta) \notin G_3$. $\square$

Now we show that the approximate solution

$$u_{\varepsilon,\delta}(t) \quad := \quad e^{-k(T-t)} J_k(t) \tag{3.1.24}$$

$$= \quad -\frac{1}{2\pi i} \int_\gamma \frac{e^{\lambda t}(1 - \varepsilon^{\frac{T-t}{T}} e^{2\lambda(T-t)})}{1 - \varepsilon e^{2\lambda T}} R_A(\lambda) x_\delta d\lambda$$

converges to the exact solution $u(t)$ as $\delta \to 0$. We write the integral in (3.1.24) as a sum of two integrals:

$$J_k(t)^1 \quad = \quad -\frac{1}{2\pi i} \int_\gamma \frac{e^{\lambda t}}{1 - \varepsilon e^{2\lambda T}} R_A(\lambda) x_\delta d\lambda,$$

$$J_k(t)^2 \quad = \quad \frac{\varepsilon^{\frac{T-t}{T}}}{2\pi i} \int_\gamma \frac{e^{2\lambda(T-t)}}{1 - \varepsilon e^{2\lambda T}} R_A(\lambda) x_\delta d\lambda$$

and choose γ in a way that the lines $\operatorname{Im}\lambda = \frac{\pi i}{T} m$ do not belong to it. Therefore the denominator is not equal to zero as $\varepsilon \to 0$.

Comparing $J_k(t)^1$ and $J_k(t)^2$ with the solution obtained by the ABC method, one can see that $J_k(t)^1$ is a solution of the boundary value problem

$$v'(\tau) = Av(\tau), \quad 0 < \tau < 2T, \quad v(0) - \varepsilon v(2T) = x_\delta, \tag{3.1.25}$$

at $\tau = t$: $J_k(t)^1 = v(t)$. Further, $J_k(t)^2$ is a solution of (3.1.25) at $\tau = 2T - t$ multiplied by a small parameter:

$$J_k(t)^2 = v(2T - t)\varepsilon^{\frac{T-t}{T}}.$$

Thus, using the results for the ABC method we obtain convergence $u_\delta(t)$ to $u(t)$ as $\delta \to 0$ and estimates for $u(t)$ satisfying the condition $\|u(2T)\| \le K$ (or on the well-posedness class $\mathcal{M} = \{u(t) \mid \|u(2T)\| \le K\}$):

$$\|u_\delta(t) - u(t)\| \le \|J_k(t)^1 - u(t)\| + \|J_k(t)^2\| \le C\delta^\gamma, \quad 0 < \gamma < 1,$$

where $\varepsilon = e^{-2kT} = c\delta^2$.

3.2 Regularization and regularized semigroups

The general theory of ill-posed problems usually treats them in the form of the equation of the first kind

$$\Psi u = f, \quad \Psi : U \to F, \tag{3.2.1}$$

where, in general, operator Ψ^{-1} either does not exist or is not bounded. Main regularization methods for (3.2.1) are variational: Ivanov's quasival-ues method, the residual method, and Tikhonov's method. In Section 3.1 we studied regularization methods for the ill-posed Cauchy problem

$$u'(t) = Au(t), \quad 0 \le t < T, \quad u(0) = x, \tag{LCP}$$

with operator $A : D(A) \subset X \to X$ that may have points of spectrum anywhere in the half-plane $\mathrm{Re}\,\lambda > \omega$.

Those methods employ the differential nature of this problem and allow the construction of a regularized solution of (LCP) without reducing it to the form (3.2.1).

In this section, we will compare the obtained regularization methods for (LCP) by the error estimates. We discuss connections between 'differential' methods and variational methods. Furthermore, we establish the connection between regularization methods for the ill-posed Cauchy problem and the C-regularized semigroup method.

3.2.1 Comparison of the ABC and the quasi-reversibility methods

The error estimates (3.1.12) for the regularizing operator $\mathbf{R}_{\varepsilon,T}$ in the quasi-reversibility method:

$$\Delta_Q \le L\delta e^{T/4\varepsilon(1-\tan^2\beta)} + L_1\varepsilon,$$

and the error estimates (3.1.21) for the regularizing operator $\tilde{\mathbf{R}}_{\varepsilon,T}$ in the ABC method:

$$\Delta_A \le \left(C_2 \varepsilon^{\frac{T}{T+\tau}} + C_3 + C_4 \frac{1}{T+\tau} \ln \frac{1}{\varepsilon} \right) \varepsilon^{\frac{\tau}{T+\tau}} + C_1(T)\frac{\delta}{\varepsilon}$$

were obtained in Section 3.1 under the following assumption: operator A belongs to class $\mathcal{A}_1$ or $\mathcal{A}_2$, respectively, and there exists an exact solution $u(T)$ of (LCP) with the property

$$\|A^2 u(T)\| \le r \quad \text{or} \quad \|u(T+\tau)\| \le r, \quad r > 0,$$

respectively. For a comparison of these methods on the basis of the error estimates, we introduce the following correctness classes:

$$\mathcal{M}_1 = \Big\{ u(T) \;\Big|\; \|A^2 u(T)\| \le r \Big\} \quad \text{and} \quad \mathcal{M}_2 = \Big\{ u(T) \;\Big|\; \|u(T+\tau)\| \le r \Big\},$$

and errors of a method on a class of operators $\mathcal{A}$ and on a correctness class $\mathcal{M}$:

$$\Delta(\mathcal{A}, \mathcal{M}, \varepsilon, \delta) = \sup_{u(T) \in \mathcal{M}, A \in \mathcal{A}} \Delta(A, u(T), \varepsilon, \delta).$$

We show that for the quasi-reversibility method there exists a class of operators $\mathcal{A}$, a correctness class $\mathcal{M}$ and a dependence between the error δ and the regularization parameter ε, such that for

$$\Delta_Q(\mathcal{A}, \mathcal{M}, \varepsilon, \delta) = \sup_{u(T) \in \mathcal{M}, A \in \mathcal{A}} \Delta_Q(A, u(T), \varepsilon, \delta),$$

the exact by the order of δ logarithmic estimate holds. Here

$$
\begin{aligned}
\Delta_Q(A, u(T), \varepsilon, \delta) \\
= \quad \Delta_Q &= \sup_{\|x - x_\delta\| \le \delta} \|\mathbf{R}_{\varepsilon, T} x_\delta - u(T)\| \\
\le \quad &\|\mathbf{R}_{\varepsilon, T} x - u(T)\| + \sup_{\|x - x_\delta\| \le \delta} \|\mathbf{R}_{\varepsilon, T} x_\delta - \mathbf{R}_{\varepsilon, T} x\| \\
= \quad &\Delta_{1_Q} + \Delta_{2_Q}.
\end{aligned}
$$

For the error in the ABC method

$$\Delta_A(\mathcal{A}, \mathcal{M}, \varepsilon, \delta) = \sup_{u(T) \in \mathcal{M}, A \in \mathcal{A}} \Delta_A(A, u(T), \varepsilon, \delta),$$

the exact by the order of δ polynomial estimate holds, where

$$
\begin{aligned}
\Delta_A(A, u(T), \varepsilon, \delta) \\
= \quad \Delta_A &= \sup_{\|x - x_\delta\| \le \delta} \|\widetilde{\mathbf{R}}_{\varepsilon, T} x_\delta - u(T)\| \\
\le \quad &\|u_\varepsilon(T) - u(T)\| + \sup_{\|x - x_\delta\| \le \delta} \|\widetilde{\mathbf{R}}_{\varepsilon, T} x_\delta - u_\varepsilon(T)\| \\
= \quad &\Delta_{1_A} + \Delta_{2_A}.
\end{aligned}
$$

Definition 3.2.1 *The error estimate* $\Delta(\mathcal{A}, \mathcal{M}, \varepsilon, \delta) \le \phi(\varepsilon, \delta)$ *is called exact by the order of* ε, δ *if*
$$\exists C > 0, \; A \in \mathcal{A}, \; u(T) \in \mathcal{M} :$$

$$\Delta(A, u(T), \varepsilon, \delta) \ge C\phi(\varepsilon, \delta).$$

Theorem 3.2.1 *For the quasi-reversibility method the exact by the order of ε, δ error estimate*

$$\Delta_Q(\mathcal{A}_1, \mathcal{M}_1, \varepsilon, \delta) \leq \delta L e^{T/4\varepsilon(1-\tan^2 \beta)} + L_1 \varepsilon \qquad (3.2.2)$$

and the exact by the order of δ error estimate

$$\Delta_Q \left(\mathcal{A}_1, \mathcal{M}_1, \frac{C}{|\ln \delta|}, \delta \right) \leq C|\ln \delta|^{-1}$$

hold. For the ABC method the corresponding estimates are

$$\Delta_A(\mathcal{A}_2, \mathcal{M}_2, \varepsilon, \delta) \qquad (3.2.3)$$
$$\leq C_1(T) \frac{\delta}{\varepsilon} + \left(C_2 \varepsilon^{\frac{\tau}{T+\tau}} + C_3 + C_4 \frac{1}{T+\tau} \ln \frac{1}{\varepsilon} \right) \varepsilon^{\frac{\tau}{T+\tau}}$$

and

$$\Delta(\mathcal{A}_2, \mathcal{M}_2, \varepsilon, \delta) \leq C r^\gamma, \quad \gamma < \frac{\tau}{T+2\tau}.$$

Proof Since constants L and L_1 in the estimate (3.1.12) do not depend on $A \in \mathcal{A}_1$ and the solution $u(T)$ is from $\mathcal{M}_1$, the estimate (3.1.12) implies (3.2.2). Similarly, (3.1.21) implies (3.2.3).

Now we show that (3.2.2) is exact by the order of ε, δ. From definitions of Δ (both Δ_Q and Δ_A) and Δ_1, Δ_2 we have the inequalities

$$\Delta_1 \leq \Delta, \quad \Delta_2 - \Delta_1 \leq \Delta, \quad \Delta \leq \frac{1}{2}\Delta_2. \qquad (3.2.4)$$

Hence, it is sufficient to prove that the estimates for Δ_1 and Δ_2 are exact by the order of ε, δ. We show that if the spectrum of operator $A \in \mathcal{A}_1$ contains sufficiently large (by the absolute value) points, then the error estimate (3.1.7) for the quasi-reversibility method:

$$\Delta_2 = \sup_{\|x-x_\delta\| \leq \delta} \|V_{A_\varepsilon}(T)(x - x_\delta)\| \leq L\delta e^{T/4\varepsilon(1-\tan^2 \beta)}$$

is exact by the order of ε, δ. An example of such an operator is

$$A = -\frac{d^2}{dx^2}, \quad D(A) = \left\{ u \in C[0,l] \mid u''_{xx} \in C[0,l], \ u(0) = u(l) = 0 \right\},$$

Let λ_n be the sequence of eigenvalues of operator A corresponding to the eigenvectors α_n such that

$$\lim_{n \to \infty} \operatorname{Re} \lambda_n = +\infty \quad \text{and} \quad \lim_{n \to \infty} \left| \frac{\operatorname{Im} \lambda_n}{\operatorname{Re} \lambda_n} \right| < 1.$$

Then, for sufficiently large n, we have $|\arg \lambda_n| < \frac{\pi}{4}$, the sequence $\mu_n = \lambda_n - \varepsilon \lambda_n^2$ is the sequence of eigenvalues for operator $A_\varepsilon = A - \varepsilon A^2$, $\varepsilon > 0$, and

$$V_{A_\varepsilon}(T)\alpha_n = e^{\mu_n T}\alpha_n = e^{(\lambda_n - \varepsilon \lambda_n^2)T}\alpha_n.$$

If we take $\varepsilon(n) = 1/(2\mathrm{Re}\,\lambda_n)$, then

$$
\begin{aligned}
\|V_{A_\varepsilon}(T)\alpha_n\| &= \left|e^{(\lambda_n - \varepsilon(n)\lambda_n^2)T}\right|\|\alpha_n\| = e^{T\mathrm{Re}\,(\lambda_n - \varepsilon(n)\lambda_n^2)}\|\alpha_n\| \\
&= e^{T\mathrm{Re}\,\lambda_n - T\varepsilon(n)[(\mathrm{Re}\,\lambda_n)^2 - (\mathrm{Im}\,\lambda_n)^2]}\|\alpha_n\| \\
&= e^{\frac{T}{2}\mathrm{Re}\,\lambda_n + T(\mathrm{Im}\,\lambda_n)^2}\|\alpha_n\| \geq e^{\frac{T}{4\varepsilon(n)}}\|\alpha_n\|.
\end{aligned}
$$

Therefore, the estimate for Δ_2 is exact by the order of ε. Now consider the estimate (3.1.10)

$$\Delta_1 = \|V_{-A^2}(\varepsilon T)u(T) - u(T)\| = \left\|\int_0^{\varepsilon T} V_{-A^2}(s)A^2 u(T)ds\right\| \leq L_1\varepsilon.$$

Since $V_{-A^2}(t)$ strongly converges to the identity operator as $t \to 0$, there exist $C > 0$, $A \in \mathcal{A}_1$ and $u(T)$ with $A^2 u(T) \neq 0$ such that

$$\Delta_1 = \left\|\int_0^{\varepsilon T} V_{-A^2}(s)A^2 u(T)ds\right\| \geq C\varepsilon. \qquad (3.2.5)$$

Hence, the estimate for Δ_1 is exact by the order of ε. From (3.2.4) and the obtained estimates for Δ_1, Δ_2 with the dependence $\varepsilon = C/|\ln\delta|$ we obtain the exact by the order of δ estimate

$$\Delta_Q\left(\mathcal{A}_1, \mathcal{M}_1, \frac{C}{|\ln\delta|}, \delta\right) \leq C|\ln\delta|^{-1}.$$

For the ABC method from the estimate (3.1.21) with the dependence $\varepsilon = \delta^{\frac{T+\tau}{T+2\tau}}$ we obtain the polynomial estimate:

$$
\begin{aligned}
\Delta(\mathcal{A}_2, \mathcal{M}_2, \varepsilon, \delta) & \\
&\leq (C_1 + C_3)\delta^{\frac{\tau}{T+2\tau}} + C_2\delta^{\frac{T+\tau}{T+2\tau}} + \frac{C_4}{T+2\tau}|\ln\delta|\delta^{\frac{\tau}{T+2\tau}} \\
&\leq C\delta^\gamma, \qquad \gamma < \frac{\tau}{T+2\tau}. \quad \square
\end{aligned}
$$

Thus it is proved that the obtained estimates for the quasi-reversibility and ABC methods hold on the class of operators $\mathcal{A}_1$ and $\mathcal{A}_2$, respectively. To compare these methods we obtain for the quasi-reversibility method an error estimate on $\mathcal{A}_1 \cap \mathcal{A}_2$, $\mathcal{M}_1 \cap \mathcal{M}_2$.

Theorem 3.2.2 *The error estimate (3.2.12) for the quasi-reversibility method is exact by the order of ε, δ on the class of operators $\mathcal{A}_1 \cap \mathcal{A}_2$ and the correctness class $\mathcal{M}_1 \cap \mathcal{M}_2 = \mathcal{M}_2$.*

Proof In the proof of the estimate (3.1.2) (Theorem 3.2.1) for an operator $A \in \mathcal{A}_1$ we can take $\lambda_n \in \mathbb{R}$. Then A belongs to the class $\mathcal{A}_2$ as well, and the estimate of Δ_2 is exact by the order of ε, δ on this class. Hence, Δ_2 is exact by the order ε, δ on the class of operators $\mathcal{A}_1 \cap \mathcal{A}_2$. Since in $\mathcal{M}_2$ there exists $u(T)$ such that $A^2 u(T) \neq 0$, the proof that the estimate for Δ_1 is exact by the order ε, δ is valid on the correctness class $\mathcal{M}_1 \cap \mathcal{M}_2$. The equality $\mathcal{M}_1 \cap \mathcal{M}_2 = \mathcal{M}_2$ on the class of operators $\mathcal{A}_1 \cap \mathcal{A}_2$ follows from boundedness of operator $A^2 V_{-A}(\tau)$ for $\tau > 0$:

$$\|A^2 u(T)\| = \|A^2 V_{-A}(\tau) u(T + \tau)\| \leq r\|A^2 V_{-A}(\tau)\|,$$

where constant r is from the definition of $\mathcal{M}_2$. Thus, the estimate (3.2.12) on $\mathcal{A}_1 \cap \mathcal{A}_2$, $\mathcal{M}_2$ is exact by the order of ε, δ. $\square$

We note here that if $A \in \mathcal{A}_1 \cap \mathcal{A}_2$ and $u(T) \in \mathcal{M}_2$ it is preferable to use the ABC method.

3.2.2   'Differential' and variational methods of regularization

The main variational methods for the regularization of the ill-posed problem (3.2.1)
$$\Psi u = f, \quad \Psi : U \to F,$$

are Ivanov's quasivalues method, Tikhonov's regularization method and the residual method. As an approximate solution of (3.2.1) one takes a solution $u \in U$ of one of the following variational problems:

$$\inf \left\{ \|\Psi u - f\| \; \middle| \; \|\Phi u\| \leq r \right\} \qquad \text{(Ivanov's quasivalues method)}$$

$$\inf \left\{ \|\Psi u - f\| + \varepsilon\|\Phi u\|, \quad \varepsilon > 0 \right\} \qquad \text{(Tikhonov's regularization)}$$

$$\inf \left\{ \|\Phi u\| \; \middle| \; \|\Psi u - f\| :\leq \delta \right\} \qquad \text{(residual method)}$$

It is proved in [129] that if $\Psi : U \to F$, $\Phi : U \to V$ are closed linear operators, $\overline{D(\Psi)} = U$, and U, F are reflexive spaces, V is a Banach space, then these methods are equivalent. That is there exists a relation between positive parameters ε, δ, and r such that solutions $u(r), u(\varepsilon)$ and $u(\delta)$ of these variational problems coincide.

Now, using the technique of Hilbert spaces, we establish that there exists a connection between the regularization parameters such that the

ABC method for ill-posed (LCP) coincides with the quasivalues method for the corresponding equation (3.2.1).

By $\mathcal{H}_+$ we denote the class of self-adjoint operators whose eigenvectors $\{\alpha_k\}$, corresponding to eigenvalues $-\infty < \lambda_1 \leq \lambda_2 \leq \ldots$, form an orthonormal basis. If $A \in \mathcal{H}_+$, then $A \in \mathcal{A}_1 \cap \mathcal{A}_2$. Consider the ill-posed Cauchy problem (LCP) with operator $A \in \mathcal{H}_+$. For (LCP) with such operators the error estimates are obtained for both methods on correctness classes $\mathcal{M}_1$ and $\mathcal{M}_2$. If there exists a solution of (LCP), then it can be written in the form of the Fourier series

$$u(t) = \sum_{k=1}^{\infty} e^{\lambda_k t} x_k \alpha_k, \quad x_k = (x, \alpha_k), \ 0 \leq t \leq T. \tag{3.2.6}$$

The solution operators $\mathbf{R}_{\varepsilon,T}$ and $\tilde{\mathbf{R}}_{\varepsilon,T}$ at $t = T$, corresponding to problems obtained by the quasi-reversibility method (3.1.2):

$$u'_{\varepsilon,\delta}(t) = (A - \varepsilon A^2) u_{\varepsilon,\delta}(t) =: A_\varepsilon u_{\varepsilon,\delta}(t), \quad 0 < t \leq T, \ \varepsilon > 0,$$
$$u_{\varepsilon,\delta}(0) = x_\delta,$$

and by the ABC method (3.1.14):

$$\tilde{u}'_{\varepsilon,\delta}(t) = A\tilde{u}_{\varepsilon,\delta}(t), \quad 0 < t \leq T + \tau, \ \tau > 0, \ \varepsilon > 0,$$
$$\tilde{u}_{\varepsilon,\delta}(0) + \varepsilon \tilde{u}_{\varepsilon,\delta}(T + \tau) = x^\delta,$$

also can be written in the form of series:

$$\mathbf{R}_{\varepsilon,T} x_\delta = u_{\varepsilon,\delta}(T) = \sum_{k=1}^{\infty} e^{(\lambda_k - \varepsilon \lambda_k^2)T} (x_\delta)_k \alpha_k, \quad (x_\delta)_k = (x_\delta, \alpha_k),$$

$$\tilde{\mathbf{R}}_{\varepsilon,T} x_\delta = \tilde{u}_{\varepsilon,\delta}(T) = \sum_{k=1}^{\infty} \frac{e^{\lambda_k T}}{1 + \varepsilon e^{\lambda_k(T+\tau)}} (x_\delta)_k \alpha_k. \tag{3.2.7}$$

On the other hand, the assumption $A \in \mathcal{H}_+$ guarantees that $-A$ generates a C_0-semigroup $\{U_{-A}(t), \ t \geq 0\}$ and the Cauchy problem

$$v'(t) = -Av(t), \quad 0 \leq t \leq T, \quad v(0) = \tilde{x}, \tag{3.2.8}$$

is uniformly well-posed. Its solution is $v(\cdot) = U_{-A}(\cdot)\tilde{x}$. Problem (3.2.8) is equivalent to the inverse Cauchy problem

$$\tilde{v}(t) = A\tilde{v}(t), \quad 0 \leq t < T, \quad v(T) = \tilde{x}, \quad \tilde{v}(t) = v(T - t). \tag{3.2.9}$$

Writing down the solution of (3.2.9) at $t = T$, with $\tilde{x} = u(T)$, we reduce the ill-posed Cauchy problem to the operator equation of the first kind (3.2.1):

$$\Psi u := \sum_{k=1}^{\infty} u_k e^{-\lambda_k T} \alpha_k = f, \tag{3.2.10}$$

where $u = u(T)$, $u_k = (u(T), \alpha_k)$, $f = x = u(0) = \tilde{v}(0)$, and Ψ is a compact linear operator on H. Regularization of (3.2.10) by the quasivalues method leads us to the following optimization problem

$$\inf_{u \in \mathcal{M}} \|\Psi u - f_\delta\|, \quad f_\delta = x_\delta$$

on the compact set

$$\mathcal{M} :=$$
$$\left\{ u = \sum_{k=1}^{\infty} u_k \alpha_k \in H, \ u_k = (u, \alpha_k) \ \Big| \ \sum_{k=1}^{\infty} \gamma_k^2 u_k^2 \le r^2, \ \lim_{k \to \infty} \gamma_k = \infty \right\}.$$

For $\gamma_k = \lambda_k^2$, set $\mathcal{M}$ coincides with the correctness class $\mathcal{M}_1$, for $\gamma_k = e^{\lambda_k(T+\tau)}$ with $\mathcal{M}_2$. Using the Lagrange multipliers method:

$$\inf \left[\sum_{k=1}^{\infty} (u_k e^{-\lambda_k T} - (f_\delta)_k)^2 + \lambda \left(r^2 - \sum_{k=1}^{\infty} \gamma_k^2 (u_k^T)^2 \right) \right],$$
$$(f_\delta)_k = (x_\delta, \alpha_k),$$

we obtain the solution of the optimization problem

$$\sum_{k=1}^{\infty} u_k \alpha_k = \sum_{k=1}^{\infty} \frac{e^{\lambda_k T}}{1 + \lambda \gamma_k^2 e^{2\lambda_k T}} (f_\delta)_k \alpha_k =: u_{\lambda, \delta}(T). \quad (3.2.11)$$

Comparing (3.2.11) and (3.2.7) we see that if $\varepsilon = \lambda$ and $\gamma_k = e^{\lambda_k T}, \tau = 3T$, then the quasivalues method coincides with the ABC method.

3.2.3 Regularizing operators and local C-regularized semigroups

In this subsection, we establish a connection between the existence of regularizing operators for (LCP) and the existence of a C-regularized semigroups with the generator A and operator $C = C_\varepsilon$ depending on the regularization parameter ε.

Theorem 3.2.3 *Suppose that operator $-A$ generates a C_0-semigroup on a Banach space X. Then the following statements are equivalent:*

(I) *A is the generator of a local C_ε-regularized semigroup $\{S(t), \ 0 \le t < T\}$ with operator C_ε strongly convergent to the identity operator as $\varepsilon \to 0$;*

(II) *for (LCP) there exists a bounded linear regularizing operator $\mathbf{R}_{\varepsilon,t}$, $0 < t < T$. It is invertible and commutes with A.*

Proof (II) $\Longrightarrow$ (I). By $U(t)$ we denote the (unbounded in general) solution operator of (LCP), $U(t)x := u(t)$. The assumption that $-A$ generates a C_0-semigroup $U_{-A}(\cdot)$ guarantees the uniform well-posedness of the Cauchy problem (3.2.8) and the uniform well-posedness of the equivalent to (3.2.8) inverse Cauchy problem (3.2.9). Since the problem (3.2.8) is well-posed, there exists an inverse operator to $U_{-A}(t)$ and we have

$$[U_{-A}(t)]^{-1} = U(t).$$

Commutativity of $\mathbf{R}_{\varepsilon,t}$ with A implies commutativity of $\mathbf{R}_{\varepsilon,t}$ with $U_{-A}(s)$ for $s \geq 0$. We define

$$
\begin{aligned}
S(t) &:= \mathbf{R}_{\varepsilon,\tau} U_{-A}(\tau - t)x = \mathbf{R}_{\varepsilon,\tau} U(t) U_{-A}(t) U_{-A}(\tau - t) \\
&= \mathbf{R}_{\varepsilon,\tau} U(t) U_{-A}(\tau), \qquad t \leq \tau < T,
\end{aligned}
$$

and

$$C_\varepsilon = S(0) = \mathbf{R}_{\varepsilon,\tau} U_{-A}(\tau).$$

We claim that $S(t)$ satisfies relations (LC1) – (LC2) from Definition 1.4.5. The equality

$$S(t + s)C_\varepsilon = S(t)S(s), \quad 0 \leq t, s, t + s < T,$$

follows from the commutativity of $\mathbf{R}_{\varepsilon,\tau}$ with $U_{-A}(t)$ and the semigroup property of $U_{-A}(t)$:

$$
\begin{aligned}
S(t + s)C_\varepsilon &= \mathbf{R}_{\varepsilon,\tau} U_{-A}(\tau - t - s)\mathbf{R}_{\varepsilon,\tau} U_{-A}(\tau) \\
&= \mathbf{R}_{\varepsilon,\tau} U_{-A}(\tau - t)\mathbf{R}_{\varepsilon,\tau} U_{-A}(\tau - s) \\
&= S(t)S(s).
\end{aligned}
$$

The operator $\mathbf{R}_{\varepsilon,\tau}$ is defined on the whole space X and is bounded. Furthermore, $U_{-A}(\tau - t)$ is strongly continuous in t, where $\tau - t \geq 0$. Therefore, $S(t)x = \mathbf{R}_{\varepsilon,\tau} U_{-A}(\tau - t)x$ is continuous in t for $0 \leq t < \tau$ and for all $x \in X$. Thus $\{S(t), 0 \leq t < T\}$ is a local C_ε-regularized semigroup.

Now we show that operator C_ε converges strongly to the identity operator as $\varepsilon \to 0$. For $x \in D(A)$ we have

$$
\begin{aligned}
\|C_\varepsilon x - x\| &= \|\mathbf{R}_{\varepsilon,\tau} U_{-A}(\tau)x - U(\tau)U_{-A}(\tau)x\| \\
&= \|\mathbf{R}_{\varepsilon,\tau} y - U(\tau)y\|, \qquad\qquad (3.2.12)
\end{aligned}
$$

where $y = U_{-A}(\tau)x$ is the solution of (3.2.8) at $t = \tau$, corresponding to the initial value x. Formula (3.2.12) and condition 2) from Definition 3.1.1 imply that $\|C_\varepsilon x - x\| \to_{\varepsilon \to 0} 0$ for $x \in D(A)$. Taking into account density of a generator of a C_0-semigroup, we have $\overline{D(-A)} = \overline{D(A)} = X$, and $\|C_\varepsilon x - x\| \to_{\varepsilon \to 0} 0$ for any $x \in X$. Now, we establish that for the complete infinitesimal generator $\overline{G}$ and the generator Z of the constructed C_ε-regularized semigroup the inclusion $\overline{G} \subset A \subset Z$ holds and implies $A = Z$. By definition of the complete infinitesimal generator, for $x \in D(G)$ we have

$$
\begin{aligned}
Gx &= \lim_{h \to 0} h^{-1} \left[C_\varepsilon^{-1} S(h)x - x \right] = \lim_{h \to 0} h^{-1} \left[U(h)x - x \right] \\
&= U'(0)x = AU(0)x = Ax,
\end{aligned}
$$

therefore, $D(G) \subseteq D(A)$ and $A|_{D(G)} = G$. This and the closedness of A imply $\overline{G} \subset A$. Now we show that $D(A) \subseteq D(Z)$ and that $Z|_{D(A)} = A$. For $x \in D(A)$ we have

$$
\begin{aligned}
C_\varepsilon A x &= AC_\varepsilon x = A\mathbf{R}_{\varepsilon,\tau} U_{-A}(\tau)x = \mathbf{R}_{\varepsilon,\tau} AU_{-A}(\tau)x \\
&= \mathbf{R}_{\varepsilon,\tau} \lim_{t \to 0} t^{-1} \left[U_{-A}(\tau - t) - U_{-A}(\tau) \right] x \\
&= \lim_{t \to 0} t^{-1} \left[S(t)x - C_\varepsilon x \right] = C_\varepsilon Z x,
\end{aligned}
$$

i.e., $x \in D(Z)$ and $Ax = Zx$ for $x \in D(A)$. Hence $A \subset Z$. From the definition of the generator and the complete infinitesimal generator for a C-regularized semigroup it follows that $C^{-1}\overline{G}Cx = Zx$ for $x \in D(Z)$. This and the inclusion $\overline{G} \subset A$ imply equalities

$$
ACx = CZx, \quad C^{-1}ACx = Zx, \quad x \in D(Z).
$$

Under the assumption $\rho(A) \neq \emptyset$ (which holds for $-A$, the generator of a C_0-semigroup, and hence for A), it follows $A = Z$.

(I) $\Longrightarrow$ (II). Suppose that for any $\tau < T$, A is the generator of a local C_ε-regularized semigroup $\{S(t), t \in [0, \tau)\}$ with C_ε strongly convergent to I as $\varepsilon \to 0$. We show that $\mathbf{R}_{\varepsilon,t} := S(t)$, $0 < t < T$, is a regularizing operator for (LCP). By our construction, the linear operator $\mathbf{R}_{\varepsilon,t}$ is defined on the whole X, is bounded, and therefore is continuous. Suppose that for some $x \in D(A)$ there exists a solution of (LCP) $u(\cdot)$. For fixed $t \in (0, T)$ we consider the error

$$
\|\mathbf{R}_{\varepsilon,t}x - u(t)\| = \|S(t)C_\varepsilon^{-1} C_\varepsilon x - u(t)\|.
$$

Since C_ε^{-1} and $S(t)$ commute on $R(C_\varepsilon)$, we have

$$
S(t)C_\varepsilon^{-1} C_\varepsilon x = C_\varepsilon^{-1} S(t) C_\varepsilon x.
$$

Consider $y = C_\varepsilon x \in C_\varepsilon D(A)$. Then by Theorem 1.4.6,

$$
v(t) = \mathbf{R}_{\varepsilon,t}x = C_\varepsilon^{-1} S(t) C_\varepsilon x = C_\varepsilon^{-1} S(t) y
$$

is the unique solution of the Cauchy problem with the generator of a C_ε-regularized semigroup:

$$
v'(t) = Av(t), \quad 0 < t < T, \ v(0) = y.
$$

On the other hand, $C_\varepsilon u(t)$ is also a solution of (LCP) with the initial value $C_\varepsilon u(0) = C_\varepsilon x = y$, therefore $\mathbf{R}_{\varepsilon,t} x = S(t)x = C_\varepsilon u(t)$. Hence, if for an initial value x there exists a solution $u(\cdot)$, then

$$\|\mathbf{R}_{\varepsilon,t} x - u(t)\| = \|C_\varepsilon u(t) - u(t)\| \underset{\varepsilon \to 0}{\longrightarrow} 0, \quad 0 \le t < T,$$

i.e., $\mathbf{R}_{\varepsilon,t}$ is a regularizing operator for (LCP). $\square$

For $A \in \mathcal{H}_+$ we now consider the regularizing operators obtained by the modified quasi-reversibility and ABC methods:

$$\mathbf{R}_{\varepsilon,t}^p x = \sum_{k=1}^{\infty} e^{\lambda_k t - \varepsilon \lambda_k^p T} x_k \alpha_k, \quad p > 1, \tag{3.2.13}$$

and

$$\widetilde{\mathbf{R}}_{\varepsilon,t}^p x = \sum_{k=1}^{\infty} \frac{e^{\lambda_k t}}{1 + \varepsilon e^{\lambda_k^p T}} x_k \alpha_k, \quad p \ge 1. \tag{3.2.14}$$

They are C_ε-regularized semigroups, where

$$C_\varepsilon x = \sum_{k=1}^{\infty} e^{-\varepsilon \lambda_k^p T} x_k \alpha_k, \qquad C_\varepsilon x = \sum_{k=1}^{\infty} \frac{1}{1 + \varepsilon e^{\lambda_k^p T}} x_k \alpha_k,$$

respectively. In [185] it is proved that the classes of regularizing operators (3.2.13), (3.2.14) have error estimates similar to (3.2.2), (3.2.3), respectively.

3.2.4 Regularization of 'slightly' ill-posed problems

In this subsection, we discuss the regularization of the Cauchy problem (CP) in the cases when A is the generator of an integrated semigroup or of a κ-convoluted semigroup.

Let X be a Banach space. Consider the Cauchy problem

$$u'(t) = Au(t), \quad t \ge 0, \; u(0) = x, \tag{CP}$$

with a densely defined linear operator A generating an exponentially bounded integrated semigroup $\{V(t), \; t \ge 0\}$. Suppose that for some exactly given initial value x there exists a solution of (CP) $u(\cdot)$ and the initial value x_δ is given with an error $\delta : \|x_\delta - x\| \le \delta$. If $x \in D(A^2)$, then by Theorem 1.2.4 the problem (CP) is $(1,\omega)$-well-posed and its solution

$$u(t) = V'(t)x = V(t)Ax + x, \quad t \ge 0, \tag{3.2.15}$$

satisfies the estimate

$$\|u(t)\| \le K \, e^{\omega t}(\|Ax\| + \|x\|).$$

The solution operators $V'(t)$ are defined and (in general) unbounded on $D(A)$. From the representation (3.2.15), it is clear that in order to regularize $V'(t)$, one has to neutralize the action of the unbounded operator A. An obvious candidate for this role is the resolvent $R_A(\lambda) : X \to D(A)$. We recall that if A is the generator of an exponentially bounded integrated semigroup satisfying property (1.2.6), then

$$\lambda R_A(\lambda)x \xrightarrow[\lambda \to \infty]{} x, \quad x \in D(A), \tag{3.2.16}$$

(see Proposition 1.5.1). Now for $t \geq 0$ and $\lambda \geq \gamma > a = \max\{\omega, 0\}$ we define the operator

$$\mathbf{R}_{\frac{1}{\lambda},t} := \lambda V'(t) R_A(\lambda)$$

on X. From the relation

$$AR_A(\lambda)x = \big[A \pm \lambda I\big] R_A(\lambda)x = \lambda R_A(\lambda)x - x, \quad x \in X,$$

we have

$$\mathbf{R}_{\frac{1}{\lambda},t} = \lambda V(t) \Big[\lambda R_A(\lambda) - I\Big] + \lambda R_A(\lambda).$$

Hence the operator $\mathbf{R}_{\frac{1}{\lambda},t}$ is bounded on X for any $t \geq 0$ and $\lambda \geq \gamma$.

Theorem 3.2.4 *Suppose that a densely defined operator A is the generator of an exponentially bounded integrated semigroup $\{V(t),\ t \geq 0\}$ satisfying property (1.2.6). Then the operator*

$$\mathbf{R}_{\varepsilon,t} = \frac{1}{\varepsilon}\Big[V(t) AR_A(1/\varepsilon) + R_A(1/\varepsilon)\Big], \quad \varepsilon \in (0, 1/\gamma],\ t > 0,$$

is the regularizing operator for (CP).

Proof Let $u(\cdot)$ be an exact solution of (CP) with the initial value $x \in D(A^2)$. Consider its approximation $u_{\lambda,\delta}(t) = \mathbf{R}_{\frac{1}{\lambda},t} x_\delta,\ t > 0,\ x_\delta \in X$, where $\lambda = 1/\varepsilon$. We now estimate the error

$$
\begin{aligned}
\big\|u_{\lambda,\delta}(t) - u(t)\big\| &= \Big\|\lambda V'(t) R_A(\lambda) x_\delta - V'(t)x \pm \lambda V'(t) R_A(\lambda)x\Big\| \\
&\leq \Big\|\mathbf{R}_{\frac{1}{\lambda},t}(x_\delta - x)\Big\| + \Big\|V'(t)\big[\lambda R_A(\lambda) - I\big]x\Big\|.
\end{aligned}
$$

Since $\|R_A(\lambda)\| \leq K$ and $\|V(t)\| \leq K\, e^{\omega t}$, we have

$$\Big\|\mathbf{R}_{\frac{1}{\lambda},t}x\Big\| = \Big\|\lambda^2 V(t) R_A(\lambda)x - \lambda V(t)x + \lambda R_A(\lambda)x\Big\| \leq \lambda^2 K\, e^{\omega t}\|x\|$$

and

$$
\begin{aligned}
\Big\|V'(t)\big[\lambda R_A(\lambda) - I\big]x\Big\| &= \Big\|V'(t)\big[\lambda R_A(\lambda) - I\big] R_A(\lambda_0)y\Big\| \\
&= \Big\|\big[V(t) AR_A(\lambda_0) + R_A(\lambda_0)\big]\big[\lambda R_A(\lambda) - I\big]y\Big\| \\
&\leq K\, e^{\omega t}\Big\|\big[\lambda R_A(\lambda) - I\big]y\Big\|,
\end{aligned}
$$

where $y := \lambda_0 x - Ax \in D(A)$. Choosing $\lambda = o(\delta^{-\nu/2})$ with some $\nu \in (0,1)$, we obtain that

$$\left\| \mathbf{R}_{\frac{1}{\lambda},t} x_\delta - u(t) \right\| \xrightarrow[\delta \to 0]{} 0.$$

Hence $\mathbf{R}_{\varepsilon,t}$ is the regularizing operator with the regularization parameter $\varepsilon = 1/\lambda \to 0$. $\square$

If A is the generator of an exponentially bounded n-integrated semigroup $\{V(t),\ t \geq 0\}$ satisfying property (1.2.6), then (3.2.16) holds for all $x \in D(A^n)$. In this case the operator

$$\mathbf{R}_{\varepsilon,t} := V^{(n)}(t) \left(\frac{1}{\varepsilon} R_A(1/\varepsilon) \right)^n, \quad \varepsilon \in (0, 1/\gamma],\ t > 0,$$

is the regularizing operator for (CP).

We now consider the Cauchy problem (CP) with an operator A generating a κ-convoluted semigroup $\{S_\kappa(t),\ t \geq 0\}$. By Definition 1.3.1, we have that $S_\kappa(t)Ax = AS_\kappa(t)x$ for $x \in D(A)$, and for $x \in X$

$$S_\kappa(t)x = A \int_0^t S_\kappa(s)x\,ds + \Theta(t)x, \quad \Theta(t) := \int_0^t \kappa(s)\,ds, \quad t \geq 0.$$

We recall that if for some x there exists a solution $u(\cdot)$ of (CP), then $v = \Theta * u$ is a solution of the corresponding convoluted Cauchy problem:

$$v'(t) = Av(t) + \Theta(t)x, \quad t \geq 0,\ v(0) = 0. \tag{3.2.17}$$

Suppose now that we are given $v_\delta(t) = \int_0^t S_\kappa(s)x_\delta,\ t \geq 0$, a solution of the convoluted problem (3.2.17) with $x = x_\delta \in X$. In order to find u, we need to regularize the following convolution equation of the first kind:

$$\Psi u := \Theta * u = \int_0^t \Theta(t-s)u(s)\,ds = v(t) \tag{3.2.18}$$

given v_δ such that $\|v - v_\delta\| \leq K\delta$. Let Λ be the region from Theorem 1.3.1. Applying the Laplace transform to (3.2.18) we obtain

$$\mathcal{L}(\Theta, \lambda) \cdot \mathcal{L}(u, \lambda) = \mathcal{L}(v, \lambda), \quad \lambda \in \Lambda \subset \mathbb{C},$$

and

$$u = \mathcal{L}^{-1} \frac{\mathcal{L}(v, \lambda)}{\mathcal{L}(\Theta, \lambda)} = \Psi^{-1} v.$$

Note that the operator Ψ^{-1} is unbounded since $\mathcal{L}(\Theta, \lambda) \to 0$ as $\lambda \to \infty$ (see Theorem 1.3.1). Using Tikhonov-Arsenin's regularization [258] we define the regularizing operator

$$\mathbf{R}_{\varepsilon,t}x := \mathcal{L}^{-1} \left(\frac{\mathcal{L}(v, \lambda) f(\lambda, \varepsilon)}{\mathcal{L}(\Theta, \lambda)}, t \right)$$

for all $x \in X$, $\varepsilon \in (0, \varepsilon_0]$ and $t > 0$. Here $f(\lambda, \varepsilon) \longrightarrow_{\varepsilon \to 0} 1$ uniformly in $\lambda \in \Lambda$ and

$$f(\lambda, \varepsilon) \underset{\lambda \to \infty}{=} \mathcal{O}\left(\mathcal{L}(\Theta, \lambda)\right), \quad \varepsilon \in (0, \varepsilon_0].$$

For example, one can take

$$f(\lambda, \varepsilon) = \frac{\mathcal{L}(\Theta, \lambda)}{(\mathcal{L}^2(\Theta, \lambda) + \varepsilon)^{1/2}}.$$

Finally, we note that the same procedure can be used in the case where A is the generator of a C-regularized semigroup. In this case one has to find a solution of the Cauchy problem

$$v_\delta'(t) = A v_\delta(t), \quad v_\delta(0) = C x_\delta,$$

and then to regularize the operator equation of the first kind: $Cu = v$ with given v_δ.

Bibliographic Remark

We would like to emphasize here that our book is not a survey in any sense. Each of the three approaches (semigroups, distributions and regularizations) to treating the abstract Cauchy problem deserves a separate volume in order to cover all aspects of the theory and the applications. Our aim was to present a concept of some of the connections between these approaches. For this reason our bibliographic notes are very minimal, and we indicate only the sources that were directly used in writing this book. The Bibliography also includes publications that are closely related and can be recommended for further reading.

Chapter 0 is inspired by R. Dautray and J.-L. Lions [51], R. Lattès and J.-L. Lions [145], A. Pazy [219], E. Zeidler [284], H. R. Thieme [256].

Chapter 1

Section 1.1 is based on H. O. Fattorini [84], S. G. Kreĭn [141], T. Kato [130], E. B. Davies and M. M. H. Pang [52], A. V. Balakrishnan [14]. We also recommend books by E. Hille and R. S. Phillips [120], K. Yosida [281], R. Dautray and J.-L. Lions [51], A. Pazy [219], J. A. Goldstein [106], K. Engel and R. Nagel [73].

Section 1.2 is based on W. Arendt [5], F. Neubrander [209], H. R. Thieme [256], H. Kellerman and M. Hieber [131], N. Tanaka and N. Okazawa [255], E. B. Davies and M. M. H. Pang [52], I. Cioranescu [36].

Section 1.3 is based on I. Cioranescu and G. Lumer [39, 38], I. V. Melnikova, U. A. Anufrieva and A. Filinkov [177].

Section 1.4 is based on E. B. Davies and M. M. H. Pang [52], N. Tanaka and N. Okazawa [255], I. Miyadera and N. Tanaka [201].

Section 1.5 is based on I. V. Melnikova and M. A. Alshansky [174, 176].

Section 1.6 is based on A. Yagi [280], I. V. Melnikova and A. V. Gladchenko [187], I. V. Melnikova [170, 171, 173], W. Arendt, O. El-Mennaoui and V. Kéyantuo [9], H. R. Thieme [256], A. Favini and A. Yagi [102].

Section 1.7 is based on I. V. Melnikova [171], I. V. Melnikova and A. Filinkov [186, 182].

Chapter 2

Section 2.1 is based on H. O. Fattorini [84], J.-L. Lions [152], J. Chazarain [30], I. V. Melnikova and M. A. Alshansky [175], I. V. Melnikova [169, 172].

Section 2.2 is based on I. V. Melnikova [169], I. V. Melnikova, U. A. Anufrieva and V. Yu. Ushkov [178].

Section 2.3 is based on H. Komatsu [135], J. Chazarain [30], E. Magenes and J.-L. Lions [153], I. V. Melnikova, U. A. Anufrieva and A. Filinkov [177], V. K. Ivanov and I. V. Melnikova [124], I. V. Melnikova [167], I. V. Melnikova and A. Filinkov [185].

Chapter 3

Sections 3.1 and 3.2 are based on I. V. Melnikova [167, 165, 166], I. V. Melnikova and A. Filinkov [185], R. Lattès and J.-L. Lions [145], V. K. Ivanov, I. V. Melnikova and A. Filinkov [128], V. K. Ivanov, V. V. Vasin and V. P. Tanana [129], M. M. Lavrentev [146], I. V. Melnikova and S. V. Bochkareva [180], A. Carasso [23].

Bibliography

[1] K. A. Ames, G. W. Clark, J. F. Epperson, and S. F. Oppenheimer. A comparison of regularizations for an ill-posed problem. *Math. Comp.*, 67(224):1451–1471, 1998.

[2] K. A. Ames and J. F. Epperson. A kernel-based method for the approximate solution of backward parabolic problems. *SIAM J. Numer. Anal.*, 34(4):1357–1390, 1997.

[3] K. A. Ames and L. E. Payne. Asymptotic behavior for two regularizations of the Cauchy problem for the backward heat equation. *Math. Models Methods Appl. Sci.*, 8(1):187–202, 1998.

[4] W. Arendt. Resolvent positive operators. *Proc. London Math. Soc. (3)*, 54(2):321–349, 1987.

[5] W. Arendt. Vector-valued Laplace transforms and Cauchy problems. *Israel J. Math.*, 59(3):327–352, 1987.

[6] W. Arendt. Sobolev imbeddings and integrated semigroups. In *Semigroup Theory and Evolution Equations (Delft, 1989)*, pages 29–40. Dekker, New York, 1991.

[7] W. Arendt and C. J. K. Batty. Almost periodic solutions of first- and second-order Cauchy problems. *J. Differential Equations*, 137(2):363–383, 1997.

[8] W. Arendt and A. V. Bukhvalov. Integral representations of resolvents and semigroups. *Forum Math.*, 6(1):111–135, 1994.

[9] W. Arendt, O. El-Mennaoui, and V. Kéyantuo. Local integrated semigroups: evolution with jumps of regularity. *J. Math. Anal. Appl.*, 186(2):572–595, 1994.

[10] W. Arendt and A. Favini. Integrated solutions to implicit differential equations. *Rend. Sem. Mat. Univ. Politec. Torino*, 51(4):315–329 (1994), 1993. Partial differential equations, I (Turin, 1993).

[11] W. Arendt and H. Kellermann. Integrated solutions of Volterra inte-grodifferential equations and applications. In *Volterra Integrodifferential Equations in Banach Spaces and Applications (Trento, 1987)*, pages 21–51. Longman Sci. Tech., Harlow, 1989.

[12] W. Arendt, F. Neubrander, and U. Schlotterbeck. Interpolation of semigroups and integrated semigroups. *Semigroup Forum*, 45(1):26–37, 1992.

[13] W. Arendt and A. Rhandi. Perturbation of positive semigroups. *Arch. Math. (Basel)*, 56(2):107–119, 1991.

[14] A. V. Balakrishnan. *Applied Functional Analysis*. Springer-Verlag, New York, second edition, 1981.

[15] J. Banasiak. Generation results for B-bounded semigroups. *Ann. Mat. Pura Appl. (4)*, 175:307–326, 1998.

[16] V. Barbu and A. Favini. On some degenerate parabolic problems. *Ricerche Mat.*, 46(1):77–86, 1997.

[17] V. Barbu and A. Favini. The analytic semigroup generated by a second order degenerate differential operator in $C[0,1]$. In *Proceedings of the Third International Conference on Functional Analysis and Approximation Theory, Vol. I (Acquafredda di Maratea, 1996)*, number 52, Vol. I, pages 23–42, 1998.

[18] V. Barbu, A. Favini, and S. Romanelli. Degenerate evolution equations and regularity of their associated semigroups. *Funkcial. Ekvac.*, 39(3):421–448, 1996.

[19] B. Bäumer, G. Lumer, and F. Neubrander. Convolution kernels and generalized functions. In *Generalized Functions, Operator Theory, and Dynamical Systems (Brussels, 1997)*, pages 68–78. Chapman & Hall/CRC, Boca Raton, FL, 1999.

[20] B. Bäumer and F. Neubrander. Laplace transform methods for evolution equations. *Confer. Sem. Mat. Univ. Bari*, (258-260):27–60, 1994. Swabian-Apulian Meeting on Operator Semigroups and Evolution Equations (Italian) (Ruvo di Puglia, 1994).

[21] R. Beals. On the abstract Cauchy problem. *J. Functional Analysis*, 10:281–299, 1972.

[22] H. Bremermann. *Distributions, Complex Variables, and Fourier Transforms*. Addison-Wesley Publishing Co., Inc., Reading, Mass.-London, 1965.

[23] A. Carasso. The abstract backward beam equation. *SIAM J. Math. Anal.*, 2:193 212, 1971.

[24] A. Carasso. The backward beam equation and the numerical computation of dissipative equations backwards in time. In *Improperly Posed Boundary Value Problems (Conf., Univ. New Mexico, Albuquerque, N.M., 1974)*, pages 124 157. Res. Notes in Math., No. 1. Pitman, London, 1975.

[25] A. Carasso. Error bounds in the final value problem for the heat equation. *SIAM J. Math. Anal.*, 7(2):195 199, 1976.

[26] R. W. Carroll and R. E. Showalter. *Singular and Degenerate Cauchy Problems*. Academic Press [Harcourt Brace Jovanovich Publishers], New York, 1976. Mathematics in Science and Engineering, Vol. 127.

[27] J. Chazarain. Problèmes de Cauchy au sens des distributions vectorielles et applications. *C. R. Acad. Sci. Paris Sér. A-B*, 266:A10 A13, 1968.

[28] J. Chazarain. Problèmes de Cauchy dans des espaces d'ultra-distributions. *C. R. Acad. Sci. Paris Sér. A-B*, 266:A564 A566, 1968.

[29] J. Chazarain. Problèmes de Cauchy abstraits et applications. In *Symposia Mathematica, Vol. VII (Convegno sulle Problemi di Evoluzione, INDAM, Rome, 1970)*, pages 129 134. Academic Press, London, 1971.

[30] J. Chazarain. Problèmes de Cauchy abstraits et applications à quelques problèmes mixtes. *J. Functional Analysis*, 7:386 446, 1971.

[31] I. Cioranescu. Operator-valued ultradistributions in the spectral theory. *Math. Ann.*, 223(1):1 12, 1976.

[32] I. Cioranescu. Abstract Beurling spaces of class (M_p) and ultradistribution semi-groups. *Bull. Sci. Math. (2)*, 102(2):167 192, 1978.

[33] I. Cioranescu. On the second order Cauchy problem associated with a linear operator. *J. Math. Anal. Appl.*, 154(1):238 242, 1991.

[34] I. Cioranescu. A generation result for C-regularized semigroups. In *Semigroups of Linear and Nonlinear Operations and Applications (Curaçao, 1992)*, pages 121 128. Kluwer Acad. Publ., Dordrecht, 1993.

[35] I. Cioranescu. On a class of C-regularized semigroups. In *Evolution Equations, Control Theory, and Biomathematics (Han sur Lesse, 1991)*, pages 45 50. Dekker, New York, 1994.

[36] I. Cioranescu. Local convoluted semigroups. In *Evolution Equations (Baton Rouge, LA, 1992)*, pages 107–122. Dekker, New York, 1995.

[37] I. Cioranescu and V. Keyantuo. C-semigroups: generation and analyticity. *Integral Transform. Spec. Funct.*, 6(1-4):15–25, 1998. Generalized functions—linear and nonlinear problems (Novi Sad, 1996).

[38] I. Cioranescu and G. Lumer. Problèmes d'évolution régularisés par un noyau général $K(t)$. Formule de Duhamel, prolongements, théorèmes de génération. *C. R. Acad. Sci. Paris Sér. I Math.*, 319(12):1273–1278, 1994.

[39] I. Cioranescu and G. Lumer. On $K(t)$-convoluted semigroups. In *Recent Developments in Evolution Equations (Glasgow, 1994)*, pages 86–93. Longman Sci. Tech., Harlow, 1995.

[40] G. W. Clark and S. F. Oppenheimer. Quasireversibility methods for non-well-posed problems. *Electron. J. Differential Equations*, pages No. 08, approx. 9 pp. (electronic), 1994.

[41] Ph. Clément, O. Diekmann, M. Gyllenberg, H. J. A. M. Heijmans, and H. R. Thieme. A Hille-Yosida theorem for a class of weakly * continuous semigroups. *Semigroup Forum*, 38(2):157–178, 1989. Semigroups and differential operators (Oberwolfach, 1988).

[42] Ph. Clément and S. Guerre-Delabrière. On the regularity of abstract Cauchy problems and boundary value problems. *Atti Accad. Naz. Lincei Cl. Sci. Fis. Mat. Natur. Rend. Lincei (9) Mat. Appl.*, 9(4):245–266 (1999), 1998.

[43] Ph. Clément, H. J. A. M. Heijmans, S. Angenent, C. J. van Duijn, and B. de Pagter. *One-Parameter Semigroups*. North-Holland Publishing Co., Amsterdam, 1987.

[44] Ph. Clément and C. A. Timmermans. On C_0-semigroups generated by differential operators satisfying Ventcel's boundary conditions. *Nederl. Akad. Wetensch. Indag. Math.*, 48(4):379–387, 1986.

[45] W. L. Conradie and N. Sauer. Empathy, C-semigroups, and integrated semigroups. In *Evolution Equations (Baton Rouge, LA, 1992)*, pages 123–132. Dekker, New York, 1995.

[46] G. Da Prato. Semigruppi regolarizzabili. *Ricerche Mat.*, 15:223–248, 1966.

[47] G. Da Prato. Weak solutions for linear abstract differential equations in Banach spaces. *Advances in Math.*, 5:181–245 (1970), 1970.

[48] G. Da Prato and E. Giusti. Una caratterizzazione dei generatori di funzioni coseno astratte. *Boll. Un. Mat. Ital. (3)*, 22:357–362, 1967.

[49] G. Da Prato and U. Mosco. Semigruppi distribuzioni analitici. *Ann. Scuola Norm. Sup. Pisa (3)*, 19:367–396, 1965.

[50] G. Da Prato and E. Sinestrari. Differential operators with nondense domain. *Ann. Scuola Norm. Sup. Pisa Cl. Sci. (4)*, 14(2):285–344 (1988), 1987.

[51] R. Dautray and J.-L. Lions. *Mathematical Analysis and Numerical Methods for Science and Technology. Vol. 1–6.* Springer-Verlag, Berlin, 1988–1993.

[52] E. B. Davies and M. M. H. Pang. The Cauchy problem and a generalization of the Hille-Yosida theorem. *Proc. London Math. Soc. (3)*, 55(1):181–208, 1987.

[53] R. deLaubenfels. Polynomials of generators of integrated semigroups. *Proc. Amer. Math. Soc.*, 107(1):197–204, 1989.

[54] R. deLaubenfels. Integrated semigroups and integrodifferential equations. *Math. Z.*, 204(4):501–514, 1990.

[55] R. deLaubenfels. Integrated semigroups, C-semigroups and the abstract Cauchy problem. *Semigroup Forum*, 41(1):83–95, 1990.

[56] R. deLaubenfels. C-existence families. In *Semigroup Theory and Evolution Equations (Delft, 1989)*, pages 295–309. Dekker, New York, 1991.

[57] R. deLaubenfels. Entire solutions of the abstract Cauchy problem. *Semigroup Forum*, 42(1):83–105, 1991.

[58] R. deLaubenfels. Existence and uniqueness families for the abstract Cauchy problem. *J. London Math. Soc. (2)*, (2):310–338, 1991.

[59] R. deLaubenfels. Unbounded well-bounded operators, strongly continuous semigroups and the Laplace transform. *Studia Math.*, 103(2):143–159, 1992.

[60] R. deLaubenfels. C-semigroups and strongly continuous semigroups. *Israel J. Math.*, 81(1-2):227–255, 1993.

[61] R. deLaubenfels. C-semigroups and the Cauchy problem. *J. Funct. Anal.*, 111(1):44–61, 1993.

[62] R. deLaubenfels. Matrices of operators and regularized semigroups. *Math. Z.*, 212(4):619–629, 1993.

[63] R. deLaubenfels. *Existence Families, Functional Calculi and Evolution Equations.* Springer-Verlag, Berlin, 1994.

[64] R. deLaubenfels. Regularized functional calculi and evolution equations. In *Evolution Equations (Baton Rouge, LA, 1992)*, pages 141–152. Dekker, New York, 1995.

[65] R. deLaubenfels. Functional calculi, semigroups of operators, and Hille-Yosida operators. *Houston J. Math.*, 22(4):787–805, 1996.

[66] R. deLaubenfels, Z. Huang, S. Wang, and Y. Wang. Laplace transforms of polynomially bounded vector-valued functions and semigroups of operators. *Israel J. Math.*, 98:189–207, 1997.

[67] R. deLaubenfels and M. Jazar. Functional calculi, regularized semigroups and integrated semigroups. *Studia Math.*, 132(2):151–172, 1999.

[68] R. deLaubenfels, G. Z. Sun, and S. W. Wang. Regularized semigroups, existence families and the abstract Cauchy problem. *Differential Integral Equations*, 8(6):1477–1496, 1995.

[69] G. V. Demidenko and S. V. Uspenskiĭ. *Uravneniya i Sistemyi Nerazreshennyie Otnositel'no Starshei Proizvodnoi.* Nauchnaya Kniga, Novosibirsk, 1998.

[70] A. A. Dezin. *Partial Differential Equations.* Springer-Verlag, Berlin, 1987.

[71] O. Diekmann, M. Gyllenberg, and H. R. Thieme. Semigroups and renewal equations on dual Banach spaces with applications to population dynamics. In *Delay Differential Equations and Dynamical Systems (Claremont, CA, 1990)*, pages 116–129. Springer, Berlin, 1991.

[72] N. Dunford and J. T. Schwartz. *Linear Operators. Parts I–III.* John Wiley & Sons Inc., New York, 1988.

[73] K. Engel and R. Nagel. *One-Parameter Semigroups for Linear Evolution Equations.* Springer-Verlag, New York, 2000. With contributions by S. Brendle, M. Campiti, T. Hahn, G. Metafune, G. Nickel, D. Pallara, C. Perazzoli, A. Rhandi, S. Romanelli and R. Schnaubelt.

[74] M. V. Falaleev. *Generalized Functions.* Irkutsk University, Irkutsk, 1996. Preprint.

[75] M. V. Falaleev. The Cauchy problem for degenerate integrodifferential equations in Banach spaces. *Vestnik Chelyabinsk. Univ. Ser. 3 Mat. Mekh.*, 2(5):126–136, 1999.

[76] H. O. Fattorini. Ordinary differential equations in linear topological spaces. I. *J. Differential Equations*, 5:72–105, 1969.

[77] H. O. Fattorini. Ordinary differential equations in linear topological spaces. II. *J. Differential Equations*, 6:50–70, 1969.

[78] H. O. Fattorini. On a class of differential equations for vector-valued distributions. *Pacific J. Math.*, 32:79–104, 1970.

[79] H. O. Fattorini. A representation theorem for distribution semigroups. *J. Functional Analysis*, 6:83–96, 1970.

[80] H. O. Fattorini. Uniformly bounded cosine functions in Hilbert space. *Indiana Univ. Math. J.*, 20:411–425, 1970/1971.

[81] H. O. Fattorini. Some remarks on convolution equations for vector-valued distributions. *Pacific J. Math.*, 66(2):347–371, 1976.

[82] H. O. Fattorini. Structure theorems for vector-valued ultradistributions. *J. Funct. Anal.*, 39(3):381–407, 1980.

[83] H. O. Fattorini. Some remarks on second order abstract Cauchy problems. *Funkcial. Ekvac.*, 24(3):331–344, 1981.

[84] H. O. Fattorini. *The Cauchy Problem*. Addison-Wesley Publishing Co., Reading, Mass., 1983.

[85] H. O. Fattorini. *Second Order Linear Differential Equations in Banach Spaces*. North-Holland Publishing Co., Amsterdam, 1985.

[86] A. Favini. Laplace transform method for a class of degenerate evolution problems. *Rend. Mat. (6)*, 12(3-4):511–536 (1980), 1979.

[87] A. Favini. Degenerate and singular evolution equations in Banach space. *Math. Ann.*, 273(1):17–44, 1985.

[88] A. Favini. An operational method for abstract degenerate evolution equations of hyperbolic type. *J. Funct. Anal.*, 76(2):432–456, 1988.

[89] A. Favini and M. Fuhrman. Approximation results for semigroups generated by multivalued linear operators and applications. *Differential Integral Equations*, 11(5):781–805, 1998.

[90] A. Favini, G. R. Goldstein, Je.A. Goldstein, and S. Romanelli. C_0-semigroups generated by second order differential operators with general Wentzell boundary conditions. *Proc. Amer. Math. Soc.*, 128(7):1981–1989, 2000.

[91] A. Favini, J. A. Goldstein, and S. Romanelli. An analytic semigroup associated to a degenerate evolution equation. In *Stochastic Processes and Functional Analysis (Riverside, CA, 1994)*, pages 85–100. Dekker, New York, 1997.

[92] A. Favini, J. A. Goldstein, and S. Romanelli. Analytic semigroups on $L_w^p(0,1)$ and on $L^p(0,1)$ generated by some classes of second order differential operators. *Taiwanese J. Math.*, 3(2):181–210, 1999.

[93] A. Favini and E. Obrecht. Conditions for parabolicity of second order abstract differential equations. *Differential Integral Equations*, 4(5):1005–1022, 1991.

[94] A. Favini and P. Plazzi. On some abstract degenerate problems of parabolic type. I. The linear case. *Nonlinear Anal.*, 12(10):1017–1027, 1988.

[95] A. Favini and P. Plazzi. On some abstract degenerate problems of parabolic type. II. The nonlinear case. *Nonlinear Anal.*, 13(1):23–31, 1989.

[96] A. Favini and S. Romanelli. Degenerate second order operators as generators of analytic semigroups on $C[0, +\infty]$ or on $L_{\alpha-1/2}^p(0, +\infty)$. In *Approximation and Optimization, Vol. II (Cluj-Napoca, 1996)*, pages 93–100. Transilvania, Cluj-Napoca, 1997.

[97] A. Favini and S. Romanelli. Analytic semigroups on $C[0, 1]$ generated by some classes of second order differential operators. *Semigroup Forum*, 56(3):362–372, 1998.

[98] A. Favini and H. Tanabe. Linear parabolic differential equations of higher order in time. In *Differential Equations in Banach Spaces (Bologna, 1991)*, pages 85–92. Dekker, New York, 1993.

[99] A. Favini and A. Yagi. Space and time regularity for degenerate evolution equations. *J. Math. Soc. Japan*, 44(2):331–350, 1992.

[100] A. Favini and A. Yagi. Multivalued linear operators and degenerate evolution equations. *Ann. Mat. Pura Appl. (4)*, 163:353–384, 1993.

[101] A. Favini and A. Yagi. Abstract second order differential equations with applications. *Funkcial. Ekvac.*, 38(1):81–99, 1995.

[102] A. Favini and A. Yagi. *Degenerate Differential Equations in Banach Spaces*. Marcel Dekker Inc., New York, 1999.

[103] C. Foiaş. Remarques sur les semi-groupes distributions d'opérateurs normaux. *Portugal. Math.*, 19:227–242, 1960.

[104] D. Fujiwara. A characterisation of exponential distribution semi-groups. *J. Math. Soc. Japan*, 18:267 274, 1966.

[105] I. M. Gelfand and G. E. Shilov. *Generalized Functions. Vol. 1 3*. Academic Press [Harcourt Brace Jovanovich Publishers], New York, 1964 1967 [1977].

[106] J. A. Goldstein. *Semigroups of Linear Operators and Applications*. The Clarendon Press Oxford University Press, New York, 1985.

[107] J. A. Goldstein. A survey of semigroups of linear operators and applications. In *Semigroups of Linear and Nonlinear Operations and Applications (Curaçao, 1992)*, pages 9 57. Kluwer Acad. Publ., Dordrecht, 1993.

[108] J. A. Goldstein, R. deLaubenfels, and J. T. Sandefur, Jr. Regularized semigroups, iterated Cauchy problems and equipartition of energy. *Monatsh. Math.*, 115(1-2):47 66, 1993.

[109] J. A. Goldstein and C. Y. Lin. An L^p semigroup approach to degenerate parabolic boundary value problems. *Ann. Mat. Pura Appl. (4)*, 159:211 227, 1991.

[110] M. Hieber. Integrated semigroups and differential operators on L^p spaces. *Math. Ann.*, 291(1):1 16, 1991.

[111] M. Hieber. Integrated semigroups and the Cauchy problem for systems in L^p spaces. *J. Math. Anal. Appl.*, 162(1):300 308, 1991.

[112] M. Hieber. Laplace transforms and α-times integrated semigroups. *Forum Math.*, 3(6):595 612, 1991.

[113] M. Hieber. On strongly elliptic differential operators on $L^1(R^n)$. In *Semigroups of Linear and Nonlinear Operations and Applications (Curaçao, 1992)*, pages 177 183. Kluwer Acad. Publ., Dordrecht, 1993.

[114] M. Hieber. On the L^p theory of systems of linear partial differential equations. In *Evolution Equations, Control Theory, and Biomathematics (Han sur Lesse, 1991)*, pages 275 283. Dekker, New York, 1994.

[115] M. Hieber. Spectral theory and Cauchy problems on L^p spaces. *Math. Z.*, 216(4):613 628, 1994.

[116] M. Hieber. Spectral theory of positive semigroups generated by differential operators. *Arch. Math. (Basel)*, 63(4):333 340, 1994.

[117] M. Hieber, A. Holderrieth, and F. Neubrander. Regularized semi-groups and systems of linear partial differential equations. *Ann. Scuola Norm. Sup. Pisa Cl. Sci. (4)*, 19(3):363-379, 1992.

[118] M. Hieber and F. Räbiger. A remark on the abstract Cauchy problem on spaces of Hölder continuous functions. *Proc. Amer. Math. Soc.*, 115(2):431-434, 1992.

[119] E. Hille. *Functional Analysis and Semi-Groups.* American Mathematical Society, New York, 1948. American Mathematical Society Colloquium Publications, vol. 31.

[120] E. Hille and R. S. Phillips. *Functional Analysis and Semi-Groups.* American Mathematical Society, Providence, R. I., 1974.

[121] L. Hörmander. Estimates for translation invariant operators in L^p spaces. *Acta Math.*, 104:93-140, 1960.

[122] V. K. Ivanov. Ill-posed problems and divergent processes. *Uspekhi Mat. Nauk*, 40(4(244)):165-166, 1985.

[123] V. K. Ivanov. Hadamard conditions for well-posedness in spaces of generalized functions. *Sibirsk. Mat. Zh.*, 28(6):53-59, 218, 1987.

[124] V. K. Ivanov and I. V. Melnikova. A general scheme for the elimination of divergences of different types. *Sibirsk. Mat. Zh.*, 29(6):66-73, 1988.

[125] V. K. Ivanov and I. V. Melnikova. Construction of quasivalues for weakly well-posed problems. *Dokl. Akad. Nauk SSSR*, 306(3):530-535, 1989.

[126] V. K. Ivanov and I. V. Melnikova. New generalized functions and weak well-posedness of operator problems. *Dokl. Akad. Nauk SSSR*, 317(1):22-26, 1991.

[127] V. K. Ivanov and I. V. Melnikova. New generalized functions and weak well-posedness of operator problems. In *Complex Analysis and Generalized Functions (Varna, 1991)*, pages 141-147. Publ. Bulgar. Acad. Sci., Sofia, 1993.

[128] V. K. Ivanov, I. V. Melnikova, and A. I. Filinkov. *Operator-Differential Equations and Ill-Posed Problems.* Fizmatlit "Nauka", Moscow, 1995.

[129] V. K. Ivanov, V. V. Vasin, and V. P. Tanana. *Theory of Linear Ill-Posed Problems and its Applications.* "Nauka", Moscow, 1978.

[130] T. Kato. *Perturbation Theory for Linear Operators*. Springer-Verlag, Berlin, 1995. Reprint of the 1980 edition.

[131] H. Kellerman and M. Hieber. Integrated semigroups. *J. Funct. Anal.*, 84(1):160–180, 1989.

[132] C. Knuckles and F. Neubrander. Remarks on the Cauchy problem for multi-valued linear operators. In *Partial Differential Equations (Han-sur-Lesse, 1993)*, pages 174–187. Akademie-Verlag, Berlin, 1994.

[133] H. Komatsu. Semi-groups of operators in locally convex spaces. *J. Math. Soc. Japan*, 16:230–262, 1964.

[134] H. Komatsu. Ultradistributions, hyperfunctions and linear differential equations. In *Colloque International CNRS sur les Équations aux Dérivées Partielles Linéaires (Univ. Paris-Sud, Orsay, 1972)*, pages 252–271. Astérisque, 2 et 3. Soc. Math. France, Paris, 1973.

[135] H. Komatsu. Ultradistributions. I. Structure theorems and a characterization. *J. Fac. Sci. Univ. Tokyo Sect. IA Math.*, 20:25–105, 1973.

[136] H. Komatsu. Operational calculus, hyperfunctions and ultradistributions. In *Algebraic Analysis, Vol. I*, pages 357–372. Academic Press, Boston, MA, 1988.

[137] H. Komatsu. Operational calculus and semi-groups of operators. In *Functional Analysis and Related Topics, 1991 (Kyoto)*, pages 213–234. Springer, Berlin, 1993.

[138] T. Kōmura. Semigroups of operators in locally convex spaces. *J. Functional Analysis*, 2:258–296, 1968.

[139] R. Kravarušić, M. Mijatović, and S. Pilipović. Integrated semigroup of unbounded linear operators—Cauchy problem. II. *Novi Sad J. Math.*, 28(1):107–122, 1998.

[140] R. Kravarušić, M. Mijatović, and S. Pilipović. Integrated semigroups of unbounded linear operators in Banach spaces. I. *Bull. Cl. Sci. Math. Nat. Sci. Math.*, (23):45–62, 1998.

[141] S. G. Kreĭn. *Linear Differential Equations in Banach Space*. American Mathematical Society, Providence, R.I., 1971. Translations of Mathematical Monographs, Vol. 29.

[142] S. G. Kreĭn. *Linear Equations in Banach Spaces*. Birkhäuser Boston, Mass., 1982.

[143] S. G. Kreĭn and K.I. Chernyishev. *Singular-Pertubed Differential Equations in a Banach Space.* Rossiĭskaya Akademiya Nauk. Sibirskoe Otdelenie. Institut Matematiki im. S. L. Soboleva., Novosibirsk, 1979. Preprint.

[144] S. G. Kreĭn and M. I. Khazan. Differential equations in a Banach space. In *Mathematical Analysis, Vol. 21*, pages 130–264. Akad. Nauk SSSR Vsesoyuz. Inst. Nauchn. i Tekhn. Inform., Moscow, 1983.

[145] R. Lattès and J.-L. Lions. *The Method of Quasi-Reversibility. Applications to Partial Differential Equations.* American Elsevier Publishing Co., Inc., New York, 1969.

[146] M. M. Lavrentev. *Conditionally Well-Posed Problems for Differential Equations.* Novosibirsk State University, Novosibirsk, 1973. Preprint.

[147] M. M. Lavrentev, V. G. Romanov, and S. P. Shishatsky. *Ill-Posed Problems of Mathematical Physics and Analysis.* American Mathematical Society, Providence, R.I., 1986.

[148] M. M. Lavrentev and L. Ya. Savelev. *Linear Operators and Ill-Posed Problems.* Consultants Bureau, New York, 1995.

[149] Y. S. Lei and Q. Zheng. The application of C-semigroups to differential operators in $L^p(R^n)$. *J. Math. Anal. Appl.*, 188(3):809–818, 1994.

[150] J. Liang and T.-J. Xiao. Integrated semigroups and higher order abstract equations. *J. Math. Anal. Appl.*, 222(1):110–125, 1998.

[151] J. Liang and T.-J. Xiao. Wellposedness results for certain classes of higher order abstract Cauchy problems connected with integrated semigroups. *Semigroup Forum*, 56(1):84–103, 1998.

[152] J.-L. Lions. Les semi groupes distributions. *Portugaliae Mathematica*, 19:141–164, 1960.

[153] J.-L. Lions and E. Magenes. *Non-Homogeneous Boundary Value Problems and Applications. Vol. I–III.* Springer-Verlag, New York, 1972–1973.

[154] Ju. I. Ljubic. The classical and local Laplace transform in the abstract Cauchy problem. *Uspehi Mat. Nauk*, 21(3 (129)):3–51, 1966.

[155] G. Lumer. Examples and results concerning the behavior of generalized solutions, integrated semigroups, and dissipative evolution problems. In *Semigroup Theory and Evolution Equations (Delft, 1989)*, pages 347–356. Dekker, New York, 1991.

[156] G. Lumer. Generalized evolution operators and (generalized) C-semigroups. In *Semigroup Theory and Evolution Equations (Delft, 1989)*, pages 337–345. Dekker, New York, 1991.

[157] G. Lumer. Evolution equations. Solutions for irregular evolution problems via generalized solutions and generalized initial values. Applications to periodic shocks models. *Ann. Univ. Sarav. Ser. Math.*, 5(1):i–iv and 1–102, 1994.

[158] G. Lumer. Singular problems, generalized solutions, and stability properties. In *Partial Differential Equations (Han-sur-Lesse, 1993)*, pages 204–216. Akademie-Verlag, Berlin, 1994.

[159] G. Lumer. Singular evolution problems, regularization, and applications to physics, engineering, and biology. In *Linear Operators (Warsaw, 1994)*, pages 205–216. Polish Acad. Sci., Warsaw, 1997.

[160] G. Lumer and F. Neubrander. Asymptotic Laplace transforms and evolution equations. In *Evolution Equations, Feshbach Resonances, Singular Hodge Theory*, pages 37–57. Wiley-VCH, Berlin, 1999.

[161] T. Matsumoto, S. Oharu, and H. R. Thieme. Nonlinear perturbations of a class of integrated semigroups. *Hiroshima Math. J.*, 26(3):433–473, 1996.

[162] I. V. Melnikova. The family M, N of operator functions and second-order equations in Banach space. *Izv. Vyssh. Uchebn. Zaved. Mat.*, 2:45–52, 87, 1985.

[163] I. V. Melnikova. A theorem of Miyadera-Feller-Phillips type for a complete second-order equation in a Banach space. *Izv. Vyssh. Uchebn. Zaved. Mat.*, 4:34–40, 86, 1985.

[164] I. V. Melnikova. Conditions for the solvability of abstract boundary value problems. *Izv. Vyssh. Uchebn. Zaved. Mat.*, 10:17–24, 1990.

[165] I. V. Melnikova. Regularization methods for ill-posed boundary problems. In *Ill-Posed Problems in Natural Sciences (Moscow, 1991)*, pages 93–103. VSP, Utrecht, 1992.

[166] I. V. Melnikova. Regularization of ill-posed differential problems. *Sibirsk. Mat. Zh.*, 33(2):125–134, 221, 1992.

[167] I. V. Melnikova. General theory of the ill-posed Cauchy problem. *J. Inverse Ill-Posed Probl.*, 3(2):149–171, 1995.

[168] I. V. Melnikova. Properties of an abstract pseudoresolvent and well-posedness of the degenerate Cauchy problem. In *Generalizations of Complex Analysis and Their Applications in Physics (Warsaw/Rynia, 1994)*, pages 151–157. Polish Acad. Sci., Warsaw, 1996.

[169] I. V. Melnikova. Properties of the Lions d-semigroups, and generalized well-posedness of the Cauchy problem. *Funktsional. Anal. i Prilozhen.*, 31(3):23–34, 95, 1997.

[170] I. V. Melnikova. The degenerate Cauchy problem in Banach spaces. *Izv. Ural. Gos. Univ. Mat. Mekh.*, (1):147–160, 183, 1998.

[171] I. V. Melnikova. The method of integrated semigroups for Cauchy problems in Banach spaces. *Sibirsk. Mat. Zh.*, 40(1):119–129, iii, 1999.

[172] I. V. Melnikova. Well-posedness of differential-operator problems. I. The Cauchy problem in spaces of distributions. *J. Math. Sci. (New York)*, 93(1):1–21, 1999. Functional analysis, 2.

[173] I. V. Melnikova. Abstract well-posed and ill-posed Cauchy problems for inclusions. In *Semigroups of Operators*, pages 203–212. Birkhäuser Boston, Mass., 2000.

[174] I. V. Melnikova and M. A. Alshansky. Well-posedness of the degenerate Cauchy problem in a Banach space. *Dokl. Akad. Nauk*, 336(1):17–20, 1994.

[175] I. V. Melnikova and M. A. Alshansky. Generalized well-posedness of the Cauchy problem, and integrated semigroups. *Dokl. Akad. Nauk*, 343(4):448–451, 1995.

[176] I. V. Melnikova and M. A. Alshansky. Well-posedness of the Cauchy problem in a Banach space: regular and degenerate cases. *J. Math. Sci. (New York)*, 87(4):3732–3780, 1997. Analysis, 9.

[177] I. V. Melnikova, U. A. Anufrieva, and A. I. Filinkov. Laplace transform of K-semigroups and well-posedness of Cauchy problems. *Integral Transform. Spec. Funct.*, 9(1):37–56, 2000.

[178] I. V. Melnikova, U. A. Anufrieva, and V. Yu. Ushkov. Degenerate distribution semigroups and well-posedness of the Cauchy problem. *Integral Transform. Spec. Funct.*, 6(1-4):247–256, 1998. Generalized functions—linear and nonlinear problems (Novi Sad, 1996).

[179] I. V. Melnikova and S. V. Bochkareva. The Maslov property for the quasireversibility method and the method of boundary value problems. In *Investigations in the Theory of Approximations (Russian)*, pages 46–50, 121–122. Ural. Gos. Univ., Sverdlovsk, 1988.

[180] I. V. Melnikova and S. V. Bochkareva. C-semigroups and the regularization of an ill-posed Cauchy problem. *Dokl. Akad. Nauk*, 329(3):270–273, 1993.

[181] I. V. Melnikova and A. I. Filinkov. Classification and well-posedness of the Cauchy problem for second-order equations in Banach space. *Dokl. Akad. Nauk SSSR*, 276(5):1066–1071, 1984.

[182] I. V. Melnikova and A. I. Filinkov. A connection between the well-posedness of the Cauchy problem for an equation and for a system in a Banach space. *Dokl. Akad. Nauk SSSR*, 300(2):280–284, 1988.

[183] I. V. Melnikova and A. I. Filinkov. Classification with respect to boundary value problems of a complete second-order equation in a Banach space. *Izv. Vyssh. Uchebn. Zaved. Mat.*, 6:39–45, 1990.

[184] I. V. Melnikova and A. I. Filinkov. An integrated family of M and N functions. *Dokl. Akad. Nauk*, 326(1):35–39, 1992.

[185] I. V. Melnikova and A. I. Filinkov. Integrated semigroups and C-semigroups. Well - posedness and regularization of operator-differential problems. *Uspekhi Mat. Nauk*, 49(6(300)):111–150, 1994.

[186] I. V. Melnikova and A. I. Filinkov. Well-posedness of differential-operator problems. II. The Cauchy problem for complete second-order equations in Banach spaces. *J. Math. Sci. (New York)*, 93(1):22–41, 1999. Functional analysis, 2.

[187] I. V. Melnikova and A. V. Gladchenko. Well-posedness of the Cauchy problem for inclusions in Banach spaces. *Dokl. Akad. Nauk*, 361(6):736–739, 1998.

[188] M. Mijatović and S. Pilipović. Integrated semigroups, relations with generators. *Novi Sad J. Math.*, 27(2):65–75, 1997.

[189] M. Mijatović and S. Pilipović. Integrated semigroups and distribution semigroups—Cauchy problem. *Math. Montisnigri*, 11:43–65, 1999.

[190] M. Mijatović, S. Pilipović, and F. Vajzović. α-times integrated semigroups ($\alpha \in R^+$). *J. Math. Anal. Appl.*, 210(2):790–803, 1997.

[191] I. Miyadera. On one-parameter semi-group of operators. *J. Math. Tokyo*, 1:23–26, 1951.

[192] I. Miyadera. Generation of a strongly continuous semi-group operators. *Tohoku Math. J. (2)*, 4:109–114, 1952.

[193] I. Miyadera. On the generation of semi-groups of linear operators. *Tôhoku Math. J. (2)*, 24:251–261, 1972. Collection of articles dedicated to Gen-ichirô Sunouchi on his sixtieth birthday.

[194] I. Miyadera. On the generators of exponentially bounded C-semigroups. *Proc. Japan Acad. Ser. A Math. Sci.*, 62(7):239–242, 1986.

[195] I. Miyadera. A generalization of the Hille-Yosida theorem. *Proc. Japan Acad. Ser. A Math. Sci.*, 64(7):223–226, 1988.

[196] I. Miyadera. C-semigroups, semigroups and n-times integrated semigroups. In *Differential Equations and Control Theory (Iaşi, 1990)*, pages 193–207. Longman Sci. Tech., Harlow, 1991.

[197] I. Miyadera. C-semigroups and semigroups of linear operators. In *Differential Equations (Plovdiv, 1991)*, pages 133–143. World Sci. Publishing, River Edge, NJ, 1992.

[198] I. Miyadera, S. Oharu, and N. Okazawa. Generation theorems of semi-groups of linear operators. *Publ. Res. Inst. Math. Sci.*, 8:509–555, 1972/73.

[199] I. Miyadera, M. Okubo, and N. Tanaka. α-times integrated semigroups and abstract Cauchy problems. *Mem. School Sci. Engrg. Waseda Univ.*, (57):267–289 (1994), 1993.

[200] I. Miyadera, M. Okubo, and N. Tanaka. On integrated semigroups which are not exponentially bounded. *Proc. Japan Acad. Ser. A Math. Sci.*, 69(6):199–204, 1993.

[201] I. Miyadera and N. Tanaka. Exponentially bounded C-semigroups and generation of semigroups. *J. Math. Anal. Appl.*, 143(2):358–378, 1989.

[202] I. Miyadera and N. Tanaka. A remark on exponentially bounded C-semigroups. *Proc. Japan Acad. Ser. A Math. Sci.*, 66(2):31–34, 1990.

[203] I. Miyadera and N. Tanaka. Generalization of the Hille-Yosida theorem. In *Semigroup Theory and Evolution Equations (Delft, 1989)*, pages 371–381. Dekker, New York, 1991.

[204] V. A. Morozov. *Methods for Solving Incorrectly Posed Problems*. Springer-Verlag, New York, 1984.

[205] F. Neubrander. Well-posedness of abstract Cauchy problems. *Semigroup Forum*, 29(1-2):75–85, 1984.

[206] F. Neubrander. Laplace transform and asymptotic behavior of strongly continuous semigroups. *Houston J. Math.*, (4):549–561, 1986.

[207] F. Neubrander. Well-posedness of higher order abstract Cauchy problems. *Trans. Amer. Math. Soc.*, 295(1):257–290, 1986.

[208] F. Neubrander. On the relation between the semigroup and its infinitesimal generator. *Proc. Amer. Math. Soc.*, 100(1):104–108, 1987.

[209] F. Neubrander. Integrated semigroups and their applications to the abstract Cauchy problem. *Pacific J. Math.*, 135(1):111–155, 1988.

[210] F. Neubrander. Integrated semigroups and their application to complete second order Cauchy problems. *Semigroup Forum*, 38(2):233–251, 1989. Semigroups and differential operators (Oberwolfach, 1988).

[211] F. Neubrander. The Laplace-Stieltjes transform in Banach spaces and abstract Cauchy problems. In *Evolution Equations, Control Theory, and Biomathematics (Han sur Lesse, 1991)*, pages 417–431. Dekker, New York, 1994.

[212] F. Neubrander and A. Rhandi. Degenerate abstract cauchy problems. In *Seminar notes in Functional Analalysis and Partial Differential Equations*, pages 1–12. Louisiana State University, 1992/1993.

[213] S. Ōharu. Semigroups of linear operators in a Banach space. *Publ. Res. Inst. Math. Sci.*, 7:205–260, 1971/72.

[214] N. Okazawa. A generation theorem for semigroups of growth order α. *Tôhoku Math. J. (2)*, 26:39–51, 1974.

[215] N. Okazawa. A remark on infinitesimal generators of C-semigroups. *SUT J. Math.*, 25(2):123–127, 1989.

[216] S. Ōuchi. Semi-groups of operators in locally convex spaces. *J. Math. Soc. Japan*, 25:265–276, 1973.

[217] M. M. H. Pang. Resolvent estimates for Schrödinger operators in $L^p(R^N)$ and the theory of exponentially bounded C-semigroups. *Semigroup Forum*, 41(1):97–114, 1990.

[218] E. Pap and S. Pilipović. Semigroups of operators on the space of generalized functions $\exp \mathcal{A}'$. *J. Math. Anal. Appl.*, 126(2):501–515, 1987.

[219] A. Pazy. *Semigroups of Linear Operators and Applications to Partial Differential Equations*. Springer-Verlag, New York, 1983.

[220] A. Pazy. Semigroups of operators in Banach spaces. In *Equadiff 82 (Würzburg, 1982)*, pages 508–524. Springer, Berlin, 1983.

[221] S. Pilipović. Generalization of Zemanian spaces of generalized functions which have orthonormal series expansions. *SIAM J. Math. Anal.*, 17(2):477–484, 1986.

[222] A. G. Rutkas. The Cauchy problem for the equation $Ax'(t) + Bx(t) = f(t)$. *Differentsialnye Uravnenija*, 11(11):1996–2010, 2108, 1975.

[223] J. T. Sandefur. Higher order abstract Cauchy problems. *J. Math. Anal. Appl.*, 60(3):728–742, 1977.

[224] J. T. Sandefur, Jr. Convergence of solutions of a second-order Cauchy problem. *J. Math. Anal. Appl.*, 100(2):470–477, 1984.

[225] N. Sanekata. A note on the abstract Cauchy problem in a Banach space. *Proc. Japan Acad.*, 49:510–513, 1973.

[226] N. Sanekata. Some remarks on the abstract Cauchy problem. *Publ. Res. Inst. Math. Sci.*, 11(1):51–65, 1975/76.

[227] L. Schwartz. Théorie des distributions à valeurs vectorielles. I. *Ann. Inst. Fourier, Grenoble*, 7:1–141, 1957.

[228] L. Schwartz. Théorie des distributions à valeurs vectorielles. II. *Ann. Inst. Fourier. Grenoble*, 8:1–209, 1958.

[229] L. Schwartz. Some applications of the theory of distributions. In *Lectures on Modern Mathematics, Vol. I*, pages 23–58. Wiley, New York, 1963.

[230] L. Schwartz. *Mathematics for the Physical Sciences*. Hermann, Paris, 1966.

[231] M. Shimizu and I. Miyadera. Perturbation theory for cosine families on Banach spaces. *Tokyo J. Math.*, 1(2):333–343, 1978.

[232] R. E. Showalter. Quasi-reversibility of first and second order parabolic evolution equations. In *Improperly Posed Boundary Value Problems (Conf., Univ. New Mexico, Albuquerque, N.M., 1974)*, pages 76–84. Res. Notes in Math., No. 1. Pitman, London, 1975.

[233] R. E. Showalter. *Hilbert Space Methods for Partial Differential Equations*. Pitman, London, 1977. Monographs and Studies in Mathematics, Vol. 1.

[234] N. A. Sidorov and E. B. Blagodatskaya. Differential equations with a Fredholm operator in the main differential expression. *Dokl. Akad. Nauk SSSR*, 319(5):1087–1090, 1991.

[235] N. A. Sidorov and M. V. Falaleev. Generalized solutions of degenerate differential and integral equations in Banach spaces. In *The Method of Lyapunov Functions in the Analysis of the Dynamics of Systems (Irkutsk, 1985) (Russian)*, pages 308–318, 328. "Nauka" Sibirsk. Otdel., Novosibirsk, 1987.

[236] Yu. T. Silchenko and P. E. Sobolevskiĭ. Solvability of the Cauchy problem for an evolution equation in a Banach space with a nondensely given operator coefficient which generates a semigroup with a singularity. *Sibirsk. Mat. Zh.*, 27(4):93–104, 214, 1986.

[237] P. E. Sobolevskiĭ. Semigroups of growth α. *Dokl. Akad. Nauk SSSR*, 196:535–537, 1971.

[238] P. E. Sobolevskiĭ and Yu. T. Silchenko. Semigroups of bounded linear operators. In *Mathematics Today '93, No. 8 (Ukrainian)*, pages 34–63. "Vishcha Shkola", Kiev, 1993.

[239] M. Sova. Cosine operator functions. *Rozprawy Mat.*, 49:47, 1966.

[240] M. Sova. Semigroups and cosine functions of normal operators in Hilbert spaces. *Časopis Pěst. Mat.*, 93:437–458, 480, 1968.

[241] M. Sova. Linear differential equations in Banach spaces. *Rozpravy Československé Akad. Věd Řada Mat. Přírod. Věd*, 85(sesit 6):82, 1975.

[242] M. Sova. Abstract Cauchy problem. In *Equadiff IV (Proc. Czechoslovak Conf. Differential Equations and their Applications, Prague, 1977)*, pages 384–396. Springer, Berlin, 1979.

[243] M. Sova. Concerning the characterization of generators of distribution semigroups. *Časopis Pěst. Mat.*, 105(4):329–340, 409, 1980.

[244] M. Sova. On a fundamental theorem of the Laplace transform theory. *Časopis Pěst. Mat.*, 106(3):231–242, 316, 1981.

[245] G. A. Sviridyuk. On the general theory of operator semigroups. *Uspekhi Mat. Nauk*, 49(4(298)):47–74, 1994.

[246] T. Takenaka and N. Okazawa. A Phillips-Miyadera type perturbation theorem for cosine functions of operators. *Tôhoku Math. J. (2)*, 30(1):107–115, 1978.

[247] T. Takenaka and N. Okazawa. Abstract Cauchy problems for second order linear differential equations in a Banach space. *Hiroshima Math. J.*, 17(3):591–612, 1987.

[248] T. Takenaka and N. Okazawa. Wellposedness of abstract Cauchy problems for second order differential equations. *Israel J. Math.*, 69(3):257–288, 1990.

[249] N. Tanaka. On the exponentially bounded C-semigroups. *Tokyo J. Math.*, 10(1):107–117, 1987.

[250] N. Tanaka. Holomorphic C-semigroups and holomorphic semigroups. *Semigroup Forum*, 38(2):253–261, 1989. Semigroups and differential operators (Oberwolfach, 1988).

[251] N. Tanaka. On perturbation theory for exponentially bounded C-semigroups. *Semigroup Forum*, 41(2):215–236, 1990.

[252] N. Tanaka and I. Miyadera. Some remarks on C-semigroups and integrated semigroups. *Proc. Japan Acad. Ser. A Math. Sci.*, 63(5):139–142, 1987.

[253] N. Tanaka and I. Miyadera. Exponentially bounded C-semigroups and integrated semigroups. *Tokyo J. Math.*, (1):99–115, 1989.

[254] N. Tanaka and I. Miyadera. C-semigroups and the abstract Cauchy problem. *J. Math. Anal. Appl.*, 170(1):196–206, 1992.

[255] N. Tanaka and N. Okazawa. Local C-semigroups and local integrated semigroups. *Proc. London Math. Soc. (3)*, 61(1):63–90, 1990.

[256] H. R. Thieme. "Integrated semigroups" and integrated solutions to abstract Cauchy problems. *J. Math. Anal. Appl.*, 152(2):416–447, 1990.

[257] H. R. Thieme. Positive perturbations of dual and integrated semigroups. *Adv. Math. Sci. Appl.*, 6(2):445–507, 1996.

[258] A. N. Tikhonov and V. Y. Arsenin. *Solutions of Ill-Posed Problems*. V. H. Winston & Sons, Washington, D.C.: John Wiley & Sons, New York, 1977.

[259] C. C. Travis and G. F. Webb. Compactness, regularity, and uniform continuity properties of strongly continuous cosine families. *Houston J. Math.*, 3(4):555–567, 1977.

[260] C. C. Travis and G. F. Webb. Cosine families and abstract nonlinear second order differential equations. *Acta Math. Acad. Sci. Hungar.*, 32(1-2):75–96, 1978.

[261] C. C. Travis and G. F. Webb. Second order differential equations in Banach space. In *Nonlinear Equations in Abstract Spaces (Proc. Internat. Sympos., Univ. Texas, Arlington, Tex., 1977)*, pages 331–361. Academic Press, New York, 1978.

[262] C. C. Travis and G. F. Webb. An abstract second-order semilinear Volterra integro-differential equation. *SIAM J. Math. Anal.*, 10(2):412–424, 1979.

[263] C. C. Travis and G. F. Webb. Perturbation of strongly continuous cosine family generators. *Colloq. Math.*, 45(2):277–285, 1981.

[264] T. Ushijima. Some properties of regular distribution semi-groups. *Proc. Japan Acad.*, 45:224–227, 1969.

[265] T. Ushijima. On the strong continuity of distribution semi-groups. *J. Fac. Sci. Univ. Tokyo Sect. I*, 17:363–372, 1970.

[266] T. Ushijima. On the abstract Cauchy problems and semi-groups of linear operators in locally convex spaces. *Sci. Papers College Gen. Ed. Univ. Tokyo*, 21:93–122, 1971.

[267] T. Ushijima. On the generation and smoothness of semi-groups of linear operators. *J. Fac. Sci. Univ. Tokyo Sect. IA Math.*, 19:65–127, 1972.

[268] T. Ushijima. Approximation theory for semi-groups of linear operators and its application to approximation of wave equations. *Japan. J. Math. (N.S.)*, 1(1):185–224, 1975/76.

[269] V. V. Vasilev, S. G. Kreĭn, and S. I. Piskarëv. Operator semigroups, cosine operator functions, and linear differential equations. In *Mathematical Analysis, Vol. 28 (Russian)*, pages 87–202, 204. Akad. Nauk SSSR Vsesoyuz. Inst. Nauchn. i Tekhn. Inform., Moscow, 1990. Translated in J. Soviet Math. **54** (1991), no. 4, 1042–1129.

[270] D.V. Widder. *An Introduction to Laplace Transform Theory*. Academic Press, New York and London, 1971.

[271] T.-J. Xiao and J. Liang. Integrated semigroups, cosine families and higher order abstract Cauchy problems. In *Functional Analysis in China*, pages 351–365. Kluwer Acad. Publ., Dordrecht, 1996.

[272] T.-J. Xiao and J. Liang. Widder-Arendt theorem and integrated semigroups in locally convex space. *Sci. China Ser. A*, 39(11):1121–1130, 1996.

[273] T.-J. Xiao and J. Liang. Laplace transforms and integrated, regularized semigroups in locally convex spaces. *J. Funct. Anal.*, 148(2):448–479, 1997.

[274] T.-J. Xiao and J. Liang. Semigroups arising from elastic systems with dissipation. *Comput. Math. Appl.*, 33(10):1–9, 1997.

[275] T.-J. Xiao and J. Liang. *The Cauchy Problem for Higher-Order Abstract Differential Equations.* Springer-Verlag, Berlin, 1998.

[276] T.-J. Xiao and J. Liang. Differential operators and C-wellposedness of complete second order abstract Cauchy problems. *Pacific J. Math.*, 186(1):167–200, 1998.

[277] T.-J. Xiao and J. Liang. Approximations of Laplace transforms and integrated semigroups. *J. Funct. Anal.*, 172(1):202–220, 2000.

[278] T.-J. Xio and J. Liang. On complete second order linear differential equations in Banach spaces. *Pacific J. Math.*, 142(1):175–195, 1990.

[279] A. Yagi. Abstract quasilinear evolution equations of parabolic type in Banach spaces. *Boll. Un. Mat. Ital. B (7)*, 5(2):341–368, 1991.

[280] A. Yagi. Generation theorem of semigroup for multivalued linear operators. *Osaka J. Math.*, 28(2):385–410, 1991.

[281] K. Yosida. *Functional Analysis.* Springer-Verlag, Berlin, 1995. Reprint of the sixth (1980) edition.

[282] S. Zaidman. Well-posed Cauchy problem and related semigroups of operators for the equation $Bu'(t) = Au(t)$, $t \geq 0$, in Banach spaces. *Libertas Math.*, 12:147–159, 1992.

[283] S. Zaidman. *Functional Analysis and Differential Equations in Abstract Spaces.* Chapman & Hall/CRC, Boca Raton, FL, 1999.

[284] E. Zeidler. *Applied Functional Analysis.* Springer-Verlag, New York, 1995. Applications to mathematical physics.

[285] A. H. Zemanian. *Generalized Integral Transformations.* Dover Publications Inc., New York, second edition, 1987.

[286] J. Zhang and Q. Zheng. On α-times integrated cosine functions. *Math. Japon.*, 50(3):401–408, 1999.

[287] Q. Zheng. Solvability and well-posedness of higher-order abstract Cauchy problems. *Sci. China Ser. A*, 34(2):147–156, 1991.

[288] Q. Zheng. Applications of integrated semigroups to higher order abstract Cauchy problems. *Systems Sci. Math. Sci.*, 5(4):316 327, 1992.

[289] Q. Zheng. A Hille-Yosida theorem for the higher-order abstract Cauchy problem. *Bull. London Math. Soc.*, 24(6):531 539, 1992.

[290] Q. Zheng. Higher order abstract Cauchy problems with N-closedness. *Math. Japon.*, 38(3):531 539, 1993.

[291] Q. Zheng. Perturbations and approximations of integrated semigroups. *Acta Math. Sinica (N.S.)*, 9(3):252 260, 1993.

[292] Q. Zheng. Strongly continuous M,N-families of bounded operators. *Integral Equations Operator Theory*, 19(1):105 119, 1994.

[293] Q. Zheng. Integrated cosine functions. *Internat. J. Math. Math. Sci.*, 19(3):575 580, 1996.

[294] Q. Zheng and Y. S. Lei. Exponentially bounded C-semigroup and integrated semigroup with nondensely defined generators. I. Approximation. *Acta Math. Sci. (English Ed.)*, 13(3):251 260, 1993.

[295] Q. Zheng and Y. S. Lei. Exponentially bounded C-semigroups and integrated semigroups with non-densely defined generators. III. Analyticity. *Acta Math. Sci. (English Ed.)*, 14(1):107 119, 1994.

[296] Q. Zheng and Y. Li. Abstract parabolic systems and regularized semigroups. *Pacific J. Math.*, 182(1):183 199, 1998.

[297] S. Zubova and K. Černyšov. A linear differential equation with a Fredholm operator acting on the derivative. *Differentsialnye Uravnenija i Primenen. Trudy Sem. Processy Optimal. Upravlenija I Sektsija*, (Vyp. 14):21 39, 107, 1976.

Glossary of Notation

Generalities

:=	equality by definition
$\Longrightarrow$	logical implication
$\forall$	for all
$\exists$	there exist
$\emptyset$	the empty set
$\square$	the end of proof
v.p.	Cauchy principal value (of an integral)

Sets, functions and numbers

$\{x \in E \mid P\}$	the subset of E whose elements possess property P
$E \times F$	the cartesian product of sets E and F
$E \hookrightarrow F$	the set E is contained in F with continuous injection
I_E	the identity mapping in E
$f\|_\Omega$	the restriction of function $f : E \to F$ to the subset Ω of E
$\{a_k\}$	the sequence $a_1, \dots, a_k, \dots$
$\overline{E}$	the closure of E
∂E	the boundary of E
$\mathbb{N}, \mathbb{R}, \mathbb{C}$	the sets of natural numbers, real numbers and complex numbers
$\mathbb{R}^N$	the set of all real N-tupels $(r_1, \dots, r_n)$
2^E	the set of all subsets of E
$f(x) =_{x \to a} \mathcal{O}(g(x))$	$\|f(x)\| \le \text{const } \|g(x)\|$ in a neighborhood of a
$o(x)$	a function satisfying $\|o(x)/x\| \to 0$ as $x \to 0$
$\operatorname{Re} z$	the real part of $z \in \mathbb{C}$
$\operatorname{Im} z$	the imaginary part of $z \in \mathbb{C}$
$\arg z$	the argument of $z \in \mathbb{C}$
$[\![x]\!]$	the integer part of $x \in \mathbb{R}$

δ or $\delta(t)$	the Dirac distribution
δ_a or $\delta(t-a)$	the Dirac distribution concentrated at a
$H(\cdot)$	the Heaviside function
$f * g$	the convolution of f and g
supp f	the support of f

Operators and spaces

$X \oplus Y$	the direct sum of vector spaces X and Y
X/Y	the quotient space of X by Y if $Y \subset X$
$\|x\|_X$ or $\|x\|$	norm of $x \in X$
I_X	the identity operator in X
$D(A)$	the domain of an operator $A : X \to Y$
ran A	the range (image) of A
ker A	the kernel of A
$\rho(A)$	the resolvent set of A
$\rho_B(A)$	the B-resolvent set of A
$\sigma(A)$	the spectrum of A
A^{-1}	the inverse of A
A^*	the adjoint of A
$\overline{A}$	the closure of A
$\|A\|$	the norm of A
$A\|_F$	the restriction of A to the subspace $F \subset X$
$R_A(\lambda)$	the resolvent of A:

$$R_A(\lambda) := (\lambda I - A)^{-1} \text{ for } \lambda \in \rho(A)$$

$[D(A)]$	the Banach space

$$\left\{ D(A), \ \|x\|_A = \|x\| + \|Ax\| \right\}$$

$[D(A^n)]$	the Banach space

$$\left\{ D(A^n), \ \|x\|_{A^n} = \|x\| + \|Ax\| + \ldots + \|A^n x\| \right\}$$

$\mathcal{L}(X,Y)$	the set of all continuous (or bounded) linear mappings of X with values in Y. If X and Y are Banach spaces, then so is $\mathcal{L}(X,Y)$ equipped with the norm:

$$\|A\|_{\mathcal{L}(X,Y)} = \sup\nolimits_{\|x\|_X} \|Ax\|$$

$\mathcal{L}(X)$ $\mathcal{L}(X) := \mathcal{L}(X, X)$

X' the dual of X, i.e. $X' = \mathcal{L}(X, \mathbb{C})$

$\langle x, y \rangle$ the bracket of duality between $x \in X$ and $y \in X'$

l_2 the space of sequences $\{x_k\} \subset \mathbb{C}$ with $\sum_{k=1}^{\infty} |x_k|^2 < \infty$

Spaces of functions and distributions

Let Ω be an open set in $\mathbb{R}^N$ and K be a compact set in $\mathbb{R}^N$.

$C(K)$ the space of functions continuous on K with $\|f\|_{C(K)} = \sup_{t \in K} |f(t)|$

$C_0(\mathbb{R}^N)$ the space of continuous functions on $\mathbb{R}$ vanishing at infinity

$C^k(K)$ the space of k-times continuously differentiable functions equipped with the norm:

$$\|f\|_{C^k(K)} = \sum_{|j|=0}^{k} \|D^j f\|_{C(K)},$$

where

$$D^j = \frac{\partial^{j_1 + \ldots + j_N}}{\partial x_1^{j_1} \ldots \partial x_N^{j_N}}, \quad |j| = j_1 + \ldots + j_N$$

$\mathcal{D}(\Omega)$ the space of Schwartz test functions

$\mathcal{D}'(\Omega)$ the space of distributions on Ω, i.e.

$$\mathcal{D}'(\Omega) = \mathcal{L}(\mathcal{D}(\Omega), \mathbb{C})$$

$L^p(\Omega)$ the space of classes of measurable functions on Ω equipped with the norm:

$$\|f\|_{L^p} = \left(\int_{\Omega} |f(x)|^p \, dx \right)^{1/p}, \quad 1 \le p < \infty$$

$H^k(\Omega)$ Sobolev space:

$$:= \left\{ f \in L^2(\Omega) \mid D^\alpha f \in L^2(\Omega), \, \forall \alpha : |\alpha| \le k \right\},$$

where

$$\alpha = (\alpha_1, \ldots, \alpha_N), \quad |\alpha| = \alpha_1 + \cdots + \alpha_N, \quad D^\alpha f = \frac{\partial^{|\alpha|} f}{\partial x^\alpha}$$

$H_0^k(\Omega)$ the closure of $\mathcal{D}(\Omega)$ in $H^k(\Omega)$
$H^{-k}(\Omega)$ the dual space of $H_0^k(\Omega)$

$W^{k,p}(\Omega)$ $:= \left\{ f \in L^p(\Omega) \mid D^\alpha f \in L^p(\Omega),\ \forall \alpha :\ |\alpha| \le k \right\}$

$C\{E, X\}$ the space of functions continuous on E,
                  with values in a Banach space X
$C^k\{E, X\}$ the space of k-times continuously differentiable functions
                  on E, with values in a Banach space X
$L^2\{E, X\}$ the space of square integrable functions on E,
                  with values in a Banach space X

Index